THE FIGHTING TANKS
SINCE 1916

The
FIGHTING TANKS
SINCE 1916

By

RALPH E. JONES
Major, Infantry, United States Army

GEORGE H. RAREY
Captain, Infantry, United States Army

ROBERT J. ICKS
First Lieutenant, Infantry Reserve, United States Army

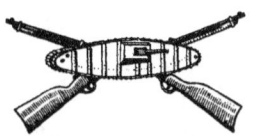

COACHWHIP PUBLICATIONS

Landisville, Pennsylvania

The Fighting Tanks Since 1916, by Ralph Jones, George Rarey, and Robert Icks
Copyright © 2012 Coachwhip Publications
First published 1933.
No claims made on public domain material.

ISBN 1-61646-138-1
ISBN-13 978-1-61646-138-6

Front cover: Tank heading towards German line, near Saint Michel, France (U.S. Signal Corps photo). Back cover: Tank advancing with Infantry near Vimy Ridge, France, April 1917 (Canada Dept. National Defence. Libraries and Archives Canada.)

CoachwhipBooks.com

All Rights Reserved. No part of this publication may be reproduced, stored in a retrieval system or transmitted in any form or by any means—electronic, mechanical, photocopy, recording or any other—except for brief quotations in printed reviews, without the prior permission of the author or publisher.

FOREWORD

The combination of mobility, fire power, crushing force, and protection, variously embodied in those classes of military vehicles ranging from tanks and combat cars to cross country carriers, is still in a state of flux. We cannot today state categorically either their potential capacities or their ultimate limitations. It is a natural consequence of this situation that the doctrines for their employment are subject to continuous and at times fundamental changes. Accordingly a complete appraisal of the value and use of these instruments of war is more than welcome at this time. It is a task to be undertaken only with judgment, discrimination and courage. This volume furnishes a solid point of departure from which the individual can go forward with the certainty that he possesses full and accurate knowledge of the facts as they exist today.

*General,
Chief of Staff, U. S. Army.*

COLONEL H. L. COOPER

Colonel Cooper, the President of the Infantry Association, the Commanding Officer of a tank regiment, and the last Commandant of the Tank School, has devoted his time to tank activities since September, 1924. He is one of the outstanding tank officers of our Army. Colonel Cooper was awarded the Silver Star Citation for gallantry in action against hostile Moros in the Philippines in 1903 and the Distinguished Service Medal for exceptionally meritorious and distinguished service while in command of the 2d Corps School at Chatillon, France, in 1918 and 1919.

THE AUTHORS

Each of the three officers, by whose joint effort this work has been produced, is an expert in the specialty dealt with herein.

Major Jones served as the Instructor in Tactics at the Tank School, and later for three years as a member of the Tank Board, at which time he had much to do with the preparation of the tank volume of the *Infantry Field Manual*. Afterward he commanded the First Battalion of the First Tank Regiment.

Captain Rarey served as an instructor in mechanical subjects at the Tank School for nearly six years, part of that time being in charge of the Design Department and the Tank Section of the School. He has initiated and advanced numerous tank projects, the air-cooled improvement in the Six Ton Tank being almost wholly due to his efforts. He is at present working with tanks in the Department of Experiment at the Infantry School, U. S. Army.

Lieutenant Icks, an Infantry Reserve Officer, residing in Minneapolis, is a graduate of Ripon College with the degree of Ph.B., and has made the study of tanks his particular hobby for about seven years. He possesses one of the most complete individual tank libraries in the world.

THE PUBLISHERS.

Washington, D. C.
June, 1933.

PREFACE

In treating of the details of tanks of the various nationalities, of their design, use, etc., this book is the first of its kind in the English language. While efforts have been made toward a reasonable degree of thoroughness, this work is not intended to be an exhaustive treatise as to all of the subjects with which it deals. The scope is wide and it has not been considered necessary or desirable to follow many of the ramifications down to their last details. In the hope of making the book useful for reference and other purposes, efforts have been made to include all fundamentals, to facilitate vision by the liberal use of illustrations, to systematize the organization of facts, and to restrict the text to a concise presentation of essentials. Where it appeared practicable, comparative facts and viewpoints of the several countries have been set forth with the idea of giving the reader an opportunity to judge these and form his own opinions. In the matter of mechanization, the authors are not conservatives; nor do they believe themselves to be radical enthusiasts. It has been their desire to maintain themselves in the middle of the road where the judicially inclined seek open-minded, common-sense judgments.

The subject is a new one. Many of the matters involved are in a formative and controversial stage. The authors have in consequence realized that it would be impossible so to discuss them as to meet with universal concurrence by those experienced in tank matters. It should not, of course, be inferred that an idea constitutes officially approved doctrine merely because it is favorably considered in this volume.

The authors do not support the view, apparently held by some, that tanks can displace other arms. It is believed that tanks will always be auxiliary to the troops on the ground. In their most common and most valuable role they are auxiliary to infantry foot troops. They are also expected to be auxiliary to cavalry. By some they are expected to be auxiliary to armies. Tanks are weapons of opportunity. Behind the protection of other troops, they are prepared for action. When the situation, the terrain, and other essential conditions are favorable, they assist in striking a decisive blow.

The collection of data regarding particular tanks, as set forth in this book, is brief, but methodically arranged. It is intended to be sufficiently complete to enable the reader to make interesting comparisons between similar vehicles as produced in the different countries. Due to the systematic arrangement, the deficiencies in some instances are very apparent. Not all of the information desired has been obtainable and not all of that obtainable has been available for publication.

To those who have assisted us in the details of this compilation, we extend our sincere thanks. Special acknowledgment is due to:

The Ordnance Department, U. S. A., for permission to use photographs, for the verification of certain information and other assistance; The Signal Corps, U. S. A., for permission to use certain photographs and for accompanying information; The Tank School, U. S. A., for permission to use photographs; Generalleutnant C. von Altrock for access to some original German sources; Major General J. F. C. Fuller and the E. P. Dutton Company, Inc., New York, for permission to quote from *Tanks in the Great War, 1914-1918;* Lieutenant Colonel Sir Albert Stern and Hodder and Stoughton, Ltd., London, for permission to quote from *Tanks 1914-1918, The Log Book of a Pioneer;* Lieutenant Colonel G. LeQ. Martel and Sifton Praed and Company, Ltd., London, for permission to quote from *In the Wake of the Tank* and to reproduce illustrations; Major Fritz Heigl (now deceased) and J. F. Lehmanns Verlag, Munich, for permission to reproduce a number of illustrations and to quote from *Taschenbuch der Tanks;* Major Clough and Mrs. A. Williams-Ellis for permission to quote from *The Tank Corps;* Major L. A. Codd, Editor, *Army Ordnance,* for permission to reproduce illustrations that have appeared in that magazine; Major Robert Sears for cooperation in the procurement of data; Major Allen F. Kingman for **several** French translations; Majors C. C. Benson and Sereno E. Brett for cooperation in the procurement of information; Captain B. H. Liddell-Hart and the *Royal Tank Corps Journal* for permission to quote from *Accurate Shooting from Moving Tanks;* Captain R. P. Butler and the *Royal Tank Corps Journal* for permission to quote from *The Tank Museum;* Captain Charles H. Unger for assistance in the procurement of data and photographs; Captain E. J. M. FitzGerald for translating accounts of German tank actions; First Lieutenant Nathan A. Smith for translating accounts of a large number of tank actions; Lieutenant Charles R. Pinkerton for cooperation in the procurement of data; Sir William Tritton and William Foster and Company, Ltd., Lincoln, England, for permission to reproduce illustrations and for preparing from the original blueprints a drawing of the "Flying Elephant"; Imprimerie et Librairie Berger-Levrault, Nancy, France, for permission to quote from *Les Chars d' Assaut* by Captain L. Dutil; Librairie Charles-Lavauzelle, Paris, for permission to quote from *l'Artillerie d' Assaut, de 1916 a 1918* by Lieut. Col. Lafitte; Vickers Armstrongs, Ltd., for furnishing illustrations of several vehicles manufactured by them; O. H. Hacker, Chief Engineer of Austro-Daimler, for furnishing photographs and information; William Beardmore and Company, Ltd., Glasgow, for furnishing a photograph of the Beardmore gun; E. S. Mittler and Son for permission to quote from *Die Deutschen Kampfwagen im Weltkrieg* by Volckheim; *Militär-Wochenblatt* for information obtained from its review of German tank actions; and Richard Carl Schmidt for permission to quote and use illustrations from *Tanks* by Krüger.

THE FIGHTING TANKS SINCE 1916

TABLE OF CONTENTS

	PAGE
FOREWORD	III
THE AUTHORS	V
PREFACE	VI

CHAPTER I
Introduction

Purpose	1
Definitions	1
Mechanization	1
The Value of Mobility	2
Classification of Vehicles	2
Classification of Tanks	2
Developments Leading to the Tank	3
Tank Combat History	4

CHAPTER II
Tank Combat History. British

Organization During the War	5
THE MARK I TANK	5
Somme, Sept. 15, 1916	5
Second Attack on Thiepval, Sept. 26, 1916	9
Beaumont-Hamel, Nov. 16, 1916	9
THE MARK II AND MARK III TANKS	9
Arras, Apl. 9, 1917	10
Bullecourt, Apl. 11, 1917	12
Monchy, Apl. 11, 1917	13
Neuville Vitasse, Apl. 11, 1917	13
Tanks in Palestine, 1917	13
Second Battle of Gaza, Apl. 17, 1917	14
Third Battle of Gaza, Nov. 1, 1917	14

CHAPTER III
Tank Combat History. British (Continued)

THE MARK IV TANK	16
Messines, June 7, 1917	16
Third Battle of Ypres, July 31 to Oct. 9, 1917	18
Langemarck-St. Julien, Aug. 16, 1917	19
Langemarck-St. Julien, Aug, 19, 1917	19
First Battle of Cambrai, Nov. 20, 1917	20
Second Battle of the Somme, Mar. 21, 1918	26
THE MEDIUM A TANK	28
Bucquoy, Night Raid, June 22-23, 1918	30

CHAPTER IV
Tank Combat History. British (Continued)

THE MARK V TANK	31
Hamel, July 4, 1918	32
Moreuil, July 23, 1918	33
THE MARK V STAR TANK	34
Amiens, Aug. 8, 1918	34
Bapaume, Aug. 21, 1918	38
Second Battle of Arras, Aug. 25 to Sept. 3, 1918	41

CHAPTER V
Tank Combat History. British (Continued)

Epehy, Sept. 17, 1918	43
Second Battle of Cambrai, Sept. 27, 1918	45
Catelet-Bony, Sept. 27, 1918	46
Selle, Oct. 17, 1918	50
Maubeuge, Nov. 2, 1918	52
Tanks in Russia, 1918	53
Tanks in Silesia, 1921	54

CHAPTER VI
Tank Combat History. French

	PAGE
Organization During the War	55
THE SCHNEIDER TANK	55
Chemin des Dames, Apl. 16, 1917	56
THE ST. CHAMOND TANK	59
Laffaux Mill, May 5, 1917	60
Malmaison, Oct. 23, 1917	61
Adelpare Farm, Apl. 5, 1918	62
Grivesnes Park	62
Bois Sencat, Du Gros Hetre, Castel, Apl. 8, 1918	62
Cantigny, May 28, 1918	62

CHAPTER VII
Tank Combat History. French (Continued)

THE RENAULT TANK	63
Defense of the Foret de Retz, May 31 to June 18, 1918	64
Ploissy-Chazelle, May 31, 1918	64
Faverolles-Corcy, June 2, 1918	65
Foret de Retz, June 2 to 12, 1918	66
Counter Attack on Mery-Belloy-Lataule-St. Maur, June 11, 1918	67
Coeuvres, June 15, 1918	68
Chafosse, June 18, 1918	68
Cutry-St. Pierre-Aigle, June 28, 1918	69
Porte and Loges Farms, July 9, 1918	70
Counter Attack South of the Marne, July 15 to 17, 1918	71
Oeuilly, July 18, 1918	71

CHAPTER VIII
Tank Combat History. French (Continued)

Soissons, July 18 to 26, 1918	72
Marfaux, Connaitreuil, Bullin Farm, Espilly, Savarts Farm, July 23, 1918	76
Bois des Dix Hommes and Bois de Reims, July 24, 1918	76
Bois de Fleury, July 26, 1918	76
Beugneux and Hill 205, Aug. 1, 1918	76
Hangest-en-Santerre, Aug. 8 to 10, 1918	76
Ressons-sur-Matz, Aug. 10, 1918	77
Tillaloy, Aug. 16, 1918	77
Roye, Aug. 17, 1918	77
Nouvron-Vingre, Aug. 20, 1918	77
Lombray, Le Fresne, Nampcel, Nouvron-Vingre, Aug. 20, 1918	78
Crecy-au-Mont—Crony, Aug. 28, 1918	78
Mennejean and Colombe Farms, Sept. 14, 1918	80
Champagne, Sept. 26 to Oct. 3, 1918	80

CHAPTER IX
Tank Combat History. French (Continued)

Flanders Offensive, Hooglede-Thielt, Sept. 26 to Oct. 20, 1918	83
Seboncourt, Petit-Verly Region, Oct. 17 to 20, 1918	84
Offensive of the First French Army South of Guise, Oct. 26 to 30, 1918	87
Attack on the Hundung Stellung, Oct. 25 to 30, 1918	88
Advance to the Escaut, Oct. 31 to Nov. 2, 1918	88
Renault Tanks in Morocco, 1925	89
Renault Tanks in Syria, 1920	94

CHAPTER X
Tank Combat History. German

THE A 7 V TANK	95
St. Quentin, Mar. 21, 1918	96
Villers-Bretonneux, Apl. 24, 1918	97
Soissons, June 1, 1918	100
Reims, June 1, 1918	102
On the Matz between Montdidier and Noyon, June 9, 1918	102
Fermicourt, Sept. 1918	104
North of Cambrai, Oct. 11, 1918	104
The German Tank Situation	104

CHAPTER XI
Tank Combat History. American

	PAGE
St. Mihiel, Sept. 12, 1918	106
Meuse-Argonne, Sept. 26 to Nov. 1, 1918	109

CHAPTER XII
Tanks of All Countries

BRITISH

Little Willie	115
Big Willie	115
Flying Elephant	115
Modified Medium A	117
Medium B	117
Medium C (Hornet)	117
Medium D	117
Mark IV, with Lengthened Tail	119
Other Mark IV Modifications	119
Mark V, Experimental	119
Mark V, with Sprung Tracks	119
Mark V, Double Star	119
Mark VI	121
Mark VII	121
British Mark VIII	121
Mark IX	121
Carden Loyd One Man Tank, Mark III	122
Carden Loyd, Mark V	122
Carden Loyd, Mark VI	122
Carden Loyd, Mark VII	124
(Carden Loyd) Light Tank, Mark I A	124
Vickers Carden Loyd Patrol Tank	126
Vickers Mark I	126
Vickers Mark I A	126
Vickers Mark II and Mark II A	126
Vickers Wheel and Track	128
Light Dragon Machine Gun Carrier	128
Sixteen Ton Tank	128
Independent Tanks, Mark I and Mark II	128
Vickers Convertible	129
Vickers Armstrongs Six Ton, Alternative A	130
Vickers Armstrongs Six Ton, Alternative B	131

CHAPTER XIII
Tanks of All Countries (Continued)

FRENCH

Renault, Model B. S.	132
Renault, M 1923	132
Renault with Rubber Tracks	132
Renault N. C. Model 1927	134
Chenilette St. Chamond, M 1921	135
Chenilette St. Chamond, M 1924	135
Chenilette St. Chamond, M. 1926	135
Peugeot	135
Delaunay-Belleville	135
Char Moyen	136
Char 1 A	137
Char 1 C	137
Char 2 C	137
Char 3 C	137
Schneider	138

CHAPTER XIV
Tanks of All Countries (Continued)

GERMAN

L K I	139
L K II	139
A 7 V U	139
K Vehicle	140

ITALIAN

	PAGE
Fiat, Type 3000	142
Fiat, Type 3000 B	142
Pavesi, M 1925	142
Pavesi, M 1926	144
Pavesi Tank Destroyer	144
Ansaldo	144
Fiat, Type 2000	144
Heavy Pavesi	146
G L 4	146
Italian Light (Carden Loyd Type)	146

CHAPTER XV
Tanks of All Countries (Continued)

RUSSIAN

Russian Renault	147
Light Tank	147
Eighty Ton Tank	147

JAPANESE

Vickers Medium C	147

SWEDISH

M 21	150
Landskrona	150

CZECHOSLOVAKIAN

Light Tank	150
K H 50-60-70	152

SPANISH

Trubia	152

POLISH

Cardosowitz	152

CHAPTER XVI
Tanks of All Countries (Continued)

UNITED STATES

Gas Electric	153
Steam Tank, Track Laying	153
Steam Tank, Three Wheeled	153
Skeleton Tank	155
Ford Three Ton	155
Mark I, Three Man	155
Mark VIII	157
Six Ton Tank, M 1917	157
Pilot Model, Six Ton, M 1917, A 1	157
Six Ton, M 1917 A 1	158
Medium A, M 1921	158
Medium Tank, M 1922	160
Medium Tank, T 1	160
Medium Tank, T 2	160
One Man Tank, Experimental	162
Light Tank, T 1	162
Light Tank, T 1-E 1	164
Light Tank, T 1-E 2	164
Light Tank, T 1-E 3	164
Light Tank, T 1-E 4	164
Christie Tank, M 1919	166
Christie Tank, M 1921	166
Christie Tank Chassis, M 1928	168
Christie Tank, M 1931	168
Christie Light Tank, M 1932	168
Combat Car, T 2	171

CHAPTER XVII
Tanks Possessed by the Various Countries

Table of Tank Types in Various Countries	172
Shapes and Relative Sizes	173

CHAPTER XVIII
Powers and Limitations of Tanks.
Considerations Governing Their Employment

	PAGE
POWERS	174
LIMITATIONS	176
PRINCIPAL CONSIDERATIONS GOVERNING THE EMPLOYMENT OF TANKS	177
Role of the Tank, in General	177
Specific Uses for Tanks	177
Principles of Tank Employment	178

CHAPTER XIX
Tank Types and Uses

War Types	180
Very Small Tanks	180
Light Tanks	181
Medium Tanks	182
Heavy Tanks	182
Amphibious Tanks	182

CHAPTER XX
Tank Design

Relatively Unimportant General Considerations	183
Weight	183
Size	183
Shape	183
The First Fundamental of Design	183
Tactical Purpose	183
Armament and Associated Matters	184
Limitations on Fire Power	184
Scout Tank	184
Light Combat Tank	184
Leading Tank	184
Heavy Tank	184
Nature of Antitank Weapons in Tanks	184
Mounting of Weapons	184
Antiaircraft Weapons	185
Crew	185
Speed and Associated Matters	185
Speed	185
Power Train and Transmission	186
Suspension	186
Armor and Other Protection Features	190
Armor	190
Bullet Splash	192
Speed	192
Gas Protection	193
Fire Protection	193
Heat and Fumes	193
Space Required	193
Crew Compartment	193
Engine Compartment	194
Determination of Horsepower	194
Engines and Cooling	194
Tracks	195
Grousers	196
Ability to Negotiate Obstacles	196
General Arrangement	198
General Arrangement. Cargo Carrier	201
Superstructure	201
Strength, Reliability, Lubrication	202
Accessibility	202
Fuel Distance	202
Location of Fuel Tanks	202
Maneuverability and Ease of Control	202
Silence in Operation	202
Lights	202

CHAPTER XXI
Tank Equipment and Accessories

	PAGE
Interior Communication	203
Exterior Communication	205
Armament	206
Protection	207
Equipment Pertaining to Tank Operation	214
Engine Accessories	215
Vehicular Maintenance	216
Miscellaneous Equipment	217

CHAPTER XXII
Miscellaneous Vehicles

COMMAND AND SIGNAL TANKS	218
British (Carden Loyd) Light Tank, Mark II	218
Command Post Vehicles	222
HALF-TRACK COMBAT VEHICLES	222
British, Morris-Martel	223
British, Crossley-Martel	224
French, Schneider Half-Track	224
French, Autochenille 1928-29 (Half-Track)	224
Russian, Austin Half-Track	224
ARMORED CARS	225
British, Lanchester Six Wheel	226
British, Crossley-Vickers Armstrongs Heavy	226
British, Crossley-Vickers Armstrongs Light	226
Swedish, M 1926	226
Czechoslovakian, PA 2 (M 1923)	228
U. S. Light Armored Car (Chevrolet)	228
U. S. Four Wheel Drive Armored Car (Franklin)	228
U. S. Armored Car T 4	228

CHAPTER XXIII
Miscellaneous Vehicles (Continued)

SELF-PROPELLED ARTILLERY	230
British. Gun Carrier Tank	230
British. Self-Propelled Mount for 18-Pounder, M 1926	232
British. Self-Propelled Mount for 18-Pounder, M 1929	232
British. Carden Loyd with Stokes Mortar	232
British. Carden Loyd with 47 mm Gun	232
U. S. Christie Motor Carriage for 8 Inch Howitzer	234
U. S. Christie Motor Carriage for 155 mm Gun	234
U. S. Christie Motor Carriage for 75 mm Gun or 105 mm Howitzer	234
U. S. Mark VII Motor Carriage	234
U. S. 75 mm Howitzer Motor Carriage, T 1	234
U. S. 4.2 Inch Mortar Motor Carriage, T 1	237
MILITARY TRACTORS	237
British. Dragon, Mark I	237
British. Dragon, Mark II	237
British. Dragon, Mark III	237
British. Special Carden Loyd Light Tractor	237
British. Carden Loyd Machine Gun Carrier	239
French. Citroen Kegresse Light Tractor	239
French. Light Infantry Supply Tank, Type V E	239

CHAPTER XXIV
Miscellaneous Vehicles (Continued)

CROSS-COUNTRY TRACTION AIDS	241
Removable Light Track for Rear Wheels of Six Wheeled Vehicles	241
Hipkins Traction Device	241
CARGO CARRIERS, CONVERTED FROM TANKS	243
British. Supply Tank, Mark IV	243
British. Supply Tank, Modified Gun Carrier	244
French. Supply Tank, St. Chamond	244
CROSS-COUNTRY CARGO CARRIERS	244
U. S. Light Cargo Carrier, T 1	244
U. S. Light Cargo Carrier, T 1-E 1	244
POWER CARTS	244

	PAGE
U. S. Large Power Cart, M 1922	246
U. S. Small Power Cart, M 1924-E	246
TANK CARRIERS	246
U. S. Mack Tank Carrier	246
U. S. TCSW (Tank Carrier Six Wheel)	246
U. S. TCSW, Pneumatic Tired	246
U. S. GMC 10 Ton Truck, Towing Medium T 2 Tank	246
SPECIAL MAINTENANCE VEHICLES	248
U. S. Wrecking Truck (Mechanized Force)	248
U. S. Ten Ton Tractor with Cranes	248

CHAPTER XXV
Tank Armament and Gunnery

ARMAMENT	249
GUNNERY	251

CHAPTER XXVI
Obstacles and Defense Against Tanks

Obstacles	254
Antitank Weapons	256
Other Means of Defense	256
Artillery Fire	256
Chemical Warfare	256
Aerial Bombing	256
Small Arms	256
Hand Grenades	257
Antitank Mines	257
ANTITANK DEFENSE METHODS	257
Tank vs. Tank	257
Principal Defense Considerations	258
German World War Methods	258
Pill Boxes	258
Armor Piercing Bullets	258
Bullet Splash	259
Hand Grenades	260
Antitank Rifle	260
Antitank Machine Gun	260
Light Trench Mortars	260
Automobile Antitank Gun	261
Tank Barriers	261
Wide Trenches	262
Mines	262
Minenwerfer	262
Defense in Depth	262
Organization of German Antitank Defense	264
Notes on Antitank Defense (Small British Tank)	266
Antitank Forts	266
Present German Doctrine	270
Present British Doctrine	270
French Doctrine	270
Japanese Doctrine	271
The Future of Antitank Defense	271
Late Developments in Antitank Weapons	271
ANTITANK WEAPONS (Table)	272

CHAPTER XXVII
Tank Organization

MAJOR ASPECTS	273
Great Britain	273
United States	273
France	273
Russia	275
Italy	275
Japan	275
Various Countries	275
DETAILS OF ORGANIZATION	275
United States	275
Regular Army	275

National Guard	276
Organized Reserves	276
Tank Organization, U. S. Army (Table)	277
Foreign Tank Organization	277

CHAPTER XXVIII
Some Tank Combat Principles

Leading Tanks	278
Accompanying Tanks	279
Assembly and Reservicing	279
Movements in Preparation for Combat	280
Cavalry	280
Armored and Mechanized Forces	281
Tanks in Reserve	281
Reconnaissance and Intelligence	282
Approaching Resistance	282
When to Stop Firing	283
In Various Tactical Situations	283
TANKS VERSUS TANKS	283
Preparation	284
Conduct of the Attack	284
Tank Actions Away from the Main Force	286
Tank Actions When with the Main Force	286
The Melée	287

CHAPTER XXIX
Cooperation Between Tanks and Other Elements

Importance of Cooperation	289
Infantry Foot Troops	289
Artillery	289
Air Corps	290
Engineers	290
Chemical Warfare Service	290
Armored Cars	290

CHAPTER XXX
Communication and Control

Communication	291
Formations	291
Tank Orders	292
Control from the Air	292

CHAPTER XXXI
Concealment and Camouflage

CHAPTER XXXII
Landings and Stream Crossings

Landings	299
STREAM CROSSINGS	299
Bridges, Rafts and Ferries	299
AMPHIBIOUS TANKS	301
British. Light Infantry D, Amphibious	301
U. S. Christie Amphibious Tank, First Model	301
U. S. Christie Amphibious Tank, Second Model	303
U. S. Christie Amphibious Tank, Third Model	304
British. Carden Loyd Light Amphibious	304

CHAPTER XXXIII
Tank Supply, Maintenance and Salvage

The Problem of Supply	307
Tank Supply	307
Tank Maintenance	308
Tank Salvage	309

CHAPTER XXXIV
Outstanding Conclusions

BIBLIOGRAPHY	314
INDEX	318

CHAPTER I.

INTRODUCTION

Purpose. The purpose of this book is to provide a concise but fairly comprehensive introduction to a study of fighting tanks. To accomplish the purpose, it is necessary to deal, to some extent, with the wider scope of mechanization, both from the domestic and foreign viewpoints. The emphasis however is placed upon the characteristics and employment of United States tanks. Armored cars and other vehicles associated with tanks are dealt with less completely, as are foreign vehicles generally.

Definitions.

The following definitions should be clearly comprehended because of their bearing on the scope of this work.

A motorized force is a motor-transported force that fights dismounted. *(U. S. Official)*.

An armored force is a force that consists chiefly of armored vehicles, the personnel of which fights without dismounting. *(U. S. Official)*.

A mechanized force is a fighting force that moves by motor power. It includes, or consists of, armored elements, and is capable of relatively independent action. *(U. S. Official)*.

An armored car is an armored combat motor vehicle of the wheel type, intended primarily for use on roads.

A tank is a self-propelled vehicle of the track-laying type, combining fire power, mobility, protection, and shock action. *(U. S. Official)*.

Origin of the Name "Tank". Some may wonder why such a vehicle is called a *tank*. In the first production of these instruments of war for use on the Western Front, it was obviously extremely desirable to maintain a high degree of secrecy so that, in their initial employment, surprise and effectiveness might be had to the maximum degree. The British were building rather large hulls which were difficult to conceal from the gaze of the curious. To lessen the prospect of the secret being given away, the new vehicles were called *tanks* and a number of ruses were employed to give the impression that they were really water carriers destined for shipment to Russia. Thus was the new instrument of war named.

Mechanization.

To the military mind, the word **mechanization** signifies the developments that have been or may be effected in military combat and transport vehicles due to the employment of the internal combustion engine. Many transport vehicles are being motorized and many combat vehicles are being both motorized and armored. This process involves many and far-reaching modifications in military organization. It has, at the present time, within its tentacles, the armies of the world. Its chief contribution is greatly increased mobility, which includes strategic mobility, tactical mobility and mobility of supply. We may add that, in its application to combat vehicles, it permits gunners, in the most advanced localities, in an unfatigued condition, and protected by armor plate, to operate their weapons either while stationary or while advancing; it permits such gunners to have, at all times close at hand wherever they go, a liberal supply of ammunition; and it provides this high degree of fire power with a great reduction in the personnel formerly required to transport, operate, and supply such weapons.

But, of course, there is the other side. Mechanization is initially expensive. Especially in the early experimental efforts, it may be expensive with relatively unimportant results. In some countries the mechanization process may be greatly delayed or curtailed due to lack of concerted effort, a shortage of motor fuel, or other restrictive influences of an economic nature.

The ability of the internal combustion engine to increase the efficiency of supply along roads has been recognized for a considerable number of years. Its ability to increase the efficiency of actual combat, and supply across country, will be more and more widely recognized in the future.

The Value of Mobility.

The value of superior mobility in dealing with an enemy is so obvious and so well borne out by the history of wars and universal tactical teachings that to attempt to emphasize it seems trite. Strange to say, the unusual situation existing in the World War seems to have blunted, to a considerable degree, our appreciation of the importance of mobility. If the tanks in the World War had possessed greater mechanical reliability and greater speed, our appreciation of mobility in tanks would be greater today. If the war on the Western Front had been fought upon a larger field with one or both flanks exposed, our appreciation of the armored car would be greater today. It is not within the province of this book to make any extensive effort to induce a belief in the value of superior mobility. A study of the history of Napoleon, or almost any other great military captain, is more appropriate and convincing.

Classification of Vehicles.

There is always a tendency to try to classify numerous divergent forms derived from the same base. It is so with motor vehicles. The aggregation of vehicles pertaining to military mechanization is complex. A perfect classification is not obvious, but perhaps the following will serve our purpose as well as any other:

Tactical vehicles, which include:

(1) *Combat vehicles;* (a) armored cars; (b) tanks; (c) supporting combat vehicles.

(2) *Combat carriers;* (a) cross-country or other motor vehicles designed to carry men and weapons to the points where they are to engage in combat. (This includes motorized machine guns and cannons but does not include vehicles carrying weapons that are normally used for combat without being dismounted from the vehicles); (b) communication vehicles.

Transport vehicles, which include:

(1) *Cross-country;* (a) cargo; (b) special (such as communication, motor-servicing, and salvage vehicles).

(2) *Road;* (a) automobiles; (b) cargo trucks; (c) tank carriers; (d) miscellaneous special.

Tractors and trailers for special purposes, where necessary.

Classification of Tanks.

During the past decade and a half, it has been the practice, as a rule, to classify tanks upon the basis of weight. There has been, however, a lack of uniformity among the nations concerned as to where the limiting lines are drawn. The present U. S. classification is as follows:

A light tank is a two-man tank that can be transported by tank carrier. *(U. S. Official)*.

A medium tank is one weighing not more than 25 tons[1] but too heavy or too large to be transported by tank carrier. *(U. S. Official)*.

A heavy tank is one of over 25 tons in weight. *(U. S. Official)*.

In England, France, and the United States, very small experimental tanks with room for only one man have been constructed. They are known as *one-man tanks*. Tanks weighing 80 tons or more are usually referred to as supertanks. The latter have been built in England, France, Germany and Russia.

The English minimum weight for the medium class is in the vicinity of 8 tons. Heigl of Austria (now deceased) in his publications gave light tanks a range of from 7 to 11 tons. He regarded tanks of over 22 tons as heavy.

Light tanks in England, France, and the United States weigh about 7½ tons. The British medium tank weighs about 13½ tons, the French medium about 24 tons. The United States has experimental, unstandardized tanks within the medium class ranging in weight from 10½ tons (the Christie) to 25 tons (the Medium, 1922). The heavy tank in the United States Army, the Mark VIII (obsolescent), weighs 43½ tons; the French heavy tank about 77 tons.

Since the 10½ ton U. S. Tank (Christie) would, if adopted and produced in quantity, supersede the tank weighing approximately 7½ tons, and since tank carriers are not contemplated for tanks that are to be built in the future, it seems that the 10½ ton tank might better be considered a light tank than a medium. This is particularly so since the possible need for a rather heavy medium tank can be foreseen.

Ten tons is probably close to the weight of the lightest tank of the future that will carry an antitank cannon. If, in the future, a tank is desired that is to be so small and light as to preclude its carrying an antitank weapon, it would seem best to call it a scout tank.

It is doubtful if weight is the best basis for the classification of tanks. A tank is not designed to weigh a certain amount. It is designed to fulfill a definite tactical mission. Its mission fixes its chief characteristics, which are its fire power, mobility, and armor; and these in turn determine its design. Its weight is merely a result. It would seem that the tactical mission of a tank, or its crew, or its armament would be a more appropriate basis of classification. The British have tanks that they call *close support tanks*. That seems to be a step in the desired direction. However, the basis of classification is after all of very slight importance provided we see clearly that weight is not one of the chief characteristics controlling design and that it is not a primary factor in the assignment of tactical missions to existing tanks. It is highly probable that an appropriate and satisfactory method of classification will be generally adopted sometime in the near future after the types and tactical uses shall have been more thoroughly developed and stabilized.

Developments Leading to the Tank.

The tank, brought into being by the World War, embodies the qualities, fire power, mobility, protection, and shock action (as stated in the definition)—also crushing power. But attempts to combine such qualities did not originate at that time. Throughout the centuries, countless attempts have been made to combine offensive power with mobility and protection. The earliest and simplest form is perhaps the com-

---

[1] In this book a ton means 2,000 lbs. In foreign writings such is not the case, since the metric ton is 2,205 lbs., and the British ton is 2,240 lbs. Gallons in this book are U. S. gallons.

bination of the mobility of the man, the offensive power of the spear, and the protection of the shield. In the early efforts, motive power was provided by man, horse, or elephant. In 3500 B. C., we find the Assyrian chariots; in 1200 B. C., the Chinese war carts; and in 300 B. C., protected towers mounted upon elephants. On the sea, the *Monitor* stands out due to its introduction of effective armor in combination with the usual mobility and fire power. All of these efforts have been engendered by this same desire or purpose: to close with the enemy with superior mobility, to strike him effectively at close range, and to be protected at the same time against his blows.

The early vehicles usually lacked adequate motive power. This bar to the development of the tank was removed by the internal combustion engine. But wheeled vehicles, especially if they are self-propelled, cannot be operated satisfactorily except over fairly favorable ground. This obstacle was removed by the development of an endless track for tractors by the Holt Caterpillar Company in the United States. Thus was laid the foundation for the invention of the combat tank. Its invention was spurred by the urgent need of a means to make possible a rapid and effective attack, without terrific losses of life, in the face of the many trenches, machine guns, barbed wire entanglements, and other obstacles existing in the defensive systems of the World War.

The Development of the World War Tanks. The development of the tanks employed in the World War is not narrated in this book. The tanks themselves are illustrated and described.

Tank Development Since the World War. Since the World War, there has been a good deal of unprofitable groping in the field of tank development. The tank of the War was an innovation. The modern tank is a second innovation and it has very different tactical qualities from its predecessor. The nations of the world, generally speaking, have not learned how to use it. They have not even learned much as to what types of tanks are best suited to their needs. During this post-war development, Great Britain has been decidedly in the lead, both from the mechanical and employment standpoints. In other countries, development has been retarded by economic and various other factors. The mechanical development and the employment of tanks are considered in greater detail later in this text.

Tank Combat History.

The account in the chapters which immediately follow is a brief review of each action in tank combat history. Its main purposes are to show what has been accomplished by the use of tanks, something of their tactics and technique and the principal reasons for their success or failure, as the case may be, in order that sound conclusions may be drawn from past experiences for future guidance in the development and employment of this new and powerful fighting vehicle.

In the historical narrative, near the point at which each particular vehicle makes its appearance, has been placed a photograph of the tank together with descriptive data, and a list of the actions in which the tank participated.

For the sake of brevity in a work which could easily become voluminous, it has been necessary to omit the important parts played by other branches in the actions described. The authors do not intend to indicate thereby that the tanks, alone, won the actions reviewed, except in isolated instances where only tanks were used.

CHAPTER II.
TANK COMBAT HISTORY. BRITISH.

Organization During the War. The British tank battalions at first consisted of 72 tanks, or 3 companies of 24 tanks each. The company consisted of 4 sections of 6 tanks each. Early in 1917, the battalions were reduced to 48 tanks of which 36 were fighting tanks and 12 were training tanks. The 36 tanks were divided into 3 companies of 12 each. Each company consisted of 3 sections of 4 tanks each.

THE MARK I TANK.

Mark I, Male and Female. Produced in 1916 by William Foster and Company, Ltd. Total production, 75 male, 75 female.

Crew: 8.
Armament: *Male;* Two 6 pounder (57 mm.-2.24 in.) naval guns and four machine guns. *Female;* Six machine guns (Vickers, later replaced by Hotchkiss).
Armor: 0.2 in. to 0.4 in.
Maximum speed: 3.7 mph.
Suspension: Rigid; rollers.
Tracks: Flat steel plates with single grousers.
General arrangement: Driver and tank commander in front, engine forward of center, transmission in rear of center, final drive in rear.
Dimensions: Length 26 ft. 5 in. (32 ft. 6 in. with tail), width 13 ft. 9 in., height 8 ft. 1 in.
Weight: Male, 31 tons; female, 30 tons.
Engine: Daimler, 6 cylinder, sleeve valve, 105 HP, forced water cooling.
Horsepower per ton: Male, 3.4; female, 3.5.
Tranmission: Sliding gear, two speeds and reverse in primary gear box, two-speed secondary gears for each track, giving a total of 4 speeds forward and 2 in reverse. Transmission differential could be locked for straightaway driving.
Obstacle ability: Trench 11 ft. 6 in.; slope 22 degrees; vertical wall 4 ft. 6 in.
Fuel distance: 12 miles. **Fuel capacity:** 53 gallons.
Special features: Bomb screens on top, later removed. Required four men for driving. Sponsons bolted in place and removed by hand for railway travel. Female sponsons larger than male. Gunners sat on bicycle saddles attached to guns. No muffler provided. Drive chains not protected from dirt and mud. Tail wheels hydraulically controlled and sprung with sixteen heavy coil springs, intended to aid in steering; later removed. Poor means of entrance and exit. Poor ventilation. Exhaust noisy and visible at night. Armor did not protect against armor piercing rifle fire. Vision was obtained by prismatic glass peepholes made into a short periscope. This was discarded due to the danger from broken glass. Polished metal surface mirrors were also tried but such vision was poor. These tanks were camouflaged with protective colouring but this was discontinued when later models were introduced.

The Mark I Tank. List of Actions in which Engaged.

Date	Battle	Number Employed
		Units Tanks
Sept. 15, 1916	Somme	49
Sept. 26, 1916	Thiepval	8
Nov. 16, 1916	Beaumont-Hamel	2
April 9, 1917	Arras (Mark I and Mark II)	60
April 11, 1917	Bullécourt (Mark I and Mark II)	11
April 11, 1917	Monchy (Mark I and Mark II)	6
April 11, 1917	Neuville Vitasse (Mark I and Mark II)	4
April 17, 1917	Second Battle of Gaza (Mark I and Mark II)	8
Nov. 1, 1917	Third Battle of Gaza (Mark I and Mark II)	5

Somme, September 15, 1916.

Forty-nine Mark I tanks were used by the British during the Battle of the Somme, *the first test of the tank in action.* Owing to the fact that these vehicles had been secretly designed, built and delivered to the battlefield, little was known about them even by the troops who were to use them. Many questions concerning the new weapon had to be

Plate 1.
1. Mark I, Male. Showing frame for grenade netting.
2. Mark I, Female. Grenade netting installed.
(Photo from British Imperial War Museum.)

answered and many new problems pertaining to their tactical use, control and supply were hastily solved. Naturally, considerable confusion existed concerning the methods to be used since no precedent or past experience of any kind was available to serve as a guide.

It was finally decided that the tanks should start in time to reach the first objective five minutes ahead of the infantry, that they should be employed in groups of two or three against strong points, and that the artillery barrages should leave lanes free from fire through which the tanks could advance.

No special reconnaissance was made by the tank personnel and, consequently, the tank commanders were not well informed as to the situation prior to the attack. However, this was only one of the links in the chain of circumstances which, as we look back at this first tank action, appears to have been designed to insure its certain failure. In their book, *The Tank Corps,* Major C. Williams-Ellis and A. Williams-Ellis, refer to the orders issued for the tanks: "For every three tanks only one set of orders had been issued, and only one map supplied; consequently we had to grasp these orders before we passed them on to the other two officers . . . . . . However, at 5 PM on the day before the battle these orders were cancelled and new verbal instructions substituted."

Although these first tank troops were severely handicapped, fate appears to have balanced the books by leaving to them the element of surprise, since the German troops apparently had no information concerning the tanks. This extraordinary achievement of secrecy in the development, construction and shipment of these tanks seems the more remarkable when it is remembered that the British had been working on the tank project for nineteen months during which time enemy secret service agents were very active in England and behind the Allied lines in France.

Instead of using this small number of tanks on a relatively small front, the 49 tanks were divided into four groups and assigned as follows: 17 of the tanks to the 14th Corps, 17 to the 15th Corps, 8 to the 3rd Corps, and 7 to the Fifth Army. Ten other tanks, all of which were unfit for action due to mechanical troubles, were held in GHQ reserve. This made a total of 59 tanks which were shipped to France prior to the first action. Many of these tanks had been practically worn out during training and demonstrations before leaving England.

The record of the Somme tank activities is one of partial success only. The available data is meager and only a brief summary of the results can be given. Of the 49 tanks assigned for the action, only 32 succeeded in reaching their line of departure, the other 17 becoming stuck or breaking down mechanically. Nine of the 32 tanks were held up on account of mechanical difficulties; 9 did not succeed in leaving the line of departure on time and therefore did not move out with their infantry, but did succeed in helping to mop up; 5 became stuck in the attack. Only 9 tanks fulfilled their missions.

One tank commander assisted the infantry troops in a difficult situation when they were held up by wire and machine gun fire, by moving his tank to a position where he could enfilade the trench from which the fire was coming. He then moved his tank along the trench and is credited with having caused the surrender of about 300 of the enemy troops. Another tank destroyed a 77 mm gun in Guedecourt. Later this tank was struck by a shell and caught fire. One of the most successful exploits was observed by a British airman who reported that "A tank is walking up High street in Flers with the British army cheering behind it."[1] Although Flers was known to contain a great many machine guns, it was taken by this tank and its infantry without casualties.

Very few casualties occurred among the tank personnel in the Somme action. Of the 32 tanks which reached their starting points, ten were

---

[1] *The Tank Corps,* Williams-Ellis.

put out of action for the time being and seven were damaged slightly. The latter, however, managed to return under their own power.

The experimental use of tanks on September 15th was not a great success but this test of the "tank idea" proved its feasibility, indicated the mechanical shortcomings of the vehicles themselves and from it many lessons were learned by the tank personnel, the infantry troops, and the higher commanders. When one considers the crude design of these first tanks, the ignorance of all concerned with reference to methods of employment, the fact that this was the first test of a new and complicated piece of machinery under battlefield conditions, and the change in the orders at the last moment, it is not surprising that the results were only moderately successful.

On the other hand, and notwithstanding the large percentage of the tanks which failed mechanically and for other reasons, there is no doubt about the impetus that was imparted to the attack in this instance by the use of the tanks. The reputation acquired by the few which actually functioned established in the minds of both friend and foe in this battle, and those to follow, a healthy respect for the new method of warfare. The factor of morale should probably be credited in a large measure for the gains made on September 15th and this morale attribute of the tank, with good reason, was to loom large throughout the war among the causes of tank success. When the matter is analyzed, the reasons why the morale factor in tank employment assumed such proportions, especially in the minds of the German front line troops, are not hard to find. Here, for the first time in history, the great characteristic of *mobility* had been successfully combined with protection, fire power and crushing power, thus providing a self-contained offensive unit the like of which had never before appeared on land in any war. The approach of a number of these mobile, metal-clad batteries, when viewed through the eyes of the front line soldier, presented a personal problem which had to be solved at once. The soldier had to stop the tank, surrender, or leave his position. The approaching tank presented a perfect example of men being faced by a condition, and not a theory. Under the circumstances it is not strange that many adopted the second or third alternatives.

In the first test of mechanical warfare on September 15th, the results obtained were sufficient to inspire the following expressions of opinion by the authorities indicated:

The Chief of Staff of the German Third Army Group reported that: "The enemy, in the latest fighting, have employed new engines of war as cruel as (they are) effective." [1]

In his report on the Somme action, Sir Douglas Haig stated: "Our new heavily armored cars, known as tanks, now brought into action for the first time, successfully cooperated with the infantry, and, coming as a surprise to the enemy rank and file, gave valuable help in breaking down their resistance." [1]

The French Minister of Munitions issued the following statement: "I think it is now unnecessary to labor on the imperative need for pressing forward with the construction of our offensive caterpillar machines as quickly as ever it is possible to do so. The English, by using prematurely the engines which, to their credit, they have constructed much more rapidly than ourselves, have debarred us of the use of the element of surprise, which should have enabled us easily to pierce the enemy's lines, though they have more or less rendered us the service of convincing even the most skeptical and most red-tape bound." [2]

[1] *The Tank Corps*, Williams-Ellis.
[2] *Log Book of a Pioneer*, Sir Albert G. Stern, K.B.E., CMG.

After the first battle most of the damaged tanks were salvaged and, together with those held up by mechanical trouble, were repaired and used in small numbers experimentally. Of these minor actions, one of the most interesting was the capture on September 25th, of Gird trench, near Guedecourt, together with 370 prisoners, by a tank, a few bombers and an airplane, at a total cost of 5 casualties.

Second Attack on Thiepval, September 26, 1916.

A small action occurred at this place which was significant in that a new principle, the use of tanks without a preliminary artillery preparation, was here tried for the first time. Thiepval had held out against several attacks by the artillery and other arms. In this action however, the attack came as a surprise to the Germans. The eight tanks and the infantry succeeded in crossing the space between the two lines so quickly, under the protection of artillery and tank weapon fire, that the defenders could not get their machine guns out of their dugouts in time to prevent the occupation of the area by the British. This attack resulted in the capture of many prisoners and an important stronghold at a very small cost.

Beaumont-Hamel, November 16, 1916.

The action at Beaumont-Hamel was more of a demonstration of the moral value of the tanks than of their efficiency along mechanical or tactical lines. They were directed to attack a considerable number of enemy troops who had successfully resisted previous attacks. In this small action the two tanks used operated ahead of the infantry, and, although one of them was able to cross the enemy front line trench, it became stuck very soon thereafter. The second tank became stuck before reaching this front trench. Both tanks were in a dangerous position when, upon examining the trenches around the tanks, the tank crews found that the enemy troops in both the front and support trenches were waving white cloths in token of surrender. The tank crews and the infantry troops who arrived in a short time made prisoners of the entire garrison before they realized that the tanks were stuck and practically at their mercy.

THE MARK II AND MARK III TANKS.

A few Mark II tanks were sent to France and were used with Mark I tanks in some of the actions in the early part of 1917. Although no definite information is available with reference to the combat history of the Mark III tanks, it is believed that they also were used in a few instances in 1917.

Mark II and III. Produced in 1916 by William Foster and Company, Ltd. Total production, 50 Mark II, 50 Mark III.

Crew: 8.
Armament: *Male;* Two 6-pounder (57 mm.-2.24 in.) guns and four machine guns. *Female;* Six machine guns.
Armor: Mark II 0.2 in. to 0.4 in; Mark III 0.2 in. to 0.47 in.
Maximum speed: 3.7 mph.
Suspension: Rigid, with improved cast iron track rollers.
Tracks: Flat steel plates with single grousers. Modified track plates spaced every sixth plate.
General arrangement: Driver and tank commander in front, engine forward of center, transmission in rear of center, final drive in rear.
Dimensions and weight: Practically the same as Mark I.
Engine: Daimler, 6 cylinder, sleeve valve, 105 HP, forced water cooling.
Horsepower per ton: Male 3.4; female 3.5.
Transmission: Practically the same as Mark I.
Obstacle ability: Trench 10 ft.; slope 22 degrees; vertical wall 4 ft. 6 in.
Fuel capacity: 84 gallons.
Special features: Trench crossing ability less due to elimination of tailpiece.

Sponsons smaller than Mark I and were hinged instead of bolted. Sponsons beveled at bottoms. Mark III protected against A P rifle fire. Fuel tank mounted outside and at the rear. Air pressure fuel system first used but replaced later by the vacuum system. Exit doors placed under the sponsons in the female tanks and in the rear end of sponson in male tank. On most of the Mark III, the 6-pounder guns were reduced in length. Lewis machine guns substituted for Hotchkiss.

The Mark II Tank. List of Actions in which Engaged.

Date	Battle	Number Employed Units Tanks
April 9, 1917	Arras	
April 11, 1917	Bullécourt	
April 11, 1917	Monchy	
April 11, 1917	Neuville Vitasse	
April 17, 1917	Second Battle of Gaza	
Nov. 1, 1917	Third Battle of Gaza	

For number employed see table under Mark I.

Arras, April 9, 1917.

The original plans for the Battle of Arras included provisions for employing two battalions of 48 tanks each. However, owing to the

Plate 2.
Mark II, Female.
(Photo from British Imperial War Museum.)

nonarrival in France by April of the tanks promised for January, the plans had to be modified.

Sixty of the Mark I and Mark II tanks were made available but this number was secured only by taking the training tanks from the instructional centers in France and England. On account of the small number of tanks, it was then decided to allot a proportion of them to each of the three armies for mopping up operations only. They were allotted as follows: Eight to the First Army, to operate against Vimy Heights; 40 to the Third Army, eight of which were to operate with the 17th Corps north of the river Scarpe; 32 with the 6th and 7th Corps, south of this river; and 12 to the Fifth Army, which was to drive north toward Visen-Artois. Thus all tanks were assigned to organizations and given missions, no provisions being made for reserves.

The actions in which these tanks were to participate were all considered as preliminary operations for the main purpose of opening the way for the advance of two cavalry divisions and the 18th Corps south of the Scarpe. This combined force was to break through at Monchy and move east to the Drocourt-Queant line.

Tank commanders had maps for this action and such reconnaissance was made as was possible from observation points from which they could see the ground to be crossed. Supply dumps were formed but it was found that the movement of the supplies forward from these dumps entailed a great amount of hard work since, like all other supplies, they had to be moved by hand.

Only the main incidents in this action will be mentioned. In some cases the tanks made friends but in others, where the new vehicle was poorly handled, the troops to whom they were assigned lost faith in them.

The previous destruction of the roads and the bad weather at the time made it impossible to bring up sufficient artillery, interfered with the handling of supplies, and seriously handicapped the tank operations. An unfortunate selection of a route from the tank park to the assault position prevented certain tanks from working with the infantry to which they were assigned. The route selected was across the Crinchon stream valley near Archicourt. The surface of the ground in this valley appeared to be firm but it was boggy underneath and, probably due to the lack of time, the route was not thoroughly investigated before the tanks were started over it. Six of the eight tanks broke through the surface and became stuck soon after starting, and the infantry, which had been trained with these tanks, had to go forward without them. The tanks were finally withdrawn from this valley, however, and placed in reserve. They were later used with good effect in the last days of the battle.

Owing to the badly shelled area in front of them and the bad weather, the tanks assigned to the First Army were also so handicapped that none of them were able to assist the infantry in their attack upon Vimy Ridge. These tanks were likewise withdrawn and later sent to the Fifth Army.

The four tanks which started east of Arras were more fortunate in the terrain to be covered and, although one of them was put out of action by shell fire soon after the action started, the rest of them aided their infantry to advance along the Scarpe by destroying the machine guns in this area. Also on the south side of the Scarpe, the tanks were able to render assistance to their troops, accounting for many machine guns and enemy troops near Tilloy les Mafflaines, the Harp and Telegraph Hill. They then went on and helped reduce two other objectives before night. Some of these tanks were mired in the low wet area near the fortification called the Harp. Several casualties occurred due to the penetration of the armor by the German "K" armor piercing bullets.

Thus the tank operations on the 9th were only partially successful. Wherever the terrain was suitable for them, however, they were of great value to their infantry, and army headquarters received many favorable messages from the front regarding them.

Bullecourt, April 11, 1917.

The last three actions in which the Mark I and Mark II tanks participated as fighting vehicles on the Western Front occurred on April 11, 1917. The attack on Bullecourt, which was planned for April 10, is interesting and important because of the lessons to be learned from it. The reason it did not take place on the day set was the failure of the tanks to arrive at the appointed place in time to move out at dawn as planned. Although the tanks started for this point at dark on the 9th, the approach march was made in a violent snowstorm. This storm was of such severity that many of the infantry units also lost their way and had difficulty in reaching their battle positions in time. The tanks had to move so slowly that they were unable to reach their assault positions at the appointed time and the Australian infantry, which had been assembled at this point, was withdrawn under heavy shell fire. Many casualties occurred during this withdrawal. The attack had to be postponed until the 11th and the plans changed considerably.

In the action on April 11th, the tanks were directed to supply the artillery barrage ahead of the infantry. The reason for this was that artillery was not available due to the roads and the weather which had prevented the artillery from reaching positions from which they could support the attack.

The plan for the attack provided that the eleven tanks being used were to be placed in line with 80 yard intervals, at about 800 yards from the German lines. Their task was to penetrate the Hindenburg line east of Bullecourt. Six tanks were then to turn westward and, of these, four were to attack the line northwest of Bullecourt. Three tanks were to go straight through the line and attack Hendecourt and Reincourt. Two were to move eastward down the Hindenburg line.

The snowstorm had continued and, when the attack started at 4:30 AM on April 11th, the infantry had great difficulty in wading through the deep snow except in the paths made by the wide tracks of the tanks. They followed these paths in single file and the sharp contrast between the tanks, the infantry and the snow, gave perfect artillery targets. Nine of the 11 tanks were knocked out before they could accomplish their missions.

The tanks ordered to capture Hendecourt and Reincourt were the only ones to accomplish their missions. One of these was put out of action by artillery fire. The other two entered the villages ahead of their infantry as scheduled but, upon their arrival, found that their flanks were unprotected because of the failure of the other parts of the line to advance.

Soon after these villages were taken, the Germans organized a strong counterattack, retaking the villages, and capturing the two tanks and several hundred infantrymen. Examination of the tanks showed the Germans that their small arms armor piercing bullets had been effective. This discovery resulted in the publication of the facts to the entire German army and caused an order to be issued requiring each infantryman to carry five rounds of the "K" armor piercing ammunition, and the issue of several hundred rounds to each machine gunner. Many of the German authorities believed they had found a satisfactory answer to the tank. However, the British had noted the effect of the "K" bullet, which enabled an infantryman or machine gunner to carry a weapon capable of penetrating and stopping tanks, so the next tank to be produced in quantity, the Mark IV, carried armor capable of stopping the "K" bullet.

As a result of the failure of the tanks to arrive at the railroad embankment on April 10th, and of their almost complete failure on the 11th, the Australians felt that the tanks were of little value and they would not work with them for several months following this action.

Monchy, April 11, 1917.

This attack was more successful than the Bullecourt action. Monchy was a strongly fortified locality which commanded the surrounding country on account of its elevated position on a ridge. The general plan called for a cavalry advance as soon as Monchy was occupied. Only six tanks were used in this attack, and of these, three were put out of action by artillery fire before they reached the ridge. The other three tanks were more fortunate. They reached Monchy and the ridge, and the fire of their weapons assisted the infantry to occupy the ridge. As soon as this point was occupied the cavalry moved forward, but this movement is reported by General Fuller to have been quickly stopped by the German rear guard machine gunners.

Neuville Vitasse, April 11, 1917.

Another attack down the Hindenburg line, starting from Neuville Vitasse was, for some reason, not supported by infantry. It was, however, attempted over ground that was suitable for tanks. Only four tanks were used and these four machines worked their way down the line to Heninel, driving the Germans underground with their machine gun fire and causing many casualties among them. At Heninel the four tanks turned to the northeast toward Wancourt. After being out and almost constantly in action for eight hours, fighting wherever they could find the enemy, all four tanks returned safely to the British lines.

This little action was an example, which was repeated upon several occasions during the war, of small numbers of tanks making lengthy excursions into German territory and remaining there several hours entirely unsupported by infantry. These successful trips can not be accounted for unless we assume that, in such cases, the Germans lacked artillery or the front line troops did not have good liaison with their artillery. The fact that such exploits were carried out over suitable terrain may have enabled the tanks to operate at their best speed, and thus, at times, avoid being hit, which they would not have been able to do if required to move over difficult ground. Another point which may have worked in their favor was the lack of antitank guns in the hands of the German infantry at the time and, finally, the moral factor should receive credit since the German front line troops were, in some cases, reluctant to fight vigorously when opposed by tanks.

Tanks in Palestine, 1917.

Although this brief account of these operations is not given in its chronological position, a break in the sequence appears to be justified at this point, in order to conclude the combat history of the Mark I and II tanks before introducing the next tank, and by reason of the fact that these operations were carried out in a theater far removed from the Western Front.

Desiring to have a test of the tank in Palestine and believing the new vehicle would aid in the operations against the Turks, the British decided to send 12 of the new Mark IV tanks, then in the process of manufacture, to Egypt. Instead of sending the new tanks however, old Mark I and Mark II tanks were sent by mistake, and the number reduced to eight. The tanks arrived in Egypt in January, 1917, but did not take part in any action until the date of the second battle of Gaza, April 17th.

The methods of using the tanks were similar to those employed in the actions in France, and the tanks, were, therefore, seriously handicapped by being used in small numbers, by being strung out over wide fronts, by having too many missions assigned to them and, in general, by being placed under the command of infantry officers who did not understand their capabilities and limitations.

Second Battle of Gaza, April 17, 1917.

For this action the eight tanks were divided among three divisions. Only two of the tanks came into action during the first phase of the attack. The attack was a surprise to the Turks, who abandoned their trenches without a fight on the fronts of two divisions. One of the two tanks was knocked out by a direct hit, but the other one accounted for many of the enemy while helping to capture and clear up the trenches in its zone.

During the second phase of the attack on April 19th, one of the two tanks assigned to the 53rd Division broke a track and went out of action. The other tank led the attack alone, driving the Turks from Samson Ridge and aiding the infantry to occupy it. This tank then went on to the El Arish redoubt. The infantry, however, was unable to follow owing to the great number of machine guns in the hands of the Turks along this line. After a six hour action, during which the crew of this tank expended 27,000 rounds of ammunition, and after every member of the crew was wounded, the tank was withdrawn from the field.

One of the four tanks assigned to the 52nd Division was lost when it fell into a gully. Another tank took its place and aided in taking Outpost Hill, after which this tank was put out of action by a direct hit. The machine gun fire from the Turkish lines at this time was so intense that another tank which was temporary being held in reserve was ordered forward, and this one aided the infantry in the occupation of Outpost Hill. During this assault very heavy losses occurred among the British infantry from machine gun fire. The Turks then organized a strong counter attack, forced the British infantry and the tank to withdraw, and captured the hill.

Of the two tanks originally assigned to the 54th Division, one had been lost during the first phase so there was now only one for use in the second phase. With this tank, the division attacked the great redoubt near Kirbet El Sihan and caused the surrender of its garrison. This victory was also a temporary one. Soon after the redoubt had been captured, a shell broke one of the tracks of the tank and a counter attack by the Turks resulted in the capture of the redoubt, the British garrison, and the tank.

Third Battle of Gaza, November 1, 1917.

Five of the old tanks had been repaired and, with three of the new Mark IV tanks which had arrived, eight were available for this action. Evidently the British authorities in Egypt still believed the tank capable of performing many missions in an action, for they assigned from three to six definite missions to each of the six front line tanks used. The following list of missions for one tank will serve to illustrate the numerous duties assigned to these tanks:

Tank No. 3 was directed to attack Zowaiid trench, then to attack Rafa redoubt, after which it was to rally, secure supplies and deliver them to the Rafa redoubt. It was then to capture Yunus trench and, when the infantry arrived, the tank was to proceed to the locality known as Sheikh Hasan, where it was to deliver supplies and then return to Sheikh Ajlin, where the reserve tanks were held. (Two of the four tanks assigned to one division were held in reserve.)

The details of this action will not be given but tank histories of this period state that the six tanks were of assistance to their infantry. Five tanks reached the first objective; four, the second, third, and fourth; and one tank reached its fifth objective. They are credited with assisting in the attacks on various redoubts and trenches, crushing wire for the infantry and delivering engineer stores to the captured localities. Two of the front line tanks became ditched owing to the darkness, another broke its track and was abandoned after it had completed its fifth mission, and the two reserve tanks, loaded with engineer material and while on their way to support the infantry, caught fire and were abandoned. A part of this material consisted of empty sacks, which were loaded on the roof of the tanks and which caught fire from the exhaust pipes.

While the part played by the tanks in Palestine was not a large one, they should be credited with having rendered some assistance and they did prove to the British that the new vehicle could be successfully operated in the desert regions.

CHAPTER III

TANK COMBAT HISTORY. BRITISH (Continued).

Returning to the Western Front, the account is resumed with the battle of Messines in which appeared a new British tank.

THE MARK IV TANK.

Mark IV. Produced in 1917 by William Foster and Company, Ltd., and others. Total production, 420 male, 595 female.

Crew: 8.
Armament: *Male;* Two 6-pounder (57 mm.—2.24 in.) guns and four machine guns. *Female;* Six machine guns (Lewis).
Armor: 0.2 in. to 0.47 in.
Maximum speed: 3.7 mph.
Suspension: Rigid. Upper track rollers were added at points of change in incline.
Tracks: Width 20½ in., pitch 7½ in. For improved operation in soft ground, a special track was provided having a width of 26½ in.
General arrangement: Same as Mark I.
Dimensions: Length 26 ft. 5 in., width (male) 13 ft. 6 in., (female) 10 ft. 6 in., height 8 ft. 2 in.
Weight: Male 31 tons; female 30 tons.
Engine: Daimler, 6 cylinder, sleeve valve, 125 HP, forced water cooling.
Horsepower per ton: Male 4.0; female 4.2.
Transmission: Practically same as Mark I.
Obstacle ability: Trench 10 ft.; slope 22 degrees; vertical wall 4 ft. 6 in.; tree 20 in.
Fuel distance: 15 miles. **Fuel capacity:** 75 gallons.
Special features: Idler adjustment provided. Engine cooling air drawn from interior of tank through radiator placed at rear of tank and forced out through rear louver. Doors placed on top of tank and below sponsons. Increase in horsepower secured by use of double carburetors, aluminum pistons, and higher RPM. Muffler provided. Vacuum feed used. Detachable grousers provided, at first made of steel, later of wood. Unditching gear provided, with rails for carrying beam over top. Flags, colored discs, and semaphores used for outside communication. A few female tanks were later fitted with male right sponsons. After the War, some Mark IV hulls were experimentally fitted with various types of transmissions: Williams-Janney hydraulic, British Westinghouse electric, Wilkins multiple clutch gear, Hele Shaw hydraulic, and Wilson planetary. In addition, one Mark IV was sent to France where the Crochat-Colardeau electric as used in the St. Chamond tank was installed.

The Mark IV Tank. List of Actions in which Engaged.

Date	Battle	Number Employed	
		Units	Tanks
June 7, 1917	Messines		76
July 31, 1917	Third Battle of Ypres		216
Aug. 16, 1917	Langemarck-St. Julien		12
Aug. 19, 1917	Langemarck-St. Julien		12
Nov. 1, 1917	Third Battle of Gaza		3
Nov. 20, 1917	First Battle of Cambrai		378
March 21, 1918	Second Battle of the Somme		320
June 22, 1918	Bucquoy Night Raid		5
Aug. 21, 1918	Bapaume	2 Bns	
Aug. 25, 1918	Second Battle of Arras		
Sept. 27, 1918	Second Battle of Cambrai	1 Bn	
Oct. 23, 1918	Selle	1 Bn	

Messines, June 7, 1917.

Seventy-six new Mark IV tanks were used in this action. The tanks were allotted a role of secondary importance at Messines, the main features of the battle consisting of the employment of a large amount of artillery and the explosion of twenty mines, containing over a million pounds of ammonal, placed under the German front lines on Messines Ridge.

The Germans, evidently suspecting that an attack was coming, used lachrymatory and other types of gas shells to very good effect the night prior to the attack. In this instance the use of gas at night made matters extremely difficult for the tank personnel during the approach march.

The British plan assigned missions to all but four tanks, 72 of the total available being divided between the three front line corps, of which 12 were allotted to the 10th Corps attacking the Oosttaverne line, 28 to the 9th Corps, of which 16 were to attack Messines ridge and 12 were sent against the Oosttaverne line, 32 to the 2d Anzac

Plate 3.
Mark IV.
(Photo by Ordnance Dept., U.S.A.)

Corps, 20 of which were sent against the ridge and 12 against Oosttaverne line. In this battle, the British used 12 supply tanks which the tank troops in France had improvised from earlier combat models.

It has frequently been said that the Messines action was not a tank battle. It is, nevertheless, a fact that tanks were credited with having given the infantry material assistance, especially in the second phase of the battle.

The details of this attack were as follows:

At 3:07 AM the British artillery preparation stopped, the mines were exploded, and the tanks and infantry started for the first objective, which was the ridge. The moral effect of the terrible explosion on the ridge, coupled with the effectiveness of the artillery preparation and the barrage, appears to have made it unnecessary for the infantry to call upon the tanks for much assistance during the first phase except in a few instances. One tank aided its infantry in passing through the village of Wytschaete where a large number of prisoners were secured after a number of machine gun nests were wiped out. This maneuver was repeated at or near Fanny's Farm where the infantry line was held up until the tanks settled with the machine guns.

By noon the first objective had been gained, 25 of the 36 tanks assigned to aid in the first phase having reached the designated line.

At 3:10 PM, 22 of the tanks assigned to the final objective (Oosttaverne line) started with the infantry on the second phase of the attack. Although this objective was taken without difficulty, the Germans soon started several counter attacks and it was during this part of the engagement that the tanks were able, by patrolling the area ahead of the Oosttaverne line, to assist the infantry troops while they were engaged in consolidating the new line. The tanks are credited with having broken up all counter attacks until the British artillery could be brought to bear upon the area in which these attacks were being organized.

Two of the tanks became stuck in this locality but they were so located that they could use their heavy guns to good effect and, moving their machine guns out of the tanks and into shell holes, they formed a very effective strong point which the Germans could not pass. In this part of the action they were able to supply some of the infantry units with machine gun ammunition, and these units then participated in the defense of the area.

The Germans used a great quantity of small arms armor piercing ("K") ammunition upon the Mark IV tanks in an attempt to pierce them, but Britain had profited from the experiences with the Mark I and Mark II tanks and had provided the Mark IV with armor which would successfully withstand that type of fire.

Third Battle of Ypres, July 31 to October 9, 1917.

This action supplies an impressive example of an attempt to use tanks on terrain which was totally unsuited to them. In his history of the World War, John Buchan describes the area about Ypres as follows: "The territory * * * was practically all reclaimed land, including Ypres and as far back as Omer, both of which a few hundred years ago were seaports. Agriculture in this area depended upon careful drainage, the water being carried by innumerable dykes."

Long before this action started, these dykes had all been destroyed by artillery, leaving the area a mere mass of mud and water. Notwithstanding the condition of the terrain over which tanks were expected to advance, a twenty-four day artillery preparation was ordered and carried out. This destroyed what little firm ground there might have been, making the place most difficult for all branches, but especially so for the tanks.

The Fifth British Army was furnished with 216 tanks for this action. The army assignments to the three corps participating were better planned than in some of the previous engagements, 72 being assigned to the 2d Corps, 72 to the 19th, 36 to the 18th, and 36 held in reserve. Each corps assigned its tanks to definite objectives and each corps held out a reserve.

The already serious condition of the terrain was made worse by a heavy rain storm on the day of the attack. In describing this action, General Fuller[1] states that "The Third Battle of Ypres was a complete study of how to move thirty tons of metal through a morass of mud and water."

However, in spite of all difficulties, the infantry first wave and the tanks moved forward at 3:50 AM July 31st, through the rain, mud and darkness in rear of the rolling barrage. Apparently with little effort, the German front line was taken along the entire seven mile front, its occupants withdrawing so rapidly that the British infantry and tanks had great difficulty in maintaining contact with the retreating forces.

After the first line was taken, however, many tanks were found to be unable to move on account of the mud, shell holes, etc., and these tanks were quickly put out of action by German artillery and airplanes.

---

[1] *Tanks in the Great War, 1914-1918*, Colonel (now General) J. F. C. Fuller.

In reading the accounts of this action, the thought occurs that the German plan must have been made with a view of drawing the British and their landships well into the Ypres sea of mud and there dealing with them in detail.

There were only three narrow lanes through which the 2d Corps troops could approach their objectives. Naturally, these lanes became narrow defiles and many tanks were knocked out by artillery on the front of this corps. In spite of all difficulties, the tanks on the 19th Corps front are credited with having reduced the expected number of infantry casualties, with breaking up a number of counter attacks, and with aiding materially in the capture of a number of farms. Likewise on the front of the 18th Corps, the tanks wiped out many machine gun nests, influenced the evacuation of the right bank of the Steenbeek, crushed wire obstacles and reduced several enemy pill-boxes (concrete machine gun emplacements) at St. Julien and Alberta, thereby saving many infantrymen. The story of this first phase of the Ypres action is one of many gallant attempts to use the tank on unsuitable terrain, some of which were successful, many of which failed. The tanks became stuck so often that much time was taken up in unditching, usually under fire from machine guns in airplanes or antitank guns and artillery. In deep holes, the water came up to the top of the engines, and yet the tank crews continued their efforts to unditch their vehicles until their tanks were struck and disabled.

The Germans made a strong counter attack after the second objective had been won and the British had to fall back along a part of the line until at one end of their line they were just in rear of their first objective and at the other end they held a part of the third objective. As a result of these reverses and the great number of tank casualties, the Fifth Army Commander turned in an adverse report on tanks and the question of further tank development appeared to hang in the balance, many of the authorities apparently having failed to realize the fact that in the Ypres area the tanks were as much out of place as gunboats would have been.

Langemarck-St. Julien, August 16, 1917.

Rain storms held up operations until, on August 16th, a small action was planned on this front, in which 12 of the 36 reserve tanks were to take part. The net result, as far as tanks were concerned, was that none of them was able to get to the scene in time to participate. What roads had been built up to the lines were required for other vehicles, supplies, etc., so the tanks had alternately to wade and attempt to swim to the starting point. Not having been designed for this kind of locomotion, they fell by the wayside and the infantry had to attack without them. The infantry was unable to capture a strong nest of pill-boxes to the northeast of St. Julien, which was well concealed and strongly held. These strongholds became the object of the next attack on this same front a few days later, the tanks being assigned the chief role.

Langemarck-St. Julien, August 19, 1917.

A division of the 18th Corps was ordered to attack these pill-boxes, the taking of which by infantry would, according to estimates made at the time, result in from 600 to 1000 casualties. It was decided to make this a tank attack and, at the request of the officer commanding the tanks assigned to this corps, it was arranged that no preliminary shelling would be ordered but that his tanks have the benefit of a smoke barrage. This is the second instance of the use of tanks with the artillery preparation eliminated, and the success of this relatively small action resulted in these tactics being used on a much larger scale at a later date.

In the action of August 19th, 12 tanks were used, of which four were held in reserve. The plan called for the tanks to maneuver to the rear of the pill-boxes, cutting off the retreat of the Germans and attacking the strongholds at their weakest point.

A dense cloud of smoke was put down around the pill-box area by the artillery at 4:25 AM and the infantry and tanks started according to schedule, the tanks in the lead. Following the plan, the pill-boxes were enveloped and captured with a total loss of two tank personnel killed and 10 wounded, and 15 infantrymen wounded. In each case, the garrisons of 60 to 100 men, in attempting to escape from their strongholds, were nearly all shot down by the tank guns or bayonetted by the infantry troops. Some of the Germans surrendered and, in one pill-box, a German officer was found to have been hanged by his command.

In *The Tank Corps,* the authors state that it was in some measure due to the tanks which won the little battle of St. Julien that the Tank Corps owed the opportunity of winning the Battle of Cambrai.

The record of the balance of the Third Battle of Ypres, from August 19th to October 9th, is one of many small attacks, most of which were failures as far as tanks were concerned. Both the Tank Corps and the Infantry made almost superhuman efforts and suffered useless casualties while trying to complete their missions in the Ypres area. Small successes were gained occasionally but at a prohibitive cost in men, machines, and time. The general result was failure, except that every one concerned appears to have learned a costly lesson, to wit: *Tanks must be used on suitable terrain.*

First Battle of Cambrai, November 20, 1917.

In this action, which was fought over terrain suitable to tank operations, the chance for success depended, according to General Fuller's *Tanks in the Great War,* upon the attack being a surprise, the tanks being able to cross the large trenches of the Hindenburg lines, and the infantry having sufficient confidence in the tanks to follow them.

In addition to these factors, this attack involved the passage of the Canal de l' Escaut, and what was at first thought to be an important obstacle, the Grand Ravine. Added to these obstacles were the great bands of well made wire obstacles protecting the Hindenburg trenches.

The three wide trenches of this system proved to be one of the greatest obstacles of all since they were too wide to be crossed by the Mark IV tanks unaided; hence 350 fascines, weighing about one and a half tons each, had to be built. The plan for crossing these trenches is interesting.[1] The tanks were divided into sections of three tanks, an advance guard tank and two infantry tanks, the former having the mission of protecting the last mentioned tanks and the infantry as they crossed the wire and trenches. Since there were three trenches and only three tanks to the section, the arrangement for the crossing operation involved the following maneuvers by the tanks of each section. The advance guard tank passed through the band of wire and, turning to the left without crossing the trench, used all weapons which could be brought to bear from the right side of the tank, as it moved along the trench, to protect the passage of the other tanks and the foot troops following. The first infantry tank approached the first trench, dropped its fascine from the forward part of the tank and,

---
[1] Information obtained from *Tanks in the Great War,* Fuller.

crossing the trench and turning to the left, moved down the right side of the trench and around its prescribed area. The other infantry tank crossed over the fascine of the first infantry tank and, going to the second trench, released its fascine and carried out the same maneuver. As soon as the second trench had been crossed by the last infantry tank, the advance guard tank turned around, crossed both trenches on the fascines already laid and started for the third trench with its fascine ready for this crossing.

Three details of infantry were assigned, the first to operate with the tanks in order to clear the dugouts, etc., the second to block the trenches at certain points, and the third to garrison the captured trenches and protect the approach of the rest of the troops. This part of the plans for the Cambrai action furnishes an excellent example of cooperation between tanks and infantry.

To attain the surprise feature of the general plan, there was no preliminary bombardment; counter-battery work and a barrage of smoke and H.E. was to start at zero hour; there was no change in the airplane activities; no change of troops on the front lines; no registering shots were to be fired by the artillery; all moves were to be made at night; and no reference to the coming battle over the telephones.

To give the infantry troops confidence in the ability of the tanks to cross all obstacles, the two were trained together and the infantry was invited to, and did, build severe obstacles which the tank personnel agreed to cross, and did cross, during the training period.

The cavalry was given a part in this battle and some of the tanks were equipped with grapnels for the purpose of clearing a path through the wire for the horses.

The British Third Army, assigned to the Cambrai attack, consisted of six infantry divisions, a cavalry corps, 1000 guns, and the available Tank Corps of nine battalions with 378 fighting tanks and 98 administrative vehicles, a total of 476 vehicles.

Briefly, the plan of the Third Army was: to break the Hindenburg line, seize Cambrai, Bourlon Wood and passages over the Sensée river, then to isolate the enemy south of the Sensée and west of the Canal du Nord and, finally, to exploit the success in the direction of Valenciennes. In the first phase, the infantry was expected to occupy a line Crevecourt - Masnières - Marcoing - Canal du Nord; the cavalry was then to pass through this line at Masnières and at Marcoing, capture Cambrai, cross the Sensée, capture Paillencourt and Pailluel and move with its right on Valenciennes. During this time the 3d Corps was to form a defensive flank on the right of the Third Army. The cavalry was given the mission of cutting the Valenciennes-Douai line to aid the 3d Corps in moving toward the northeast.

At 6 AM General Hugh Elles led 350 tanks forward and the prearranged artillery fire started. The element of surprise played a big part in the success of the action. The Hindenburg wire and trenches were reached and crossed as planned, much to the surprise of the defenders, and Havrincourt, Marcoing, and Masnières were captured and occupied.

While the passage of the Hindenburg trenches was being made, many interesting incidents occurred. The commander of a tank observed that the infantry appeared to be under fire, but none of the crew could locate the point from which the fire was coming. Finally, three infantry scouts advanced toward the tank by rushes. One of them reached the tank and, with his hat on his bayonet, indicated the direction of the hostile machine guns. This tank had orders to wait until the next tank dropped its fascine into the second trench before trying to cross, it having already dropped its own fascine. However, the infantry was under fire from guns which had been located, so the tank commander decided to attempt a crossing of the second line unaided. This was finally accomplished and the tank made for the machine guns near the crest of a hill. The German gunners made no move to leave their position or cease firing; they continued their fire regardless of the approaching tank until their weapons and one of the gunners were crushed by the tank.

In response to a signal from the tank commander, the British infantry now came forward without losses. Being too far ahead, the tank commander could have waited for the rest of the tanks which were coming with the infantry, but he decided to move on over the crest of the hill. As soon as he had done this, he observed four German field guns which were a short distance away and apparently prepared to fire. The German gunners seemed to be as much surprised as the tank crew. Although no doubt realizing that he was alone and unsupported, and that the guns could go into action before he could move his tank out of their field of fire, the tank commander gave orders for full speed ahead.

With its weapons firing, the tank made for the battery at about 4 m.p.h., its highest speed, and the German gunners soon went into action. The first two rounds were high, then one gun fired short, the aim of the gunners, no doubt, being somewhat influenced by the suddenness of the attack and the excitement of the moment. The tank commander was at first undecided as to whether to zigzag in his approach and thus cause the gunners to re-lay, but, as the tank drew nearer, he decided against this course and continued straight for the center of the battery. The tank was a few yards from the guns when the upper cab was struck by a shell, temporarily dazing the crew. One man was fatally wounded by a shell splinter. The crew recovered and the tank continued, much to the surprise of the artillerymen. In a moment it was among the guns, the fire from its machine guns and the case shot from its six pounders wiping out the remainder of the gun crews.

The infantry and the other tanks arrived and the partly disabled tank moved out to aid in taking the next objective. No casualties occurred among the remaining members of the crew during this part of the action although several were slightly wounded by bullet splash.

The next mission of this tank was to seize a bridge over a canal for the use of the troops following. The route passed through Mairie and, as the tank was passing through this place, retreating German artillery limbers were observed in another street making for the canal and the bridge. The tank commander followed the limbers and ordered his gunners to hold their fire, as he reasoned that the bridge would not be blown up as long as the artillery was on his side of the canal. As soon as the limbers passed across the bridge, the German officer detailed to destroy the bridge came up to see if any more German troops were to use the bridge and found the tank upon it. The tank gunners fired at him but missed as he ran under the bridge to light the fuze. Two of the tank crew quickly followed and shot him with their revolvers before he succeeded in lighting the fuse, thus saving the bridge. The tank moved forward into position to cover the approach to the bridge and await the coming of the infantry.

When the infantry and another tank arrived, preparations were made to return the tank that saved the bridge to its rallying point as the crew were by this time exhausted. Deciding to give them a little more time to rest before starting back, the tank commander withheld the order to move back. Before the order was issued he was requested to aid an infantry company which was being held up by fire from a nearby ridge. Knowing he had only enough gas to reach the rallying point and that, by this time, all other tanks had gone back and, consequently, he would have no tank support, and believing that his crew was physically incapable of the additional effort necessary to take the strong point on the ridge, the tank commander at first decided against making this additional effort. As the infantry officer who had made the request moved away to return to his company, the tank commander changed his mind and called for three volunteers from his remaining six men. The six men responded. As soon as the tank reached the hill it came under very heavy fire from all directions. The machine gun being operated by the tank commander jammed and, as the tank was now close to the German troops, he opened the front flap and fired at them with his revolver. Lead splash from bullets striking the open flap blinded the commander, but case shot from the tank six-pounders drove the German troops from the ridge. Soon thereafter three shots from a German field gun struck the tank and set it on fire. His vision having improved somewhat by this time, the tank commander moved his men from the tank and, taking charge of some of the many German guns left on the ridge, prepared to hold the position until the arrival of the infantry. With these weapons and the small tank crew, which was augmented by the arrival of an officer and three men from an infantry company, three counter attacks were stopped and the ridge was held until the rest of the infantry arrived.

Graincourt was the farthest point reached by the infantry troops during the 20th. The tanks continued on from this place but, due to the exhaustion of the foot troops by this time, no further gains could be made and held. At many points during the advance, heavy fighting took place. Among the most interesting encounters was the duel near Lateau wood between a tank and a German 5.9 inch howitzer. After the latter had, at close range, blown off a part of one sponson and before the howitzer gunners could load again, the tank struck the big gun and crushed it.

At all points along the advance the infantry and tanks cooperated as planned except in the case of one division operating near Flesquières, which, according to General Fuller, had devised an attack formation of its own. This formation prevented the desired close cooperation between the division and the tanks and, as the tanks moved forward, they came over a ridge and found themselves under direct short range artillery fire from a single gun which is said to have knocked out several tanks before it was silenced. These tanks evidently came over the ridge one at a time in plain view of the gun.

The supply tanks were advanced to their new positions, the wireless tanks reported the capture of Marcoing, and the tanks assigned to clear the wire for the cavalry opened up wide passages as directed. The tanks completed their part of the program by 4 PM. and the successful conclusion of the first day's efforts had more than justified the faith of the tank advocates in these vehicles.

As no provision had been made for tank reserves in the general plan, the best of the remaining tanks and crews were formed into companies for use on the 21st. Twenty-five tanks aided in the capture of Anneux and Bourlon Wood and 24 tanks helped capture Cantaing and in the attack on Fontaine-Notre Dame. At the latter place, 23 tanks entered the town ahead of the infantry. The Germans defended their position from the tops of houses, firing at the tanks and throwing bombs on them. The British infantry was so exhausted that it could not support the tanks and take advantage of the opportunity provided by them, so the tanks had to withdraw from the town.

Infantry and tanks captured Bourlon Wood. The tanks then went on toward the town of Bourlon nearby, but, owing to casualties among the infantry, the troops available were not sufficient to capture and hold Bourlon. Tanks and infantry attacked both Bourlon and Fontaine-Notre Dame on the 25th and 27th but did not succeed in taking and holding either place. The plans made for the employment of cavalry were not carried out.

The following incidents in the defense against tanks in Fontaine, published in *Taschenbuch der Tanks,* are given to illustrate the efficient methods used by both the offense and defense in this early instance of street fighting.

The author and leader was Lieutenant Spremberg, commander of the 5th Company, Infantry Regiment 52.[1]

My aim was the village entrance to Fontaine. Should the first tank succeed in coming out of Fontaine, our battalion was lost, since it would be subject to the flanking fire of the tank on an open field. With twenty men of my company I wheeled somewhat to the right, ran through the connecting trench on the double, in order to reach the first house before the tank arrived. My men with full packs, heavy lumps of clay on their feet, rushed after me. No one held back, as each one realized his task at this moment.

We saw the tank about 100 meters ahead of us advancing and holding the entire village street under its fire. However, we quickly sprang into and behind the yards. We had found a hand grenade dump in a previous assault on the village and tried at first to throw the hand grenades under the tracks of the tank. That succeeded. The single grenades, however were too weak in explosive ability. I then ordered that empty sand bags be brought and four hand grenades to be placed in them, with one grenade tied near the top of the bag so that only the firing spring showed. In the meantime the tank, which had stopped, was kept under steady rifle fire, particularly the eye slits, so that my assault group could work to better advantage.

Then came a favorable moment and Musicians Buttenberg and Schroeder, both storm troopers, rushed upon the firing giant and, from throwing distance, tossed two bunched charges under the tracks. A single explosion, the tracks on the left side flew in the air, and the tank stood still. At this there was a cheer from our little group. An approach was not to be thought of since the tank held everything under its fire. In a few minutes the fire ceased. Suddenly a second tank appeared, armed with cannon, and opened fire, penetrating the lower house walls so that we had to flee into the farm yard. In spite of that we placed it under rifle fire and saw to our dismay that it moved to the right side of the tank which we had disabled. Since we could not cross to the other side of the street, we could do it no harm even though it was only 10 meters from us. What happened? The second tank evacuated the crew of the first and, firing constantly, departed like a roaring lion.

Then came information from my noncommissioned officer post under 1st Sergeant Lutter whom I had placed at the east exit from Bourlon Wood, that sixteen tanks were advancing against the west exit of Fontaine from Bourlon Wood. Volunteers to report to the regiment! Noncommissioned officer Maletzki, who made his way through a heavy barrage, requested artillery fire. The tanks, some of which were seen at 10:30 were destroyed. The artillery, particularly the heavy artillery, had completely put them out. So it was with the English infantry, who, at about 2:00 in the afternoon, were on the defile of Bourlon Wood near the western village entrance of Fontaine in dense column of march. This time the heavy artillery fulfilled its obligation. The English infantry was destroyed. With that, the English attack was temporarily halted.

(After 2 o'clock) We all know that the Englishman is tough. Near 3 o'clock they organized another attack with 80 tanks deeply echeloned on a narrow front, and attacked energetically, single tanks penetrating as far as the village of Fontaine. Their watchword then was clearly, "Take Cambrai, cost what it may!" The first tank that came into Fontaine was C-47. We sat in a house and had prepared ourselves well with armor piercing ammunition and bunched grenades. "They are coming!" was heard. My orders went to each subordinate leader. We could hardly raise our heads over the lower window sills so heavy was the enemy machine gun and shell fire. It was necessary to flee to the yard since the shells fired from 10 meters easily penetrated the walls of the house. We let the tank go by and opened fire at nearly 20 meters on the eye slits in the rear walls, at leisurely but continuous fire with armor piercing ammunition. Then I saw a reservist firing with trembling hands from a window and hitting nothing. Taking his gun, the first shot cracked, and a yellow flame came out of the tank. I repeated, shot once more, and already my men were yelling, "Hurrah, Lieutenant, you hit it!" I saw two bright flames leaping from the rear. Everyone ran out, covered by the houses, behind the still-moving tank until it suddenly began to smoke and then stopped. The tank crew fired wildly at us so that none of my men could approach. After about five minutes the doors of the tank opened. Believing that the crew wished to surrender, we held our fire. But no, the crew had no thought of surrender, but continued to fire like fury, as they had only wanted fresh air. My command was repeated to direct fire against the now closed

---

[1] From *Taschenbuch der Tanks, 1927,* Dr. Fritz Heigl.

doors. This incident perhaps lasted seven minutes. All of a sudden everything in the tank was quiet; firing ceased. Carefully we ran up, ripped open the door and found truly that the entire crew had met their battle death.

New tanks were reported by 1st Sergeant Lutter. One of these monsters came along the road from Bapaume as far as the schoolhouse in spite of the fact that we continued to ply it with armor piercing ammunition. At this point on the road we had a machine gun that we had taken over from the 46th Regiment during a counter attack. Point blank fire was placed on the right side of the tank up to 5 meters, but then we had to flee to the house. The tank moved suddenly to the south part of the village and began to smoke. We followed; suddenly the crew threw smoke bombs and utilized the opportunity to escape to the cellars. We took possession of the tank but were unable to find any trace of the crew, who surely were provided with civilian clothing by the residents who remained.

Suddenly (at 4:30), behind us is heard a characteristic and well known din. We saw at the road bend toward Cambrai, and awaited with delight, two motor guns. Commanding them was a keen captain who reported to me immediately and became oriented. At once the captain placed one gun at the road bend toward Bourlon Wood and placed one, concealed, near the Bapaume road.

Soon, through the defile, as into a rat trap, from Bourlon Wood came nine tanks toward Fontaine. The gun crews stood to their guns, burning with eagerness. The captain commanded "Steady men, it will soon be time." Now the tanks are climbing out hardly 100 meters away. The command rings, "Rapid fire!" The first tank rears upward, those following halt. One direct hit after another strikes the tank company. The crews, which were left alive, fled and abandoned sound tanks. For us it was a rare fine moment. All praise to the motor guns and their personnel.

In Fontaine now were concentrated all available troops. The English appeared to have given up their desire for further advance.

In a crater on the road to Bapaume lay a tank. Members of every possible organization had, with tremendous losses, attempted to storm the colossus. Constantly, new troops of the steadily increasing units, stormed it and were regretfully required to retreat because of heavy losses. The tank crew defends itself well. They mow down in every direction with their machine guns. 30 to 40 brave field grays, some dead, some wounded, lie about the monster. My 1st Sergeant Luban and Musician Schoenwetter bend all their skill and succeed. Crawling along, using every crater for cover, they approach the monster and strike against the doors with their rifle butts. The doors were opened and a single Englishman appeared. The rest of the crew were already dead. The Englishman who had defended himself so bravely was taken prisoner. One machine gun, the only one still capable of firing was taken as a trophy.

By November 28th the tank units had become so depleted that it was decided to withdraw two of the tank brigades. This plan was carried out and, while it was in progress, the remaining tanks and exhausted infantry had to bear the brunt of the strong German counter attack, which started on November 30th, and which, due to lack of preparation by the British for such a contingency, was destined to turn the tables and practically wipe out the advantages gained in the brilliant victory of the 20th and 21st. By way of comparison of results, it was noted that an advance of 10,000 yards, from a base 13,000 yards wide, was made on the 20th in about twelve hours, while at the Third Battle of Ypres, it required three months to effect a similar advance. The tank force at Cambrai numbered about 4200 men. The casualties in the 3rd and 4th Corps during the 20th exceeded 4000.

An interesting view of the British use of tanks at Cambrai was published in the *Militär-Wochenblatt*, December 25th, 1929, and illustrates a German view of this famous tank action:

The British failed to make the proper tactical use of tanks at Cambrai in November 1917. Instead of being concentrated in the main attack they were distributed along the whole front. Each division used its tanks differently. The principal mission was that of accompanying the infantry in breaking through and rolling up the first position. After breaking through, 30 tanks waited at Marcoing for the cavalry to arrive. Four hours later the German 107th Division, advancing from Cambrai, succeeded in blocking the bridges over the canal and established a new front between the canal and Bourlon Wood. Had the British tanks continued on their advance, they would have engaged the German artillery, captured the crossing at Marcoing and struck the reserves before they could have organized for defense. The Battle of Cambrai therefore brought out the following lessons in the employment of tanks:

(1) Their action must be concentrated in the main blow, which must be directed by the higher command.

(2) The missions assigned must exploit the principal characteristics of the weapon—armor, fire power, and mobility.

(3) Their armor permits their use as battering rams to assist the infantry in penetrating and fighting its way through the hostile position.

(4) Their fire power and protection enables them to engage the hostile artillery in rear in close combat.

(5) Their speed facilitates their use, after breaking through, to block the main routes in rear and engage the hostile reserves.

(6) These three missions cannot be executed at the same time, hence tanks must be used in waves, each strong enough to carry out a specific mission which must be assigned at the jump off.

Let us see how the British should have used their tanks at Cambrai to carry out the above principles. In the first place they should have been concentrated in a main effort of 5,000 to 6,000 yards between the Bois de Couvillet and Havrincourt. The mass of tanks should have been organized in five waves, the first four containing 80 tanks each, and the last, the remaining 60 tanks.

The first two waves, without regard to the infantry advance, should have moved out as leading tanks and, rapidly penetrating the hostile position, should have pushed through to engage the hostile artillery, the first wave making the actual penetration and what is left of this wave assisting the second in the close combat with the hostile artillery. The third wave is the accompanying wave, assisting the infantry in its combat against the strong points through the hostile position.

The fourth wave is launched through the gap and through the zone of the hostile artillery to engage the hostile reserves. The fifth wave is at the disposition of the high command as the situation may require, either to roll up the flanks of the broken front, to reinforce the leading waves, or to initiate the pursuit.

Had the British thus employed their tanks at Cambrai, the battle would have had an entirely different outcome. The first and second waves advancing at 7 AM under a screen of smoke and the protection of artillery fire on the dominant terrain in the hostile zone, (would have) quickly penetrated the hostile position. Upon arrival on line at Ribécourt, the left half should have finished off the German batteries not silenced by counterbattery fire, and then pushed on to engage the German batteries west of Flesquières and at Anneux. The right half should have cleaned up the artillery being counterbatteried west of Marcoing, and then pushed on to secure the canal bridge east thereof. In the meantime, the third wave would have assisted the infantry in overrunning the hostile position before the enemy could have reorganized his defense. The fourth wave would then follow, pushing right through the center in the direction of Cantaing where it would have arrived about 11 AM just as the 107th Division, in reserve and quickly alerted at Cambrai was advancing to engage in the fight. This mass of tanks would soon have been reinforced by the fifth wave and the break-through at Cambrai would have been successful, for no other reserves were available until the next day.

The above criticism, by one who was no doubt familiar with both the British and German activities on November 20th, points out some of the recognized errors made by the British in their plans for the Cambrai action. This criticism, which was written 12 years after the date of the action, by which time the lessons of the Great War were more clearly understood by all, would, no doubt, be more acceptable to the members of the Royal Tank Corps who participated in the action, had the author referred to the failure of higher authority to provide sufficient infantry to take advantage of the many opportunities provided by the vigorous attacks of the tank personnel, and to hold the territory won by the novel, even if faulty, tank tactics; and had he referred to the failure of the cavalry divisions to carry out the missions assigned. Except as noted above, the German criticism seems to be well founded.

General Rockenbach, the Chief of the U. S. Tank Corps, reported that the salient points brought out by the successful use of tanks at Cambrai, when a penetration of nearly six miles was made on a seven mile front in 12 hours, were as follows: economy in men per weapon, in men per yard of front, in casualties, in artillery personnel, in cavalry personnel, in ammunition, manufacture, transportation and tonnage, in labor on the battlefield, in property handled, and in time.

Second Battle of the Somme, March 21, 1918.

At this time the British Tank Corps in France consisted of five brigades, or thirteen battalions. These units were equipped with 320 Mark IV and 50 medium tanks. Contrary to the request of the Tank Corps, in December 1917, that it be concentrated at Bray-sur-Somme, the Corps was directed, in January 1918, to occupy a number of stations between Roisel and Bethune, thus scattering the battalions over a frontage of sixty miles. By this order, the British high command placed the Tank Corps at a disadvantage for unity of supervision, supply, and control prior to action, and decreased the possibilities for concerted action against a major hostile offensive. The lesson of concentration had apparently not yet been learned by the authorities, as far as tanks were concerned. Hence the tank battalions were under the command of many different divisions and in widely separated localities when the German attack came on March 21st, and they were used, or withheld from action, in accordance with the ideas of the more or less inexperienced staffs of the various divisions. Frequently the tank battalion commanders had to act upon their own initiative. No records are available to show exactly how many of each of the two types of tanks were employed in the defensive actions during this great German advance. The Whippets did not come into the fight until March 26th.

The German attack was preceded by an intense bombardment placed upon the British forward lines, headquarters, artillery positions, lines of communication and supply, starting at 4:45 AM on the 21st. In this action the Germans made use of special methods of infiltration in which they had thoroughly trained the forces used.

Following their barrage, which included gas shells, small bodies of these special troops worked through the British front lines under cover of fog. These troops carried small mortars and light machine guns. Close behind these groups came field batteries supporting the advanced groups and helping to make gaps in the British lines through which the infantry troops came. A signal system of flares and rockets kept the attacking troops oriented in the fog and enabled them to cooperate with each other. When one regiment had reached its limit of operation, another was ready to take its place. So the advance continued by waves, a good example of an attack organized in depth.

The British lines on the fronts of the Third and Fifth Armies gave way, their command and supply facilities became disorganized, and their efforts at defense accomplished little during the 14 days following, while a penetration of about 40 miles was made by the Germans. British authors have stated that the advance was finally checked, not by British resistance, but because the Germans had reached the limit of their supply facilities.

During this great retreat, in the confusion and uncertainty as to the location of the various headquarters, of the lines themselves, and supply points, the tanks apparently were unable materially to affect the outcome. All they could do was temporarily to check the advance at various points by counter attacks. This they did whenever they were allowed, and so long as their supplies of fuel, oil, and ammunition held out. There were not sufficient tanks on the ground to break up the attack, so they covered the withdrawal of the infantry when they were allowed to enter the action. Major Williams-Ellis, in his book *The Tank Corps,* states that only 180 tanks, of the 370 which were in condition to fight, saw any action, and intimates that in many cases the infantry commanders held their tanks out of action, moving them from point to point until the fuel was exhausted or the tank disabled by some mechanical trouble. They then had to be blown up and abandoned without having had a chance to fire a shot.

A few instances of the use of tanks during the retreat are given: About noon on the 21st a portion of the 4th Tank Battalion, in an action near Emilie, in company with two companies of infantry, captured a battery of heavy guns. Later these tanks took part in the counter attack on Ronssoy Wood.

The 8th Tank Company assisted an infantry brigade on the Third Army front in a counter attack on the village of Doignies. The attack was not started at the hour set and it was practically dark when the objective was reached. The tanks cleared the village but, owing to the darkness, the infantry did not fully occupy the area so the tanks were obliged to withdraw.

The 2nd Tank Battalion and two companies of infantry were ordered to counterattack on the 22nd near Beugny. When the time came for the action, the infantry could not be spared, so the tanks moved out alone through an artillery concentration promptly put down by the Germans. After putting a battery of field guns out of action, they got into position along the German trenches which they enfiladed, causing many casualties and forcing the garrison to retreat.

The following is quoted from an article by Major Heigl:

> It became very disagreeable for us when a tank crossed our line a number of times and set to work against us with machine gun and artillery fire. * * * * * This beast ran systematically along our line and subjected one of our sections to heavy casualties. Four men who were standing in front of their gun were directed from out of the tank to walk alongside, whereupon it turned about, being covered in part against our fire by these wounded men.
> The Prussian Infantry Regiment 52, * * * * was obliged to withdraw under heavy losses, and this when the English accompanying infantry had not participated, as is clearly shown by the regimental history. The tank can therefore also take prisoners and did that here, alone.

Although 17 of the 30 tanks used were hit and 70% of the tank crews became casualties, this action checked the German advance at this point.

The 4th and 5th Tank Battalions were used on both sides of the Cologne river to cover the withdrawal of part of the Fifth Army. Counter attacks delivered by these tanks at Epehy, Roisel and Hervilly gained valuable time for the retreating forces. Several times during these actions the German ranks broke when the tanks appeared, and the frequency with which this occurred during the German advance indicates that the whole attack must have broken down if the British had possessed tanks in numbers appropriate to the two armies defending the line.

The 5th Battalion arrived at Brie Bridge south of Peronne on the 23rd. The bridge was struck and demolished shortly after their arrival and, as they were out of gas, all but three were blown up to prevent their capture. Of the three, one succeeded in crossing the damaged bridge and the other two escaped through Peronne. These three were lost on the 24th.

The tanks of the 8th Battalion took part in a successful action on the 24th while aiding the withdrawal of the 2d Division. They caused many casualties among the German infantry and checked it long enough to allow the 2d Division to withdraw from its difficult position. The German advance was not again taken up until the artillery had been brought up to oppose the tanks.

The Mark IV tanks were used in a great many more brief engagements during this retreat, usually in small numbers. A large percentage of those used were lost due to lack of fuel. When tanks were abandoned due to this cause, or due to mechanical trouble, the machine guns were removed with the remaining ammunition and the crews were organized into Lewis gun ground crews. These joined the infantry and saw much service in this capacity. The 5th Battalion alone sent 84 machine guns into the line, and these crews held the line at Mont Rouge during an attack in force. At one time, a mile and a half of the British line was held by machine gun crews taken from abandoned tanks and from tank organizations which did not get into the action with their tanks.

THE MEDIUM A TANK.

Medium A (Whippet). Produced in 1917 by William Foster and Company, Ltd. Total production, 200.

Crew: 3.
Armament: Four machine guns in turret, at front, rear, and each side.
Armor: 0.2 in. to 0.55 in.

Maximum speed: 8.3 mph.
Suspension: Rigid; the 6 center rollers being on ball bearings.
Tracks: Width 20½ in., pitch 7½ in.
General arrangement: Fuel and cooling system in front, engines in center, crew and final drive in rear, driver in right side of crew compartment.
Dimensions: Length 20 ft.; width 8 ft. 7 in.; height 9 ft.
Weight: 15.7 tons.
Engines: Two, Taylor, 4 cylinder, 45 HP each, forced water cooling. One engine for each track.
Horsepower per ton: 5.7
Transmission: Sliding gear, a transmission for each engine, 4 speeds forward, 1 reverse.
Obstacle ability: Trench 7 ft.; stream 3 ft.; slope 40 degrees; vertical wall 31.5 in.; tree 14 in.

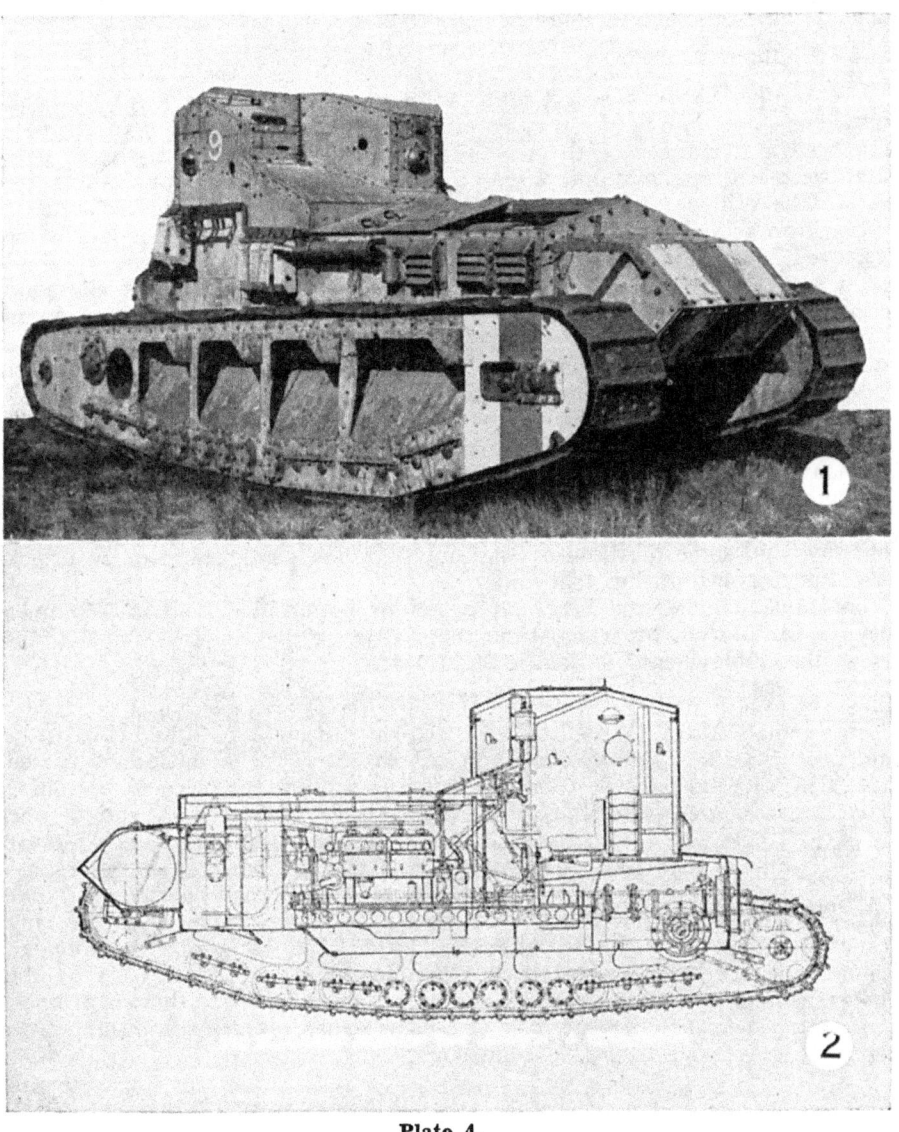

Plate 4.
1. **Medium A (Whippet).**
(Photo by F. Mitchell, Bovington Camp, England.)
2. **Medium A (Whippet).**
(From *Tanks* by R. Krüger.)

Fuel distance: 40 miles. **Fuel capacity:** 84 gallons.
Special features: Hotchkiss air cooled machine guns used. Upper portion of track supported by rollers. Steering accomplished by variations in engine speed for each track by means of steering wheel. Engines mounted side by side. Drive shafts could be independently driven or locked together. Radiator, two cooling fans, and fuel tank in front. Turret cooled by separate fans. Intended for use with cavalry.

The Medium A Tank. List of Actions in which Engaged.

Date	Battle	Units	Tanks
March 21, 1918	Second Battle of the Somme	1 Bn	50
Aug. 8, 1918	Amiens	2 Bns	96
Aug. 21, 1918	Bapaume	2 Bns	96
Aug. 25, 1918	Second Battle of Arras	1 Bn	
Sept. 27 1918	Second Battle of Cambrai	1 Bn	
Oct. 17, 1918	Selle	1 Bn	
Nov. 4, 1918	Maubeuge	1 Bn	

The Medium A or Whippet tanks were issued to the 2nd and 3rd Tank Battalions some time before they entered the Somme action. Engine trouble developed in a number of them while they were well forward where spare parts were not available and where the crews had no time for making repairs. The Whippet units answered many calls for assistance before actually getting into action and a good many of those in need of repairs had to be destroyed before they could fire a shot.

On March 26, 1918, twelve of these tanks from the 3d Tank Battalion received orders to move through Colincamps to investigate the situation in that village. At the far side of the village they met about 300 Germans en route to this place in close formation. These troops retreated in disorder under the fire of the Whippets' machine guns, some of them surrendering to the British infantry following the tanks. While moving in the direction of Serre, the Whippets met and turned back several enemy patrols. These German troops were making an enveloping movement against a point (Hebuterne) where there was a break in the British line.

The Whippets were out on patrol duty a number of times during the next few days and some of those of the 2nd Battalion participated in an attack near Bouzencourt during that time.

The tank actions were few and of minor importance from the 27th to the 31st of March, by which time the Battle of the Somme was over as far as the employment of tanks was concerned.

Bucquoy, Night Raid, June 22 to 23, 1918.

Five female Mark IV tanks were used on the night of June 22-23 in a raid near Bucquoy with five platoons of infantry. The operation started at 11:25 PM. Its purpose was to destroy or capture the German personnel in certain localities. In this action, the infantry advanced ahead of the tanks until held up in "No man's land" by machine gun and trench mortar fire. The tanks then advanced through this fire and carried out their missions, causing a number of casualties among the German garrisons. One tank was attacked by numerous Germans but they were repulsed by the tank crew with revolvers. During the action, this tank recaptured a wounded platoon commander who had been taken by the Germans. None of the tanks was damaged by the barrage and all returned safely to their own lines. This action indicated the possibility of using tanks, over short distances, in night actions, under favorable conditions.

CHAPTER IV.

TANK COMBAT HISTORY. BRITISH (Continued).

THE MARK V TANK.

Mark V. Produced in 1918 by William Foster and Company, Ltd., and Metropolitan Carriage Wagon and Finance Company, Ltd. Total production, 200 male, 200 female.

Crew: 8.
Armament: *Male;* Two 6-pounder (57 mm.-2.24 in.) guns and four machine guns. *Female;* Six machine guns.
Armor: 0.2 in. to 0.47 in.
Maximum speed: 4.6 m p h.
Suspension: Rigid; rollers.
Tracks: Flat steel plates with single grousers, width 26½ in., pitch 7¾ in.
General arrangement: Driver and tank commander in front, engine in rear of driver, armament at front, sides, and rear, final drive in rear.
Dimensions: Length 26 ft. 5 in.; width (male) 13 ft. 6 in., (female) 10 ft. 6 in.; height 8 ft. 8 in.
Weight: Male, 31.9 tons; female 31 tons.
Engine: Ricardo, 6 cylinder, 150 HP, forced water cooling.
Horsepower per ton: Male 4.7; female 4.8.
Transmissions Planetary, 4 speeds forward, and four reverse.
Obstacle ability: Trench 10 ft.; stream 3 ft.; slope 35 degrees; vertical wall 4 ft. 11 in.
Fuel distance: 25 miles. **Fuel capacity:** 108 gallons.
Special features: Hotchkiss machine guns substituted for Lewis. One machine gun placed for fire to the rear. Entrance door behind the rear cab. Air louvers on each side of tank near rear using cool air for engine radiator. Engine encased in sheet metal. Cooling fan to force warm air out through roof. Three fuel tanks mounted outside at rear, protected by armor. Vacuum feed. Unditching beam could be attached from rear cab. Beam held in position on rails by sprung brackets. In this tank the controls were improved so that one man could drive. A few female tanks were fitted with male right sponsons and called Mark V Composite.

Plate 5.

Mark V, Male.
(Photo by Ordnance Dept., U.S.A.)

The Mark V Tank. List of Actions in which Employed.

Date	Battle	Number Units	Employed Tanks
July 4, 1918	Hamel	2 Bns	60
July 23, 1918	Moreuil	1 Bn	35
Aug. 8, 1918	Amiens	8 Bns	288
Aug. 21, 1918	Bapaume	9 Bns	about 108
Aug. 25, 1918	Second Battle of Arras	2 Bns	
Sept. 17, 1918	Epehy	3 Bns	
Sept. 27, 1918	Second Battle of Cambrai	6 Bns	
Oct. 17, 1918	Selle	1 Bn	
Nov. 4, 1918	Maubeuge	2 Bns	

Hamel, July 4, 1918.

The new Mark V tanks were used for the first time during this action. There were 60 fighting tanks in all, of which 48 were employed in the first wave and 12 in what was termed a "reserve wave." In addition to their fighting missions, these tanks carried ammunition and water for the infantry. Four converted gun carriers were used as supply carriers to transport engineer and other material. Each of these carriers delivered 12,500 pounds of supplies to within 500 yards of the final objective, arriving within thirty minutes of the time that this objective was captured. Only 24 men were used with the carriers. Assuming a load of 40 pounds per man, over 1200 men would have been required to carry the same amount of supplies.

The artillery missions consisted of harassing fire prior to zero hour, a rolling barrage for the advance, smoke screens placed on three localities known to be dangerous, and a standing barrage ahead of the final objective during the consolidation thereof. The harassing fire was carried out at the same time on previous mornings to get the Germans accustomed to it. On the morning of the attack, the harassing fire, together with a barrage fire, provided a noise to mask the noise of the moving tanks for an interval of twelve minutes, during which the tanks covered their approach march of 1200 yards. In addition to these arrangements, a flight of airplanes was detailed to accompany and cooperate with the tanks.

In this action the tanks were supporting the 4th Australian Division. It was the first time the Australians had consented to work with tanks since the action at Bullecourt in April of 1917 and, remembering their experiences at that time, the Australians specified that the tanks were to follow the infantry and were to take the lead only when the infantry was held up. Prior to the action, infantry and tank officers studied the plans together and agreed upon measures for mutual assistance in the accomplishment of their missions.

The infantry and tanks being in position at 3:10 AM, the barrage on the German front line continued for four minutes and, as it lifted, the assault troops and tanks moved forward. The increased maneuverability of the new tanks and the ease with which they could be handled, together with the initiative and energy of the infantry, brought a quick and economical victory. The Germans fought gamely and, where tanks were not near enough to prevent it, German machine guns held up the Australian infantry, inflicting severe casualties. This condition was quickly reversed, however, upon the arrival of the tanks and a great many of the German weapons were run over and crushed during the action. The tanks used to support the infantry battalions that had been given the mission of passing around the Vaire and Hamel woods and the village of Hamel, guarded the flanks of these troops during this maneuver.

One tank commander, after aiding his infantry to reach its final objective, hunted out and destroyed machine guns between this objective and the protective barrage. Noticing two field guns in action against the Australians, he attacked them, running over one and capturing the other. Observing next that machine guns were firing through the barrage while the position was being consolidated, he drove his tank through the British barrage and was in action against these guns when his armor was pierced by a bullet from a German antitank gun. This damaged some of the machinery and caused this very energetic tank crew to return to its own line.

All objectives were taken on schedule. The cost in infantry casualties was 672. The tanks had 16 men wounded and 5 tanks hit. The latter were all salvaged and returned that night. During the 2500 yard advance, over 200 machine guns were destroyed and 1500 prisoners captured. This successful action demonstrated the ability of tanks, when properly used, to reduce casualties among the infantry and, from this time on, the Australians had confidence in them and desired their assistance.

Moreuil (or Sauvillers), July 23, 1918.

In this action the British 9th Battalion supported the 3rd French Infantry Division in the attack on St. Ribert wood and against the German batteries near St. Ribert. Having originally 42 Mark V tanks, the battalion went into action with 35, the remaining 7 being mechanically unfit after their move of about 16 miles under their own power from their training area.

In preparation for the action, tactical training operations were carried out by the tank troops and French infantry units, combined reconnaissance of the terrain was made, and all details of the attack were settled. The attack was preceded by an artillery preparation of short duration, which included heavy counter-battery work. A rolling barrage was arranged, which was to include smoke and high explosive shells. The tanks were used in sections of three, two in front and the third in support of the leading tanks. The infantry was organized in small groups and followed the tanks and a French infantryman who could speak English became a temporary member of each tank crew. Twelve tanks and four infantry battalions were assigned to the first objective, 24 tanks and four infantry battalions to the second, and all surviving tanks and a strong infantry force were to attack the third objective.

The advance started at 5:30 AM. Arrachis wood and Sauvillers were occupied with little difficulty and the first objective was gained, although it was necessary to wipe out several machine guns in these localities. Two tanks were put out of action by the German barrage during this part of the fight.

Sauvillers wood was too thick for the tanks to pass through, so some of them, used in the attack upon the second objective, passed around and fired broadsides into it. Other tanks advanced toward the St. Ribert wood but found that the infantry did not keep up with them, so they turned back to regain touch. While seeking their infantry, six of these tanks were put out of action by direct hits from German artillery located near the St. Ribert wood.

The second objective included the clearing of the area north of Sauvillers wood and the capture of a portion of the Harpon wood. The first part of this mission having been accomplished, an attack was made on the woods by a company of tanks and a battalion of infantry. While nearing the wood, one of the tanks located a German battery. One of the six-pounders in this tank scored a direct hit on one of the German guns and then, using its machine guns, the tank crew drove the gun crews away from the other guns, thus putting the whole battery out of action. While trying to get into position to tow one of the captured guns back, this tank was put out of action by two direct hits.

The attack upon Harpon wood was successful, however, and posts were established in this wood. A German shell put another tank out of action, and the tank commander, in order to save the following tanks from being struck by the same gun, left his tank with a number of smoke bombs with which he prepared a smoke screen in front of the following tanks. Under cover of the smoke they changed direction and escaped the enemy guns. This officer then went back to his tank, under heavy fire, and demolished it.

The third objective was captured before the action ended that evening. The French were well pleased with the cooperation and aid received from the British tanks. The French had 1891 casualties in all three divisions used in the action, of which approximately one-third occurred in the 3rd Division, which was the only one supported by tanks. This division, however, was given the largest system of enemy defense to overcome. 1858 prisoners, 5 guns, 45 trench mortars and 275 machine guns were captured.

THE MARK V STAR TANK.

Mark V Star. Produced in 1918 by various manufacturers. Total production, 642.

Crew: 8.
Armament: *Male;* Two 6-pounder (57 mm.-2.24 in.) guns and five machine guns. *Female;* Seven machine guns.
Armor: 0.24 in. to 0.59 in.
Maximum speed: 4 mph.
Suspension: Rigid; rollers.
Tracks: Flat steel plates with single grousers, width 26½ in., pitch 7¾ in.
General arrangement: Driver in front, engine in front of center, tank commander in rear of center, armament on all sides including 1 machine gun in commander's turret, final drive in rear.
Dimensions: Length 32 ft. 5 in.; width (male) 13 ft. 6 in.; (female) 10 ft. 6 in.; heighth 8 ft. 8 in.
Weight: Male 37 tons; female 36 tons.
Engine: Ricardo, 6 cylinder, 150 HP, forced water cooling.
Horsepower per ton: Male 4.1; female 4.2.
Transmission: Same as Mark V.
Obstacle ability: Trench 14 ft.; stream 3 ft. 3 in.; slope 30 degrees; vertical wall 4 ft. 10 in.; tree 22 in.
Fuel distance: 40 miles. **Fuel capacity:** 111 gallons.
Special features: Ventilation poor. Maneuver difficult due to increased length. Twenty-five additional men with weapons could be carried but 14 was the normal number. Commander communicated with driver by speaking tube and gave range and deflection to gunners by Bowden wire controls. Semaphores and flags used for outside communication. Intended as tank or as a troop carrier.

The Mark V Star Tank. List of Actions in which Engaged.

Date	Battle	Number Employed	
		Units	Tanks
Aug. 8, 1918	Amiens	2 Bns	72
Aug. 21, 1918	Bapaume	2 Bns	72
Aug. 25, 1918	Second Battle of Arras	1 Bn	
Sept. 17, 1918	Epehy	1 Bn	
Sept. 27, 1918	Second Battle of Cambrai	4 Bns	
Oct. 17, 1918	Selle	2 Bns	

Amiens, August 8, 1918.

This action was carried out by the Fourth British Army, consisting of the Canadian Corps, the Australian Corps, the 3rd Corps, and a cavalry corps of three cavalry divisions. Eight battalions of Mark V tanks, two of Mark V Star, and two of Whippet tanks were allotted to the Fourth Army. With the exception of the 1st Tank Brigade, which was still equipped with the Mark IV tanks, the entire Tank Corps (in France) was used at Amiens. The Mark V Star tank battalions had 36 tanks each, the Whippet battalions had 48 tanks each, and the Mark V battalions normally had 36 fighting and 6 training tanks each. Approximately 450 fighting tanks were thus assigned for this action, but some of these were held in mechanical reserve and one battalion was held in general reserve. In addition to the fighting tanks, 118 supply tanks were used, including 22 of the converted gun carriers.

Since tanks were to be featured in the attack, no preliminary artillery preparation was permitted. The plan called for the starting of the barrage at zero. Eighty-two brigades of field artillery, 26 brigades of heavy artillery, and 13 batteries of heavy guns and howitzers were allotted for the attack. Most of the heavy guns were assigned to counter-battery work. The field artillery was to move up and give close support to the infantry. Another mission given to the artillery was to provide a noise barrage to cover the approach march of the tanks. To aid in the rear area movements after the action started, the 2nd Tank Field Company was detailed to keep the Berteaucourt-Thennes road clear of obstacles and to provide crossings over the Luce river.

An interesting innovation was the plan to have the 1st and 15th Tank Battalions (equipped with the Mark V Star tanks) carry to the 3rd objective

Plate 6.
Mark V, Star.
(Photo by Ordnance Dept., U.S.A.)

four additional machine gun crews and their weapons in each tank. These crews were to aid the cavalry divisions in holding this objective until relieved by infantry troops. All tanks except these and the Whippets were to rally after the 2nd objective had been taken. The second mission assigned to the cavalry was to move out against the line, Roye-Chaulnes, after being relieved at the 3rd objective. In this mission, as in the first, they were to be supported by the Whippet tanks. One battalion of tanks was assigned to each of the two divisions which were to be used on these missions. The cavalry was to advance at zero plus four hours. The formation planned for the Whippet tanks was: one company, with 200-yard intervals between tanks, as a screen ahead of each cavalry division, one company in support, and one in reserve.

The 17th Armored Car Battalion, like the cavalry and the Whippet tanks, was designated for exploitation after the second objective had been taken. Foreseeing the difficulty these vehicles would have in crossing trenches and in passing obstacles on the roads, a detail of tanks was attached to the battalion to help them overcome these difficulties. The armored cars were placed under the orders of the 5th Tank Brigade, which was supporting the Australian Corps.

Another feature, in the preparation for the Amiens attack, was a plan carried out for the purpose of misleading the German forces into believing that the next attack would be made in Flanders. Canadians were ordered into the line in the First Army Sector and were identified by the Germans, tanks were trained with infantry in the rear areas occupied by this army in view of the German long distance reconnaissance aircraft, and wireless messages contained references to troop concentrations in the area.

Squadron No. 8 of the Royal Air Force, with 18 planes, was directed to provide a noise barrage for the tanks in the approach march, to act as contact and counter attack patrols, to aid tank units, and to report the location of these units to headquarters by means of dropped messages at prearranged localities.

At 4:45 AM the tanks started forward with the various divisions. The cooperation between the infantry and tanks was good, and the advance, with the exception of that on the 3rd Corps front, where this corps had been in action on August 7th, went forward according to schedule with a few minor exceptions where the tanks were hindered by the fog and difficult terrain. Many machine guns were crushed by the tanks. The attack of the 3rd Corps, to which had been allotted but one battalion of tanks, was held up and, consequently, the German artillery on Chipilly ridge, which was in the zone of this Corps, accounted for a number of the tanks assigned to the Australian Corps by firing from that flank.

The two Whippet battalions and the cavalry were in action east of Mezières, at Guillaucourt, and south of Harbonnières against determined resistance encountered at these places.

During the advance on August 8th, the armored cars were helped across the obstacles by the tank detail and passed through the British lines before the troops reached their final objective. They successfully passed through the British artillery barrage and encountered the German troops, causing many casualties as they passed through them. Continuing, the six cars in this group surprised and killed a number of the headquarters troops of a German corps. Other cars encountered many transport vehicles near Framerville. A number of mounted officers and teams, on the road, were fired upon and scattered, several being killed and wounded. Arriving at a German headquarters, the crews of the armored cars ran up the Australian flag. Soon after this an armored car sighted a German railroad train, and its locomotive was put out of action by machine gun fire from the car. The train was captured by the cavalry. Moving into Proyart, the cars fired upon German troops who were found at dinner. As they were leaving this place, the car crews observed great numbers of German troops retreating from the attack of the Australians. The cars were concealed until the Germans were within about 100 yards, at which time surprise fire was delivered. The retreating German troops changed direction toward Chuignolles, but they were again fired upon by other cars from this village. One car encountered a truck load of German troops and fired at it until the truck ran into the ditch. Many cases occurred during this action where the cars overtook German vehicles on the road and fired into them at short range. More than half of the armored cars were put out of action on the 8th, but none of the crews was seriously injured.

The Mark V Star tanks completed their mission, delivering the machine gun crews at the designated objective, but, due to the heat and fumes in the tank, these crews were in no condition to fight when they arrived at their destination. Many of them left the tanks and walked rather than be subjected to the discomforts of riding in the tank. The Mark V Star was not sufficiently ventilated to serve as a successful troop carrier.

Throughout the attack on August 8th and on succeeding days as well, the planes of the 8th Squadron cooperated effectively upon many occasions by reporting the location of various tank units, armored cars, infantry units, a German observation balloon, and adjacent French units. They reported the successful occupation of the first objective and the fact that the tanks were preparing to move forward, and their report of the location of certain tanks beyond Demuin indicated that the bridge at that place had not been damaged. They bombed the German lines and retreating transport and fired upon many troop targets. The operations during this action indicated a need for the employment of attack aviation against German field guns in order to reduce the number of tank casualties.

The penetration on August 8th at the farthest point was 7½ miles. During this day, 100 tanks were put out of action by artillery fire. Due to the heat a great many tank crews were exhausted and, as the battalion held in reserve (the 9th) had been refitting and was unable to enter the action at this time, composite companies had to be formed for the action on the 9th. With the exception of the 3rd Corps objectives on the north, all objectives were captured on August 8th and more than 16,000 prisoners and 200 guns were taken. The prisoners reported that the advance of the tanks had been so rapid that resistance was useless. The British, on the 8th, had reached the old Amiens defense line Proyart, points just west of Rainecourt, east of Vauvillers, Rosières, Meharicourt, Rouvroy and Bouchoir. The British line was joined up on the south with the French.

The action was continued on August 9th. Sixteen tanks of the 10th Battalion supported the 3rd Corps. This advance was temporarily stopped by machine gun fire coming from the Chipilly woods and, owing to the thick woods and difficult terrain in this locality, the tanks were severely handicapped. Five tanks were knocked out in this action, but the objectives were taken. Eighty-nine tanks supported the Australians and the Canadians on the 9th. Five of the tanks were put out of action near Lihons and one tank was hit near Framerville. The cooperation of the infantry with the tanks engaged near the last mentioned village prevented additional tank casualties as they kept well forward and fired upon the German field gunners who opposed the tanks.

Liaison between the Australians and the tanks was materially improved by the detail of a noncommissioned officer to ride in each tank as an observer. Cooperation between the cavalry and the Whippet tanks was not altogether satisfactory because of terrain conditions, their different speeds, capabilities and limitations.

Thirty-nine of the 145 tanks used on the 9th were put out of action by German artillery.

During August 10th, the remaining tanks took part in small actions, meeting considerable resistance. The advance had now arrived at the edge of the Somme area, which was a serious obstacle due to the condition of the ground. Old trenches and rough terrain prevented the cavalry and the Whippet tanks from accomplishing the capture of Parvillers as directed. Of the 67 tanks used, 30 were knocked out by artillery.

On the night of the 10th an attempt was made by the Australian Corps to surround and capture a salient in the Somme valley which had been created by the advance. Two columns were used in the attempt to get around the German positions on the Etinehem spur, north of the Somme, and the ridge east of Proyart, south of the Somme. Tanks were allotted to each column. Both columns were to advance eastward and then turn toward the Somme river. The north column operation was successful but the operation south of the Somme did not accomplish much, as the Germans bombed this locality and their artillery fire and machine guns delayed the attack until it had to be stopped and the forces withdrawn. In this case, darkness proved to be a handicap since liaison was more difficult and the tank crews had no way of locating the German guns except when they were fired. Two tanks were knocked out in this partially successful night attack.

On August 11th a few tanks were used in small operations of which the most important was the capture of Lihons by the Australians who were supported by 10 tanks. The remaining tanks assigned to the Australian and Canadian Corps were withdrawn from the action on the 11th. In the four days of fighting practically 70% of the tanks were out of action from various causes. However, a great advance 12 miles deep had been made, 22,000 prisoners and 400 guns had been captured, and, in aiding in this accomplishment, the tanks had prevented great losses among the infantry. Following this advance a statement was issued by the Crown Prince's Group of Armies which contained a brief account of the British success, and concluded with the following significant order: "Messages concerning tanks will have priority over all other messages or calls whatsoever."

Bapaume, August 21, 1918.

The plan was for the Third British Army to attack north of the Ancre toward Bapaume on August 21st, while the Fourth British Army attacked south of this river. August 22nd was to be used by the Third Army in preparation for a continuation of the attack of the 23rd. If success attended the efforts on the 23rd, both armies were to exploit any advantages gained, and the First Army on the north was to make an additional attack.

To the Third Army were allotted the following: The 1st Tank Brigade, consisting of one battalion of Whippets, one battalion of Mark IV tanks, one battalion of Mark V tanks, and the battalion of armored cars; the 2d Tank Brigade, consisting of one battalion of Whippets, one battalion of Mark IV tanks, and one battalion of Mark V tanks; and the 3d Tank Brigade consisting of two battalions of Mark V tanks and one battalion of Mark V Star tanks. This army was to deliver the principal attack. The Fourth Army was to be assisted by the 4th and 5th Tank Brigades, each of which had three battalions of Mark V tanks. Although 15 battalions were thus allotted for this action, these units were much reduced in numbers of tanks available, due to the casualties incurred in the Battle of Amiens, and the battalions at this time had only from 10 to 15 tanks each.

About this time the Germans adopted new tactical methods of defense which consisted mainly in holding the front line lightly, as a line of observation or outpost line, with their reserves and guns farther back, thus extending the depth of their defense. On account of this extension and the difference in maneuverability of the three types of tanks available, the Third Army arranged for the Mark IV tanks to go no farther than the 2nd objective, while the Mark V and Mark V Star tanks were to aid in the attack on the 2nd objective and go as far as the Albert-Arras railway. The Whippets were to operate beyond the railway.

In addition to Squadron No. 8 of the Royal Air Force, Squadron No. 73 was attached to the Tank Corps for use against German field guns.

Starting at 4:45 AM, the Mark IV tanks and their infantry took the first objective and moved on to the second objective which was hurriedly evacuated by the German troops.

A different situation was encountered, however, when the Mark V, Mark V Star and Whippets arrived at the railway. This line had been prepared for defense, and was strongly held by machine gun nests and by field guns placed well forward. In addition to these measures, it was found that the points on the railway which were not embanked were defended by concrete blocks and iron antitank stockades. The fog lifted about the time the tanks arrived, with the result that they came under very heavy fire. A number of tanks were put out of action at this time but the infantry, who avoided the tanks, moved forward without many casualties. One of the Whippets succeeded in getting across the railway before the fog lifted but it was then right in front of the German guns and was put out of action. The tank commander got his crew out of the tank and then went back to the point of crossing, under heavy fire, to warn the other tanks which were intending to cross at that point. As soon as the fog lifted, the planes designated for attacking field guns came forward and were of material assistance from this time on. At some points the German machine gunners and artillerymen fought to the last, while at others large groups surrendered before the tanks could use their weapons.

The crews of the Mark V and Whippet tanks suffered severely from lack of ventilation, heat, engine fumes, and gas. In some cases the entire crew of the Mark V tanks fainted and the tanks were consequently out of action temporarily due to the fault in design whereby the crew compartment was not properly ventilated. After about an hour of operation, the Whippets were almost as bad in this respect as the Mark V tanks. Even the weapons became hot, and the steering wheel, in one case, became too hot to hold.

The armored cars operated successfully after the Whippet tanks had towed them through a large hole in the road near Bucquoy. Passing through this place, they entered Achiet-le-Petit before the infantry arrived and put several machine gun nests out of action. Two cars received direct hits at this place and were knocked out.

The objectives set for this date were gained. The cost in tank casualties was 37 out of the 190 tanks used.

The following account, which describes an instance of a tank catching on fire during this action and which was written by L. A. Marrison, a member of the crew, was published in the *Royal Tank Corps Journal:*

> By August 21st we were behind Courcelles and Gommecourt, where the enemy were concentrated in force. We were warned of tank traps, antitank guns, and all sorts of other devices to entertain us. The weather was still mainly fine, which was lucky for us, for we slept at night beneath our buses. At least we were fairly safe from shelling and bombing.
>
> We went forward from the tape in an impenetrable curtain of fog. Not a leaf stirred, even the sound of the guns seemed blanketed. At eight o'clock we were in the thick of it, firing going on from all directions, the dense mist enveloping us, and none of us knowing exactly what was happening. Then all manner of accidents took place at once. The crew commander bravely got out of the bus to discover how we stood, and to get into touch, if possible, with company headquarters; a few minutes later we were surrounded by Germans; one of them got under the six-pounder and fired his automatic through the aperture; poor Morris at the other gun was shot through the spine; then the engine burst into flames. I told the gunners to blaze away like the devil with machine guns and revolvers, while the rest of us seized the Pyrene extinguishers and directed streams of acid on the burning engine. Dense smoke soon filled the cabin, and the rank stench of singeing rubber. We fought desperately, in terror lest the petrol went off. The gunners cleared the Germans, all but a persistent begger who was crawling round the bus and firing through every loophole he could find.
>
> Miraculously he was missing us by inches, but he was just a bit more than we could stand. In sheer desperation I got hold of a Mills bomb, hopped out of the cab, and ran around the back, to meet him face to face as he was returning. We both started back automatically; I suppose it would have been funny in other circumstances. For a spare moment we looked into each other's eyes. I don't know what he read in mine, except sudden funk, but the pistol slid from his fingers and he bolted round to the other side of the bus. This gave me back some courage. I pulled out the pin, lobbed the bomb over the bus and cowered down in the shelter of its near side. The Mills went off all right, and fragments sang in the air overhead.
>
> I was shaking all over. The boys had opened the sponsons and were heaving out the ammunition. Morris was on the ground, livid and speechless, his head wagging from side to side. The flames had got the upper hand, and the interior was a lurid inferno. Two stood at the doors and threw out all we could save—machine guns and ammunition. Then Muir volunteered to go back and remove the wedges from the six-pounders; which he did, although he got badly burned. Mr. Allan returned then, to find his bus blazing merrily, ammunition popping off, one of the crew dying, the rest pretty "dicky".
>
> That was the end of our jaunt that day. I found my way to Brigade with Mr. Allan's report, and we were ordered to withdraw. We saw Morris to a field dressing station and returned to company headquarters.

A new penetration of 4000 yards was effected August 22nd by the 3rd Corps of the Fourth Army on a 10,000 yard front. The Mark V tanks, which had started out in rear of the infantry, moved forward when resistance was encountered and thereafter led the infantry during the remainder of the action, aiding in the capture of Albert, Meaulte, and the western edge of Bray-sur-Somme.

Eighteen Mark V Star and eight Mark V tanks supported two infantry divisions on the 23rd in a very successful attack on the Hamelincourt-Heninel spur, and the objectives of both divisions were taken with few casualties.

On the 23rd, six Mark V Star tanks assisted in the capture of the German positions on the Usna and Tara hills. All three battalions of the 5th Tank Brigade of Mark V tanks participated in the attack on Chuignolles. After this place was in the infantry's possession, one battalion rallied and the other two were used to exploit the success north of this place. Here a determined resistance was encountered, the German machine gunners, in many cases, continued their fire until their weapons were crushed by the tanks. A good word picture of this action is contained in the account of the experiences of one of the female tanks which supported the Australians. In leading its infantry forward this tank zigzagged toward the woods at Chuignolles and there came under the fire of an antitank gun. This gun was finally silenced by a male tank which was advancing on the right of the female tank. The latter then proceeded along the side of the village, firing into the crops and brush, forcing several German machine gunners to withdraw. About this time two large shells dropped short and exploded in front of the tank and the smoke from these explosions hid the tank momentarily. As it came through the smoke the tank crew found themselves facing a German battery. Using their machine guns to good advantage they overcame the battery gunners and then maneuvered their tank to the rear of another battery where the operation was repeated. This tank was then requested to clear out some machine guns in the village. Twelve machine guns were accounted for in Chuignolles. During this time the other tanks had gone on, so this tank commander started after them. While climbing over a steep bank, a shell hit the front end of the tank, breaking a track and forcing the tank down the bank. This damage was repaired under fire. By the time the work was finished the other tanks were returning to the rallying point so this tank returned with them.

The 3rd Division of the Third Army started its attack on August 23rd at 4 AM against Gomiecourt. This was a moonlight operation. Ten Mark IV tanks were used and the village was taken without difficulty. Soon thereafter four Mark IV tanks aided the Guards Division to take Hamelincourt.

The 6th Corps, supported by 15 Whippet tanks, moved out at 11 AM in the direction of Ervillers-Behagnies-Sapignies. The infantry was held up near the latter place, but the Whippets went on. During this part of the action, the officer and NCO in one of the Whippet tanks were killed. The remaining member of the crew removed these bodies from the tank, followed the other tanks, and operated his machine gun whenever targets were located. Although this attack was not completely successful, it aided the 37th Division, and the six Mark V tanks supporting it, to capture Achiet-le-Grand and Bihucourt.

The tanks accounted for great numbers of machine guns during the day's action. One tank put over thirty machine guns out of action and several other tanks did almost as well. Armor piercing ammunition, used by the Germans, pitted the sides of the tanks and, in some cases, penetrated the armor. The tanks unquestionably saved a great many British lives by destroying a great number of machine guns in the hands of the German troops.

While the Whippets were advancing with their infantry near Courcelles, the tank commanders became aware of the fact that the artillery barrage had slowed down and they consequently had to maneuver their tanks to keep from getting into the bararge. Being under antitank fire at the time, however, seven of them passed through the barrage, attacked large numbers of Germans who had taken cover, and caused many casualties among them.

It was these tanks that cut off several hundreds of German troops near Achiet-le-Grand who had been holding up the infantry, and drove them toward the British lines, where they were captured.

Both the infantry and the air service cooperated very successfully with the tanks on this date. The former took advantage of the opportunities provided by the aggressive advance of the tanks, and the latter kept headquarters informed of the progress being made, thus enabling the supply of reserves and replacements to be effected where and when needed. The squadron designated for attacking field guns rendered excellent assistance. One report from this squadron shows that upon observing German batteries preparing for action near Behagnies, it dropped bombs upon them and fired into them with its machine guns, thus causing the horses to stampede, the limbers to be overturned, and the gun crews to be dispersed.

As in the action of Amiens, the interior of the Mark V tanks became very disagreeable due to the hot weather in which this action took place. In some cases the heat and fumes caused the crews to become delirious. The engine was located in the front part of the crew compartment, and the heat generated was not properly exhausted to the outer air. The exhaust manifolds became very hot and, the joints having expanded due to the heat, engine fumes were released inside the tank. The crews, being thus exposed to carbon gases, soon became exhausted and frequently became unconscious.

The 1st, 3rd, and 4th Tank Brigades supported the Third and Fourth Armies on the 24th. Seven Mark IV and 19 Whippet tanks supported attacks on Sapignies, Grevillers, Biefvillers, and Loupart wood, all of which were captured. After making an approach march of 10,000 yards, 11 Mark V Star tanks supported a successful attack on the line St. Leger—Henin-sur-Cojeul. This line having been taken, the attack was continued (unsuccessfully) at 2 PM supported by two of the tanks. The officer in charge of one of the tanks became unconscious when the tank received a direct hit. Later on he recovered and continued on his mission. The Germans partly surrounded the other tank and, by using phosphorus bombs, forced the crew to dismount. Before the tank commander left the tank, however, he turned it toward the British lines. He then got out and walked ahead between the front horns until the gas fumes had been cleared from the tank. He then drove it back to the British front line.

Five Mark V tanks supported an attack on the line, Mory copse-Camouflage copse, in the afternoon of the 24th. Little resistance was encountered until the Mory copse was reached. This garrison fought to the last and practically all of them were killed, one group of about sixty were accounted for by four rounds of case shot from the six-pounders. Happy Valley and the village of Bray were captured by an infantry division supported by five Mark V Star tanks.

After the close of the fighting on August 24th, the 3rd Tank Brigade was directed to report to the Canadian Corps of the 1st British Army at Achicourt, in preparation for the Second Battle of Arras. This change involved a night march of about 16½ miles.

Second Battle of Arras, August 25 to September 3, 1918.

This action was in reality a continuation of the Battle of Bapaume and consisted of numerous small actions generally initiated by the divisions concerned. Small attacks were made on August 25th on the Third Army front and, on the 26th, Mark V tanks supported an attack on Mory in a heavy fog which practically prevented cooperation between the infantry and tanks and greatly hindered the latter in maintaining direction. The tanks were bothered considerably by German antitank guns and five members of one tank crew were wounded by bullets from these weapons. This action was only partially successful.

Mark V and Mark V Star tanks supported two Canadian divisions on August 26th in the attack on Wancourt, Guemappe, and Monchy-le-Preux. At the last-mentioned place, several tanks were put out of action and the tank crews joined the infantry in stopping a counter attack. A few of these tanks were used in local operations on August 27th, but there were no more tank actions after that until the 29th, when, the Germans having evacuated Bapaume, a few Whippet tanks were used by the Third Army near Beugnatre. On the 30th, an attack was made against Ecoust, Longatte, and the trenches beyond these places, in which it was planned to use six Mark IV tanks. The orders to the tanks were delivered late the night before the attack, a four-mile approach march was necessary to reach their assault position, the only reconnaissance officer who knew the route had just been ordered away from the area, and, as a result, the tanks became lost and did not reach the front lines in time to go forward with the infantry. The infantry suffered many casualties and the attack failed.

The 1st Tank Brigade supported attacks by the 5th British and the New Zealand Divisions on Fremicourt, Beugny, Bancourt, Haplincourt and Velu wood, all of which were occupied. On the 31st, nine Mark IV and four Whippet tanks aided in the capture of Longatte trench, Moreuil Switch and Vraucourt trench. This success was exploited on September 1st when five Whippets were used with the 185th Brigade in the attack on Voux-Vraucourt. Exceptionally heavy fire from machine guns and antitank guns was encountered by the tanks in this action. Bullets removed practically all of the paint from some of the tanks. The infantry came under this fire and a large number of casualties resulted, but the objectives were successfully occupied.

On September 2nd, a combined attack on the Drocourt-Quéant line was made by various divisions which were supported by the remaining tanks of the 1st, 2nd, and 3rd Tank Brigades. The Sensée Valley again gave trouble in the assembling of the tanks for the attack. The British expected that this line would be strongly held and knew that it was protected by very strong wire obstacles, so they made careful preparations for the action. In all, 81 tanks were assigned to the attacking infantry divisions.

The Mark IV tanks of the 12th Battalion encountered very heavy field gun and antitank rifle fire near Lagnicourt. The antitank rifles caused many casualties among the tank crews during this action. One female tank used its machine guns until all but one were out of action and five of the crew were wounded. As it was being brought back by the tank commander, it was set on fire by a direct hit. In addition to Lagnicourt, Moreuil and Morchies were taken by the 6th Corps, supported by the 12th Tank Battalion and other units of the 2nd Tank Brigade. The 1st Tank Brigade supported two divisions in the capture of Beugny and Villar-au-Flos. The 3rd Tank Brigade cooperated with three divisions in an attack on the Drocourt-Quéant line and this line, in the Canadian sector, was captured about noon. The resistance at this point consisted mainly of many machine gun nests and antitank rifles. Many of the German gunners surrendered to the tanks, and one company of tanks is reported to have thereupon crushed about 70 machine guns. The armored cars participated in the attack on this line. They lost four of their machines which were struck by shells from concealed German batteries. The German gunners were prevented from capturing the armored car personnel by fire from the accompanying planes.

The German forces were now retreating rapidly and, when the Whippet tanks advanced as far as Hermies and Dermicourt on September 3rd, very little opposition was encountered. All tank units were withdrawn at this time and became a part of GHQ reserve while they were being reconditioned.

CHAPTER V.

TANK COMBAT HISTORY. BRITISH (Continued).

Epehy, September 17, 1918.

This action, carried out by the Third and Fourth Armies and in cooperation with the 1st French Army, had for its purpose the capture of the Hindenburg outpost line in order to secure observation over the main line and to permit the advancing of the artillery positions for the main attack.

Tanks were not used until September 18th, when GHQ allotted the 4th and 5th Tank Brigades to the Fourth Army. Twenty Mark V tanks of the 2nd Battalion, the first to arrive, supported the 3rd Australian Corps and 9th Corps over a wide front. Visibility was poor during the fight and the tank compasses in these tanks proved useful. On the 3rd Corps front, Epehy was taken, many Germans surrendering upon the arrival of the tanks. The village of Ronssoy was well defended by machine guns with armor piercing ammunition and by the German antitank rifles. The infantry was stopped in the attack on Fresnoy by heavy machine gun fire coming from the strong point called the Quadrilateral, a strongly fortified system of trenches and buildings which were an important part of the Fresnoy and Selency defenses. Two tanks advanced against this resistance. The first one to arrive became ditched in a sunken road near the strong point and the intense machine gun fire prevented the crew from using the unditching beam. In the meantime the driver of the second tank had been killed and the assistant driver badly wounded, so the tank commander personally drove his tank, while the rest of the crew operated the guns until the tank caught on fire. As the crew left this tank, they were surrounded and captured by the Germans. The crew of the first tank then left their tank, removing their machine guns, and, taking up a position away from the tank, held the Germans off until the infantry arrived. The diversion of many German weapons to these two tanks had, in the meantime, so lessened the opposition to the British infantry that it was able to advance without difficulty. Ronssoy and Hargicourt were also captured on this date, although the attack had not made rapid progress during the day.

The advance was not started again until September 21st, when an attack was made by troops of the 3rd Corps against the Knoll, Guillemont, and Quennemont farms. Nine tanks in all, seven Mark V and two Mark V Star supported this attack. In this action, land mines, field guns, antitank rifles and machine guns using armor-piercing ammunition opposed the tanks. There were not enough tanks to take care of the German machine gun nests and the attack failed. The two Mark V Star tanks successfully carried their load of infantry machine gunners and their weapons forward, but the tanks were under such heavy machine gun fire when they arrived at their destination that the transported troops could not be unloaded at the point designated.

The next advance was started on September 24th with a view of completing the improvement of the line, which, on account of the determined resistance up to this point, had not yet been satisfactorily effected. In the meantime the 8th, 13th, and 16th Battalions, together with the 5th Supply Company tanks had arrived. Nineteen Mark V tanks of the 13th Battalion supported two divisions in the attack on this date against Fresnoy-le-Petit and upon the Quadrilateral area, the previous attack having secured only a small portion of the latter stronghold. The tanks were moved up to the assault position by the operating crews and the fighting crews were later carried forward by truck. During the final approach march, some difficulty was had on account of the semaphores of the tanks catching on the overhead signal wires. During this part of the approach march, the Germans used gas effectively, causing the tank crews to wear their masks for two or more hours and, when they had arrived at the assault position, German planes dropped flares over the tanks and they were subjected to heavy shell fire. Antitank guns were used effectively and about half of the tanks were put out of action. Infantry troops and three tanks entered the Quadrilateral but all three tanks were knocked out by one German gun. By night the British line had been brought to a point which flanked the Quadrilateral, and certain observation points had been gained, but the full object of the attack had not been accomplished.

The following account, which describes an instance of the use of tanks to carry infantry troops and of the effect of gas on the tank crew, was written by L. A. Marrison, and appeared in the *Royal Tank Corps Journal*.

In those late September days, with the daylight fading and the first leaf falling and the winter chill advancing, a new emotion came to warm us and cement us still more closely. This was the strange, exulting sense of victory that pervaded the air. You could feel it; you thrilled to it. Our "news-room" in those days of preparation was crowded after the arrival of the official bulletins. The number of prisoners became incredible; the swiftness of the advance amazing. We girded up our loins and worked feverishly to be ready for the crowning triumph. If we could capture Cambrai and all the network on communications of which it was the center, the end would be in sight. The four years' depression was lifting at last.

But I was not to see the end. My next engagement was the last. We advanced on the Epehy front against the famous Quadrilateral near Fresnoy-le-Petit. The Quadrilateral was the pivot on which the German defense hinged in this sector. Groups of cottages had been reinforced and fortified and encircled by an elaborate system of trenches. It was infested with field guns and antitank guns, and simply bristled with machine guns. And we hadn't nearly enough tanks that day to make good our losses.

To save the storming troops till the last moment, we packed as many as we could inside our buses. Talk about sardines! The infantry didn't enjoy their ride and vowed that the Tank Corps was a vastly overrated affair when it came to a question of comfort. But when they caught a glimpse of the "cloudburst" outside they were resigned to remain where they were. We dipped and dodged, taking advantage of every bit of natural cover there was, but the bombardment found us everywhere. However, it was the gas that defeated us. There is no saying how far we might have gone if we could have had free use of our faculties. But when the gas barrage drifted toward us and began to seep into our bus, we were handicapped as much as if we had been blindfolded and handcuffed. To clamp a box-respirator over your face in a hermetically sealed cab with the temperature at a hundred degrees is a bit too thick. The sweat streams down, the damp folds cling to your face, the eye-piece dim, you can see nothing at all and you feel you are suffocating. All the while the bombardment went on, concentrating on each tank as a spearpoint of attack. Shells missed us by hair's breadths; our sides were splintered with shrapnel and machine-gun bullets; we couldn't see where to go or what to do; and the gas clung persistently round us. Again and again we pushed our hands into our respirators to sniff the air, but it was no use; the bus was permeated.

We jogged along aimlessly, dodging, dipping, zigzagging, swinging away from shell craters that were suddenly formed in front of our noses. At last it became intolerable. There we were, the engine roaring, the guns blazing, the cab stacked with explosives, seven of a crew and fourteen of the infantry with not enough air to keep an oyster alive. I couldn't stick it any longer. Besides, I couldn't see anything. I whipped off the mask and sniffed the air. It didn't seem so bad now, and the fact that I could see and hear was everything.

Not long afterwards a terrible cramp seized me, right across the stomach. The pain increased, gripping me until I doubled up in agony. Then I began to vomit, violently, endlessly. Everything seemed to be upheaving. I thought it would never stop. I coughed, spluttered, choked and retched, rolling on the floor with my knees up. I didn't care a cuss what happened to me. If the bus had blown up it would have been a blessed relief. Something did happen to the bus soon after, but not as bad as that. A shell pierced the rear sprocket and put us out of action. The crew had then a terrible fusillade to withstand before they could withdraw, during which the armor was twice penetrated by an antitank gun. Mr. Allan applied such remedies as he could find in the first-aid chest, but nothing seemed to help until the vomiting ceased. When evacuation was decided upon they wanted to take me along on a stretcher, but I had the decency to resist and hobbled back with them somehow, clinging to a couple of shoulders, as far as the field ambulance. Luckily for me they didn't inquire too closely into my case at the C.C.S., for really it was all my own fault.

By the time I was back the battalion had withdrawn to billets near Blangy, and the Armistice was signed before we were called upon again.

Second Battle of Cambrai, September 27, 1918
(Cambrai-St.Quentin).

This action was carried out by the First, Third and Fourth British Armies on a front of about 16 miles. A point of interest in the general plan was the arrangement for the First and Third Armies to start the attack on September 27th; and on the 29th, when the Germans had had time to place their reserves in front of these armies, the Fourth British Army was to attack in its sector. The 7th Battalion of Mark IV tanks and the 11th and the 15th Battalions of Mark V Star tanks were to support the attack on September 27th. For the 29th, on the Fourth Army front, the 3rd Tank Brigade, consisting of the 5th and 9th Battalions of Mark V, and the 6th Battalion of Whippet (Medium A) tanks, was allotted to the 9th Corps; the 4th Tank Brigade, consisting of the 1st and 4th Battalions of Mark V, and the 301st American Battalion of Mark V Star tanks, was allotted to the Australian and American Corps; and the 5th Tank Brigade, consisting of the 8th and 13th Battalions of Mark V and the 16th Battalion of Mark V Star tanks, was held in Army reserve. The 8th Squadron, Royal Air Force, (cooperating with tanks) and the 17th Battalion of Armored Cars were also attached to the Fourth Army. A tank supply company was attached to each tank brigade.

The First and Third Armies, on the north, were to capture Bourlon hill, operating between the Sensée river and Gouzeacourt; and the Fourth Army, on the south, was to operate against the Knoll, Guillemont and Quennemont farms. One of the obstacles in the First Army sector was the dry Canal du Nord, which had steep slopes. This canal was about 12 feet deep and approximately 50 feet wide at the bottom. To make it a better tank obstacle, the Germans cut the banks to make a vertical wall about 9 feet deep in some places. Airplane pictures were taken and a mosaic prepared to supply the latest information of this obstacle, and all other available data, such as engineering drawings and reports from prisoners and refugees, were studied. A unique plan was made as a last resort. In order to insure that the tanks would be able to cross and support the infantry, four old tanks were to be driven into the canal. Thus, it was hoped, a bridge would be formed over which the fighting tanks could cross. These four tanks were prepared with internal timber bracings to insure sufficient strength. They were to turn and form in line after reaching the bottom of the canal, and thus provide the bridge. As it turned out, the old tanks selected were unable to reach the canal, so this interesting plan was not executed. When it was reached, the canal proved to be less of an obstacle than was expected. The only tank that failed to cross was one damaged by a mine, located in a road that went through a gap in the canal.

When the advance started on September 27th, the infantry and tanks encountered strong resistance on the right of the line near Beauchamp Ridge. In this area the German troops displayed great bravery. In some cases they surrounded the tanks and tried to pull out the machine guns and six-pounders. This procedure, of course, resulted in many German casualties both by fire from the tanks and by crushing. Many of the German machine guns were also crushed. Twenty-six Mark V Star tanks supported the attacks on Flesquières and Prémy Chapel, also an attack south of Bourlon wood. The Canal du Nord was successfully crossed after some difficulty, and although eleven of the tanks were put out of action by artillery, the objectives were taken. Sixteen Mark IV tanks supported the Canadian Corps in the attack on Bourlon village and the western edge of Bourlon woods. Fifteen of these tanks succeeded in crossing the canal. One tank encountered many German troops at Silkem Chapel and Wood Switch who were poorly led, and caused many casualties among them with its six-pounders, using case shot at point blank range. In addition to the tank which was damaged by a mine before it crossed the canal, two more were lost due to shell fire from a battery close to Deligny hill.

The 15th Battalion Mark V Star tanks were engaged and, at about 4:30 PM (September 27th), when the fuel supply was getting low, they were requested to make another advance to support an infantry brigade which was about to advance again. In order to bring his brigade to the Marquion line, the brigade commander accepted responsibility for the tanks being stranded without fuel. Two tanks accordingly moved out in rear of the infantry. Heavy machine gun fire was received from Fontaine-Notre Dame, and both tanks used their heavy weapons against the resistance. While at this point, they also broke up and caused many casualties among an additional group of Germans who were approaching the Marquion line.

Fontaine-Notre Dame, Bourlon Wood, Bourlon village, and Sauchy-Lestrée were all taken on this date. Cooperation from the air was as effective as could be expected considering the distance the planes had to cover to reach the area. Important messages were delivered by the planes, and bombs and machine gun fire were used against German field guns in Bourlon wood.

On September 28th, seven Mark V Star tanks supported the 5th Corps in the capture of Villers Guislain and Gonnelieu. Six Mark IV tanks of the 7th Battalion supported a successful attack on Raillencourt and St. Olle. Epinoy and Haynécourt were captured, and all objectives assigned were occupied.

10,000 prisoners and 200 guns were captured in this first part of the Second Battle of Cambrai, but the cost in tank casualties was considerable.

Catelet-Bony, September 27, 1918.

In the meantime, on the Fourth Army front, a local action had been carried out on September 27th by the American 27th Division, supported by 12 Mark V tanks and under the direction of the Australian Corps Commander, which had for its purpose the advancement of the line at that point a distance of 1000 to 1200 yards, so that the Knoll, Guillemont and Quennemont farms could form the assault position for the larger attack which was to start on September 29th. It will be remembered that repeated attacks had been made on this part of the line previously, but without success. While this division was preparing for the attack on the 29th, orders for the attack on the 27th were received and had to be hurriedly considered. This attack (of the 27th) failed, although small parties of the 27th Division reached their objectives and maintained their positions in German territory until relieved on September 29th. The attack left the remainder of the 27th Division in a worse condition than it would have been had the attempt to straighten out the line not been made, because only one day remained in which to complete the plans for the attack on the 29th and, since part of the 27th was known to be in the area over which this attack was to be made, an artillery barrage could not be used. Consequently the 27th Division and the supporting tanks had to advance on the 29th without any artillery support.

The purpose of the action on September 29th was to break the Hindenburg line in the sector of the American 2nd Corps. The St. Quentin Canal crossed this front. This canal ran underground through a tunnel between

Bellicourt and Vendhuille. The tunnel was used as a dugout by the German troops.

It was planned that the American 2nd Corps (27th and 30th Divisions) would take the first objective east of Bony, and that the Australian Corps would then move ahead of the Americans. The attack started at 5:50 AM on September 29th, with that part of the jump-off line planned for the use of the American 2nd Corps (Knoll-Guillemont Farm-Quennemont Farm) still occupied by German troops. The barrage, as planned, came down ahead of this line, so that the 2nd American Corps, which started about 1200 yards in rear of this line, had no chance to catch up with the barrage. In addition to this handicap, a part of the 301st Battalion tanks and British tanks were routed through an area where the British forces had placed a mine field in March, 1918, as a defense against German tanks and, in going through this field, two or three of the 301st Battalion tanks and some of the British tanks were blown up.

The wind carried the smoke produced by the British barrage into the German lines until about 10:30 AM when a change in the wind caused the battlefield to be covered with smoke. This interfered seriously with the cooperation between the tanks and the infantry. A great many of the tanks received direct hits early in the day and many were ditched later when the field was covered with smoke. The Germans were so successful in making direct hits on the tanks that it was thought that they had some unusual means of determining just where the tanks were located. One of the 301st tanks stripped its gears, one had difficulty with oil pressure, and several had trouble with the vacuum systems. The tank crews were handicapped by the lack of pistols. This frequently resulted in men being captured when they had to evacuate their tanks. The Germans used antitank rifles extensively in this action, although in several cases where the bullets penetrated the armor, they did no harm. Bullet splash in these tanks was very bad.

The attack of the American corps was not successful. At the end of the day, the 30th Division had reached its objective, but this point should have been reached by 9:00 AM, according to the plan. The 27th Division had been subjected to massed machine gun fire from the start of the action and, without any aid from its artillery, did not make much headway. At the end of the day its line ran from Bony to Maguincourt trench. The 301st Battalion lost 105 killed and wounded and seven missing. The casualties among the infantry were also severe on this date, a great many of them being killed when the smoke lifted, at which time both infantry and tanks provided good targets for the German guns of all calibers.

The tanks allotted to the 9th Corps on the right made better progress, taking Nauroy and Bellicourt and passing through the Hindenburg line. However, since the attack by the 2nd Corps had not progressed, the troops of the 9th Corps were in a position threatened from the flank and rear, and tanks which had been assigned to other objectives found it necessary to intervene at this point. The great amount of wire and the numerous trenches encountered had slowed up the advance and consequently the tanks and infantry were behind the barrage schedule.

On the right of the 9th Corps, the advance moved rapidly, the infantry troops crossing the St. Quentin canal by any means available, some by wading and swimming. The tanks had to go up to the point near Bellicourt and cross at the tunnel. Then, coming down along the other side of the canal, they joined their infantry and aided in the capture of Magny. The smoke and mist raised a little after noon and the German artillery became very effective, causing many tank casualties.

On September 30th, 18 Mark V tanks of the 13th Battalion, from the reserve, moved up the Hindenburg and adjacent lines, but, as the infantry did not follow, little was accomplished. Six Mark V tanks cooperated with two Canadian divisions in the capture of Cuvilliers, Blécourt and Tilloy, wiping out many machine guns near these places. On the 31st, additional tanks from this battalion aided in the occupation of the Fonsomme line. In this action the tanks produced their own smoke clouds to blind German observation points. This method proved quite successful and was thought to have reduced the number of tank casualties.

On October 3rd, 20 tanks aided in an attack on the Sequéhart-Bony line, when Sequéhart, Ramicourt and Doon Copse were captured. On the 5th, 6 Mark V tanks and a new division of infantry made an unsuccessful attack on Beaurevoir. These troops had no previous training with tanks and did not cooperate effectively in this action. Australian troops were supported by 12 Mark V Star tanks in a very successful attack on Montbrehain on this date, cooperation between the infantry and the tanks being exceptionally good. The tanks did considerable execution with their six-pounders in this action.

On the 8th of October, 82 tanks of all types were used. On the Niergnies-La Targette line, Mark IV tanks of the 12th Battalion proved of great assistance to their infantry. A tank-against-tank action occurred near this area. The Germans used one male and three female tanks of the Mark IV type, which had been captured by them from the British. The British accounts of this action show that one of the 12th Battalion male tanks put the captured male tank out of action with a six-pounder shell; that one of the female tanks was knocked out by a shell fired by a tank officer from a captured German field gun; and that the other two female tanks retreated. Two British male tanks were struck during the fight, one of them being set on fire. A female tank was nearby, but had a leaky radiator and was about out of water. It was its crew which found and used the field gun.

Two companies of Mark V Star tanks supported two divisions in other areas. In one division area, Villers-Outreaux was captured. Before the other division could reach its objective, it had to cross a great band of wire. The troops were held up by this until the tanks crushed it for them.

On this date, the American 301st Battalion and the British 6th Battalion were assigned to the 2nd American Corps. The latter battalion was equipped with the Medium A (Whippet) tanks. These assisted materially in the attack on Fraicourt wood, one tank succeeding in overcoming three batteries of field guns in this action. The 3rd Battalion of Whippets supported an attack on the strongly defended village of Serain.

The 301st Battalion tanks supported the attack of the American 59th Brigade on Brancourt. Twenty-three tanks, including a wireless tank were used. Of these, three failed mechanically, but twenty were able to leave the assault position. During the last part of the approach march, the Germans shelled and bombed the tanks but no casualties occurred. This attack was very successful and eleven tanks succeeded in reaching the final objective. A light railway embankment was encountered which was defended by a great many machine guns. The tanks cleared up this point in advance of the infantry. Brancourt was attacked by tanks coming in from the northern and southern flanks. Many machine guns were wiped out by these tanks and one captured a German battery by approaching it from the rear. Cooperation between infantry and tanks was excellent. Each helped the other when the necessity arose. Having reached the objective, some of the tank commanders failed to rally and continued on to aid the Whippets in the exploitation. This battalion (the 301st) lost 18 killed and wounded.

Among the defects reported by the crews were: faulty vacuum systems, broken exhaust springs and leaky gear boxes. The official report recommends the carrying of ampules of ammonia for use of the crew when affected by engine fumes, the enlargement of the field of vision of the periscope, and the prevention of bullet splash entering through the sides of the six-pounder turrets. The wireless tank was used to advantage in supplying both the infantry and the tank headquarters with the latest information on the progress of the action.

A good description of what a tank action looks like from the inside of a tank is contained in a report of the personal experiences of 2d Lt. Paul S. Haimbaugh, U. S. Army, one of the tank commanders in the 301st Battalion during the action on the 8th of October, an extract from which follows:

* * * The doughboys spring to their feet and start forward. You urge your tank on until your are nosing the barrage; ahead the German distress signals flare in the sky. In a moment the enemy barrage will fall. Here it is, and it's disconcertingly close. You think maybe you better zigzag a bit, maybe you can dodge 'em. The doughboys trudge sturdily on and here and there one sags into a heap (shell splinters). One shell nearly gets you as it bursts nearby with the rending crash peculiar to high explosive. Seems like it nearly lifts your tank into the air. A dozen pneumatic hammers start playing a tattoo on the sides and front of your tank, and splashes of hot metal enter the cracks and sting your face and hands. The infantry, a hundred yards back, is prone on the ground; too hot for them.

Well, it's up to you to locate the enemy machine guns and put them out. Observation from the peep holes reveals nothing. You pop your head through the trap door and take a quick look around. There they are. A hasty command to the six-pounder and machine gunners. Crash! goes the port six-pounder, and the tank is filled with the fumes of cordite. A hit! A couple more in the same place and a belt of machine gun cartridges suffices to quiet that machine gun nest.

"Come on Infantry!" As the tank passes, you see the grey forms sprawled grotesquely around their guns. You are glad the "bus" is a male for these six-pounders certainly do the work.

Up ahead is a railway embankment and a sunken road, a likely place for machine gun nests. Tat-tat-tat! They've already begun to strafe you. Slipping from your seat you shout commands to your gunners. Picking their targets they pepper away with the machine guns and six-pounders. The noise is terrific and the tank is filled with cordite and gasoline fumes. There is a sickening smell of hot oil about. You are pretty close now, so you order case shot and the six-pounders rake the embankment and road with iron case shot, with deadly effect. The place is a shambles—grey forms sprawled in the road—huddled in gun holes—lying in position about their guns. It's war, and you had to get them first. A half a dozen Germans scramble to their feet with hands upraised and you let them pass to the rear. "Come on Infantry!"

Your tank surmounts the embankment and your hair raises for on a ridge 500 yards ahead are two 77's, sacrifice guns left to get you. Crash! A shell lands 50 yards to your left and short. Your starboard six-pounder lets go. Crash! Another one which lands close sends a shower of dirt and stones against the side of the tank. Working like mad, your gunner sends four shells after the two guns. Good work, for they are silenced. One member of the two gun crews is able to run away.

Beads of perspiration stand on your forehead. Hot work this. The combination of powder and gasoline fumes, the smell of hot oil and the exhaust begins to daze you but you pull yourself together and rumble on. The infantry swings along behind, bombing dugouts and "mopping up," assisted by your running mate, a female tank armed with machine guns only.

It's a mile to your objective now, but it's a mile of thrills. You get "shot-up," put out a half dozen machine gun nests; clean up another sunken road with machine guns placed every ten feet along it. A one-pounder in a hedge scares you with several well placed shots before it goes "west." Some German artillery observer, way back, spots you and chases you over the landscape, dropping now a shell in front and then in back of you. Here is our objective! You wait for the infantry to come up and your crew enjoys a breath of fresh air. After the infantry has dug in and consolidated its position, you turn towards the rallying point.

The 17th Battalion of armored cars worked with the cavalry corps on October 8th, but moved out ahead of it on the 9th to Maretz and beyond. These cars assisted the South African infantry on this date. By running through the German line and delivering enfilade fire upon it, they forced the machine gun crews to leave their positions. The cars killed many of the gunners and captured ten machine guns and two trench mortars. Two cars moved up the Maretz-Honnechy road in an effort to secure a railway bridge. They were observed by the Germans, who lighted a demolition fuse and then retreated. The first car got across before the bomb exploded, but the second could not cross. Approaching the village of Honnechy, the car that had crossed passed a building as a group of Germans were leaving it. A short burst was fired into them and five dropped. The car went through Honnechy, which was full of troops, using its guns effectively, and on to Maurois where the procedure was repeated, causing many casualties. Knowing of another bridge by which they could get back to the other side of the river, the crew approached this bridge with guns pointing to the probable location of the demolition crew. These men were fired upon before they could light the charge. The bridge was saved and proved of value later.

The advance was continued on October 9th. Eight Mark V tanks supported an attack east of Premont. The 17th Battalion of armored cars

Plate 7.
British Armored Car, Used in Combat in 1918.
(From *The Tank Corps* by Williams-Ellis.)

saw action at Maurois and Honnechy. The last tank action in this battle was on the 11th when five Mark V's were used to support an attack north of Riguerval wood. The Hindenburg line had been penetrated on a front of about 30 miles. Cambrai was occupied on the 9th and the air forces reported a great many troops on the roads in German territory about Le Cateau. The congested troops on these roads were attacked by the British planes.

Since August 8th, 819 tanks had been damaged and salvaged and the tank personnel had lost more than one-third in casualties. Many of the tanks were later repaired and returned for service.

Selle, October 17, 1918.

In this action the British Fourth Army and the French First Army attacked southward from Le Cateau to Vaux Andigny on a 12-mile front. The 4th Tank Brigade, supporting this action, was composed of: the 1st British Tank Battalion which supported the 9th British Corps on the right, the 301st American Battalion which supported the 2nd American Corps in the center, and the 16th British Battalion which supported the 13th British Corps on the left of the Fourth Army. The 6th British Tank Battalion was held in the Fourth Army reserve.

The Germans held the right bank of the river Selle, which crossed the ground between the two forces, and they had observation posts on the

Allied side of the stream. These posts made reconnaissance, on the part of the Allied troops, very difficult. This work was accomplished, however, and safe points of crossing were selected prior to the action. All tanks carried cribs to be used, if necessary, in crossing the river. Forty-eight tanks in all were used by the Fourth Army. The fog was so thick, when the action started at 5:30 AM on October 17th, the tanks had to depend upon compass bearings. Those tanks in which the compasses were out of order had great difficulty in keeping direction. The infantry, in several cases, had to depend upon the tanks for direction. All three battalions crossed the river safely, although it had been dammed in places to make it a more difficult obstacle for tanks. The opposition was not as severe as had been expected, although the Germans made good use of their artillery and, at certain points, their outposts resisted effectively. The attack on Demi-Lieue was held up, and Whippet tanks were requested to aid. Several Whippets responded but, due to the heavy machine gun fire, the infantry still had difficulty. When nearing the town, three Whippets were put out of action by one gun and the infantry then retreated. This village was taken later in the day.

Of the 301st Battalion tanks, 10 were allotted to the 27th Division and 15 to the 30th Division for the Le Cateau action. In the sector of the 30th Division four tanks supported the advance from the line of departure to the river, a distance of about 1000 yards, on account of the German posts known to be located on the American side of the river. The other American units were not to receive tank support until the river had been crossed. Five of the 25 tanks had mechanical trouble before the attack so only 20 started. The map of the river, prepared by the air service and which was furnished to all tank commanders, proved to be a valuable aid in this attack. It showed the river throughout this sector, and on it were marked all points where the tanks would probably have difficulty in crossing, points where the banks were marshy, points where bridges had been blown up, and other tank obstacles.

The river was crossed successfully by all but two tanks, one of which was lost due to the fog and tried to cross above the designated point. The other tank was ditched in the marshy ground near the banks. Even after the river had been passed, the fog seriously interfered with the operation of those tanks in which the compasses were out of order. Several tanks became lost on this account and, by the time the fog lifted, the fuel was exhausted and some of them were unable to continue in the action. Only three of the tanks reached the first objective but several gave assistance to the infantry at various points during the advance. The tanks assigned to aid the infantry on the right put a number of machine gun nests out of action, permitting the infantry to advance. The losses in tank personnel were ten killed and wounded. Seventeen tanks rallied but all were in need of overhaul.

On October 20th, four Mark V Star tanks cooperated with the 5th Corps in the attack against Neuvilly and Amervalles, between Le Cateau and the Scheldt canal. This advance also crossed the Selle river at a point where the river was a more serious obstacle than was encountered by the 301st Battalion. The crossing on the 20th was made possible by building a bridge just under the surface of the water. This bridge was built at night and was not visible to the German aircraft in the daytime. The tanks crossed without difficulty and all reached their objectives.

On October 23rd, 37 tanks participated in a night attack starting at 1:20 AM near Le Cateau against a line near Valenciennes. The 10th Battalion of Mark V, the 11th Battalion of Mark V Star, the 12th Battalion of Mark IV British tanks and the 301st American Battalion of Mark V and

Mark V Star tanks supported the attack. In addition to the darkness, the crews had to contend with gas. They were forced to use gas masks in some localities and the attack did not make much progress until daylight. The attack was continued on the 24th and the tanks were credited with having rendered effective assistance. In one case, a tank six-pounder shell exploded a German ammunition dump, causing many casualties among the German troops.

About 12 Mark V and Mark V Star tanks of the 301st Battalion participated in the night attack. They were divided into three sections of 4 tanks each, one tank from each section being held in reserve. The preparations for this action were not thorough. Orders were not delivered to the 301st Battalion in time to make a satisfactory reconnaissance, the maps issued were not accurate, and the tank and section commanders were not able to familiarize themselves with landmarks prior to the action. Lack of transportation for section commanders was a factor in preventing the proper dissemination of instructions to tank commanders. Two of the tanks of the left and center sections reached their objectives but had to return to pick up their infantry which had not kept in touch with them. These tanks captured 18 prisoners and turned them over to the infantry. The infantry was unsuccessful in consolidating its position, so a later attack, supported by two tanks, was made over the same ground. The objective was then taken and held. Although the right tank section was subjected to an intense barrage just before zero hour, it moved out as scheduled. One of its tanks became ditched within 300 yards of the objective. The compass in another tank became stuck and useless, and this tank, in consequence, lost direction. The third tank patrolled the objective but met with no resistance and was withdrawn.

Maubeuge, November 2, 1918.

The action on this date, although a part of the Battle of Maubeuge, was one of the frequent small actions intended to improve the line at certain localities with a view to obtaining a more suitable point for starting a more important advance. Three Mark V Star tanks supported the attack on, and capture of, positions in the vicinity of Happeharbes. These positions would have facilitated the advance planned for November 4th. However, the Germans counterattacked and forced the British out of the positions soon after they had been captured.

The last important British tank action during the war occurred on November 4th, a week before the armistice, when 37 Mark V and Whippet tanks supported the Third and Fourth Armies. The 17th Battalion of armored cars was attached to the Fourth Army. Tanks of the 10th Battalion supported the attack on, and capture of, Catillon and Happeharbes, thus assuring the British of a point for crossing the Oise canal. In this part of the battle, supply tanks, armed with one machine gun each, appear to have been well forward. Three of them, being mistaken by the Germans for fighting tanks, moved ahead of the infantry which was held up by machine gun fire on the British side of the canal near Landrecies, and, although one of the tanks was put out of action, the other two, in conjunction with the infantry that followed them, caused the surrender of the German troops on the far side of the canal.

On account of the heavy fog and the trees and brush, where the 9th and 14th Tank Battalions entered the action, vision was difficult. The infantry depended upon the tanks for direction and the tanks relied upon their compasses. The hedges provided good locations for German machine guns. These were dealt with by the tanks. Resistance by the enemy in this action was not uniform. At places the German barrage and gas were used effectively; at others the shelling was unimportant. Some machine gunners resisted while others did no firing. Many Germans were found in hiding, waiting to be captured. Field guns, placed to enfilade the hedgerows, and antitank guns were found in position, but many of these had not been fired.

One company of the 9th Tank Battalion, operating with the 5th Corps, overcame severe resistance during their advance into the Mormal forest. Tanks with the 37th Division in the attack on Jolimetz aided the infantry in the capture of about 1000 prisoners, and entered the forest. Tanks aided the New Zealand infantry, in the capture of Le Quesnoy. On this day the attack penetrated to a point five miles beyond Valenciennes. The armored cars had been assigned to work in the Mormal forest. They found the roads narrow and slippery and, operating under adverse conditions, they accomplished very little.

On November 5th, the very last tank action occurred when eight Whippets supported an attack north of the Mormal forest. The ground was broken badly by ditches, which facilitated the German rear guard action, and the infantry advanced with difficulty. Two of the tanks went out of action due to mechanical trouble. The ditches and hedges made it difficult to locate the German machine guns, so they had to be hunted out by the tanks. Considerable resistance was encountered by two tanks and two separate attempts were made to capture an objective near Buvignies. One of the tanks had mechanical trouble and the driver was wounded while repairing the difficulty. The other tank continued until it became damaged by running on to a stump and was thus compelled to stop in the German lines. The crew did not leave the tank until that night. They then blew it up and returned to the British lines.

The armored cars continued in pursuit of the retreating German troops. At different points they overtook the troops and worked their way past whole trains of ammunition, guns, trucks, etc., being withdrawn by the Germans. These cars were near the Belgian line, twelve miles ahead of the nearest English troops, on November 11th when they were overtaken by a messenger with the news of the armistice.

Tanks in Russia, 1918.

The Allies despatched forces to Archangel and Uladivostolc before the armistice in 1918 for the purpose of assisting the White Russians to overcome the Bolsheviks, to keep two German divisions from the Western Front (sent to aid in the Revolution), to keep open a sea route to Japan, and to form rallying points for counterrevolutionaries.

In March 1919, two detachments of British tanks were sent from France to Batum to aid General Denekin. Workshops were established at Ekaterinoda, and White Russians were trained in the handling of tanks, both Mark V and Medium A. In June 1919, a single tank led an attack on Tsaritsin, which had previously withstood three assaults. By July 1919, the number of tanks had increased to 57 Mark V and 17 Medium A's, manned by White troops. Morale was low due to Red propaganda and retirement was begun from Karkor, Taragog, and finally from Ekaterinoda and Novorossiisk. The tanks which had not been captured were used mainly in rear guard actions. The remaining tanks were reorganized at Sevastopol. The British Mission left for England when the downfall of the Denekin Army occurred.

In July 1919, another British tank detachment was sent to Northwest Russia for service with the White Russians and Estonians. The tanks here were of the Mark V Composite model. The detachment landed at Reval and moved to Narwa. A special battalion of tank-accompanying infantry was organized but finally had been so misused that it ceased to function. Gdov was captured by two tanks and, on October 11, 1919, three of the six available tanks assisted in the capture of Jamburg, the first step in the proposed advance on Petrograd. On October 15th, the tanks forded the river Luga and caught up with the infantry at Gatchina, whereupon they advanced with it to the outskirts of Tsarkoe Selo, a few miles from Petrograd. A salient had been formed here, due to indifference on the part of the Estonians on the flanks. The Red defense stiffened and gradually forced the Whites back to a White defensive position around Narwa. The tanks moved back to Reval on November 18th and thence to England. The armistice was signed on December 31st.

A small force of four Mark V and two Medium A tanks was sent to Archangel, arriving there on August 29, 1919. Their chief use was to cover the withdrawal of the Allied forces and, after the withdrawal, the tanks were successfully used by the Whites until the fall of Archangel. At this time, the tanks were destroyed to prevent their falling into the hands of the Reds. Only in South Russia were tanks captured by the Bolsheviks. There considerable numbers fell into their hands.

Tanks in Silesia, 1921.

Political troubles between the Poles and Germans over the boundary between the two countries caused the British to send a small division with one heavy tank company to move between the opposing forces which were in battle front across Upper Silesia. Extensive use was made of railroad facilities in this civil disorder, the tanks detraining and entraining frequently. During the winter, the troops were kept at Beuthen and, in June 1922, another flare-up was quelled with tank demonstrations. The troops were finally returned to Cologne in July 1922.

CHAPTER VI.

TANK COMBAT HISTORY. FRENCH.

Organization During the War. As originally organized, both the Schneider and the St. Chamond units were designated as batteries, groups and groupments. The battery corresponded to our platoon and had four tanks. The group comprised four batteries. The groupment included two or more groups and corresponded to our battalion. In May 1917, this organization was modified. The group was then composed of a forward echelon and a (rear) repair echelon. The forward echelon included four batteries of four tanks each. The Schneider groupment included four groups and a maintenance and supply section. The St. Chamond groupment included three groups and a maintenance and supply section.

The Renault tanks were organized into sections, companies, battalions and regiments. The section was composed basically of five tanks, the company of three sections, the battalion of three companies, and the regiment of three battalions. The company was divided into a forward and a rear echelon. The former comprised the three sections and a combat train. The latter was equipped with nine replacement combat tanks and one signal tank.

The information for this narrative of French tank history has been obtained largely from *Les Chars d' Assaut* by Captain L. Dutil, with permission from Berger-Levrault publishers; and, in part, from *l' Artillerie d' Assaut de 1916 à 1918* by Lieutenant Colonel Lafitte with permission from Charles-Lavauzelle and Company, publishers.

THE SCHNEIDER TANK

Schneider. Produced in 1916 by Schneider and Co. Total production, 400.

Crew: 6.
Armament: One short 75 mm (2.95 in.) gun and two machine guns.
Armor: 0.2 in. to 0.95 in. Double armor plates on front, sides, and top. These plates are separated having about 1.5 in. space between them.
Maximum speed: 5 mph.
Suspension: Vertical coil springs with jointed bogie frames.
Tracks: Solid plates with single grousers, width 14 in., pitch 10 in.
General arrangement: 75 mm gun in right front, driver to left of gun, engine forward and left of center, machine guns at sides, final drive in rear.
Dimensions: Length 19 ft. 8 in.; width 6 ft. 7 in.; height 7 ft. 10 in.
Weight: 14.9 tons.
Engine: Schneider, 4 cylinder, 70 HP, forced water cooling.
Horsepower per ton: 4.7
Transmission: Sliding gear, 3 speeds forward, 1 reverse.
Obstacle ability: Trench 5 ft. 10 in.; stream 2 ft. 7 in.; slope 30 degrees; vertical wall 2 ft. 7 in.; tree 16 in.
Fuel distance: 25 miles. **Fuel capacity:** 53 gallons.
Special features: Equipped with double tailpiece. Nosepiece intended as wire cutter and to assist in crossing obstacles. Unditching beam carried on right side. Dome ventilating louver on top of hull. Overhanging hull interfered with cross-country mobility.

The Schneider Tank. List of Actions in which Engaged

Date	Battle	Units	Tanks
April 16, 1917	Chemin des Dames	8 groups (about)	132
May 5, 1917	Laffaux Mill	2 groups "	32
Oct. 23, 1917	Malmaison	3 groups "	36
April 5, 1918	Adelpare		6
April 7, 1918	Grivesnes		6
April 8, 1918	Bois Senecat, DuGros-Hetre, Castel.	3 batteries	

Date	Battle	Number Employed	
		Units	Tanks
May 28, 1918	Cantigny	3 batteries	
June 11, 1918	Mery-Belloy-Lataule-St. Maur	1 groupment	"
July 9, 1918	Porte and Loges Farm	4 batteries	
July 18, 1918	Soissons	3 groupments (about)	123
Aug. 16, 1918	Tillaloy	3 groups	32
Aug. 20, 1918	Bieuxy	1 group	
Sept. 12, 1918	St. Mihiel	2 groups	24
Sept. 26, 1918	Champagne	3 groups	32
Sept. 26, 1918	Foret d' Argonne	2 groups	22

Chemin des Dames, April 16, 1917.

In placing their Schneider tanks in action for the first time during the above operation, the French believed they were working under a severe handicap and that, in a sense, their tanks were obsolete as far as their break-through mission was concerned. In their opinion this condition had been brought about by the premature use of tanks by the British; the most important result of the Somme action, according to the French view, having been the awakening of the German army to their danger from tanks and the consequent widening of German trenches so that the French Schneiders could not cross without assistance.

This view of the situation caused the French to use their tanks as assault artillery and to bring them into the action only after the infantry troops

Plate 8.
Schneider.

had advanced to the point where their supporting artillery could no longer aid them without moving forward. On account of the number of widened trenches which had to be crossed, the plans provided for pioneers to assist at the crossings. In some cases the tanks were not to deploy until they had passed the German 2d line (moving in rear of the French infantry), and in others the deployment was not to take place until the 3rd or 4th line had been passed. In carrying out this plan, a long approach march was made by all tank units in column and part of the march was in broad daylight. German aircraft discovered the tank columns soon after they started. Part of the approach was in view of German artillery and as a result many tanks were lost as they moved up.

Eight groups were assigned to the Fifth French Army, five of which, or 82 tanks, went to the 32nd Corps which was to strike the main blow between the Aisne and the Miette, and three groups, or 50 tanks, to the Fifth Corps west of the Miette.

The groupment supporting the 32nd Corps encountered many obstacles which blocked the road, including poorly made trench crossings, and these, together with artillery concentrations which put a few tanks out of commission, delayed the march. Preceding infantry troops were to have prepared the trench crossings but had failed to do so. Difficulty was had in crossing the German 1st line and the tanks were delayed here 45 minutes, during which time they wiped out two machine gun nests which the infantry had been unable to overcome. It was 10:15 AM before the first tank crossed the German 1st line. The commander of the 2nd Group deployed his tanks between the 1st and the 2nd line to expedite the crossing. The infantry which had arrived ahead of the tanks was unable to aid in preparing the crossing on account of the artillery fire, so the tank crews had to leave their tanks and do this work. At this point the groupment commander's tank was put out of action by a direct hit and this officer was killed. By noon five tanks had crossed the second line and had reached Hill 78, but the infantry hesitated to leave the shelter of the trenches in the face of heavy machine gun fire and shelling. By 3:00 PM two more tanks had come up and the seven advanced. Two of them were struck and burned soon after the start and a little later a 77 mm gun put two more out of action. The crews of the disabled tanks took possession of a German first aid station and prepared it for defense. The remaining three tanks moved back behind Hill 78. A German counter attack drove the crews from the aid station but the three tanks later dispersed this attack. The group commander was killed shortly after this and the remaining tanks of this group were taken over by Group No. 6.

Group No. 6 had followed Group No. 2 and had reached the 2nd German line without losses. It deployed at 11 AM but this movement was witnessed by a German plane observer and soon an artillery concentration was put down which stopped six of these tanks, five of which were set on fire. As a result of these reverses the infantry failed to advance on both sides of the Miette river. The remaining tanks aided in stopping the counter attack mentioned above and then retired.

Another attack was planned about 3 PM, when some large infantry detachments came forward. The tanks and infantry advanced at 5:20 PM, and Hill 78 was taken.

One of the tanks of Group No. 5 was struck and burned on the march forward. Two officers were wounded, one being the group commander. The other tanks advanced in two columns until they arrived at the Wurzburg trench, when they deployed and opened fire. The Germans retreated and the tanks moved on, losing four in this attack. The remainder continued until they reached a hollow where they were somewhat protected from the front, when they halted and signalled the infantry to advance. Although the tanks remained here from 1 to 4 PM and repeated the signal for support from time to time, the infantry, for some reason, did not move forward. Three tanks were sent on reconnaissance and all three were knocked out before they returned to the group. At 4:30 the group commander was notified that on account of losses the infantry would be unable to advance farther. The group then prepared to withdraw, but before they moved back the Germans started a counter attack. The tanks broke this up easily with their fire and the infantry requested that the tanks remain in front of them during the night, as sentinels. Fortunately, orders arrived directing the tanks to return to their deployment point.

Group No. 9 lost two tanks during the approach march. They deployed to cross the Wurzburg trench but found few crossing points, so they reformed in column for the crossing at the most favorable place. The tanks were kept at 30 meters from each other but at this distance they were so badly bunched that ten of the eleven were put out of action by the artillery fire which was brought to bear upon them. Seeing that the infantry could not advance, the remaining tank returned to the French lines.

Group No. 4 was the last in the column to leave the tank park for the approach march. It was delayed by the preceding tanks because of the obstacles encountered by them. No passages over the trenches had been prepared by the infantry, and this group did not reach the second German line until 3:30 PM. Five of the tanks had crossed this line when a German counter attack started which drove the infantry back. The tanks moved up in front of the infantry and stopped the attack, driving the Germans back. Soon thereafter, German artillery opened on the group. A small cannon, brought forward by hand also opened fire and in a few moments four of the tanks were put out of action. The rest had now come forward but the fire from the artillery and many machine guns from front and flank was so intense that the entire group was forced to retire. Of the 82 tanks which left the line of departure, 31 were put out of action by artillery fire and 13 by other causes, making a total of 44 tanks lost. Twenty-six officers and 103 men were killed and wounded during the day.

The three groups supporting the 5th Corps arrived at the point of deployment at 6:00 AM. They left this place at 6:20 with only two and a half groups, eight tanks of the 8th Group having become stuck in marshy ground during the approach march.

Group No. 3 headed the column. After they succeeded in crossing the French trenches and came out of the woods at this point, they were observed by German airmen who signalled the German artillery, which opened a heavy fire on the column. The tanks increased their speed as much as possible and increased the distance between tanks. The infantry could not follow the tanks on account of the heavy fire and dispersed in all directions seeking shelter. The group reached the last French trench at 6:50 AM, but found that the crossings had not been well prepared so had to stop and improve them. This stop cost two tanks which were hit by artillery fire. Upon their arrival at the first German line at 7:15 AM they found a serious obstacle, as the trenches were about five yards wide and very deep. The tanks here were hidden from the view of the artillery and this fire slackened somewhat. Some of the infantry troops were brought forward and aided the tanks in preparing the crossing. This work was done under intense machine gun fire which caused many casualties among the working troops. The approach of Group No. 7, following Group No. 3, started the hostile artillery fire again and many of the tanks were hit, four of them being set on fire. The tank crews took their machine guns into the trenches alongside of the infantry. Deciding that the tanks could not make the crossing successfully, the group commander directed them to withdraw. This movement again started the German artillery fire and four of the five remaining tanks were hit and burned.

Group No. 7, having arrived at a point near the Plain trench while Group No. 3 was engaged in crossing, deployed to the left to prevent forming an artillery target. This brought the tanks into the view of artillery from which they had been hidden and several batteries at once opened fire, knocking out several tanks. The group commander ordered the remaining tanks back but during the withdrawal more were struck and only 5 of the 16 were left, 7 of the 11 tanks struck being on fire.

Group No. 8 had only eight tanks left, the other eight having been mired in the marsh during the approach march. As this group reached the rise near Temple farm it came under heavy artillery fire. As it crossed the last French trench, it reached the rear of the preceding column which was held up at the crossing of the German line. When Group No. 7 deployed to the left, Group No. 8 tried to deploy to the right but found that this movement did not give protection, so the group commander ordered his tanks back to the woods in rear of this point while he went forward to find a crossing for them. Only four of the eight tanks reached the woods, the rest were knocked out on the way back.

Of the 50 tanks sent into the 5th Corps area, 32 remained on the field and 26 of this number had been put out of action by artillery. The losses in personnel were 7 officers and 44 men.

This first action showed that the Schneider tanks were defective in several respects. So many of them caught fire when struck by artillery that it was

Plate 9.
St. Chamond.
(Photo by Ordnance Dept., U.S.A.)

found necessary to give the fuel tanks better protection and change their location; interior communication, observation and ventilation were poor; the track plates were too narrow; and the vertical armor plates were not proof against the German "K" bullet.

THE ST. CHAMOND TANK.

St. Chamond. Produced in 1916 by the St. Chamond Company. Total production, 400.

Crew: 9.
Armament: One 75 mm (2.95 in.) gun and four machine guns. 75 mm gun and one machine gun in front; one machine gun at each side and at rear.
Armor: 0.2 in. to 0.67 in.
Maximum speed: 5 mph.
Suspension: Coil springs, rollers and bogies.
Tracks: Solid plates with double grousers, width 19¾ in., pitch 10 in.

General arrangement: Driver left front, engine forward of center, electric motors in rear.
Dimensions: Length 25 ft. 10 in.; 28 ft. 10 in. (including gun); width 8 ft. 9 in.; height 7 ft. 8 in.
Weight: 25.3 tons.
Engine: Panhard, 4 cylinder, 90 HP, water cooled, with dynamo, two electric motors, and storage batteries.
Horsepower per ton: 3.6
Transmission: Crochat-Colardeau, electric.
Obstacle ability: Trench 8 ft.; stream 31 in.; slope 30 degrees; vertical wall 15 in.; tree 15 in.
Fuel distance: 37 miles. **Fuel capacity:** 66 gallons.
Special features: One model flat topped forward, with two towers and armored searchlight in center; one with front top of hull jutting upward and with or without a small tower on left. Both engine and speed control rheostats controlled by foot pedal. Duplicate controls enabled the tank to be driven from either end. The tank had poor cross-country ability.

The St. Chamond Tank. List of Actions in which Engaged

Date	Battle	Number Employed Units	Tanks
May 5, 1917	Laffaux Mill	1 group	16
Oct. 23, 1917	Malmaison	2 groups	56
June 11, 1918	Mery–Belloy–Lataule–St. Maur	2 groupments	
July 18, 1918	Soissons	3 groupments	100
Aug. 20, 1918	Lombray–Camelin– LeFresne	1 group	10
Aug. 21, 1918	Camélin–LeFresne–Besme	1 group	
Aug. 22, 1918	La Jonquire	1 group	7
Sept. 12, 1918	St. Mihiel	2 groups	36
Sept. 14, 1918	Thielt–Gand	1 groupment	
Sept. 26, 1918	Foret d' Argonne	2 groups	
Sept. 26, 1918	Champagne	2 groups	

Laffaux Mill, May 5, 1917.

The new St. Chamond tanks were first used in a small action on the above date. One group of these tanks and two groups of Schneider tanks participated. The Schneiders passed through their infantry, when these troops were held up by wire and machine guns. These tanks were credited with having materially aided the infantry by destroying many machine guns and one pill box. One tank caused a number of Germans to surrender, and later the Schneider tanks quickly stopped a counter attack.

The St. Chamond tanks did not accomplish much in this action. They had to make a long approach march covering two days before the attack, during which many mechanical failures occurred. Twelve St. Chamonds were at the line of departure at zero hour on May 5th, although some were unable to advance. During the advance the infantry was held up by wire until two of the tanks arrived. They promptly crushed the wire and fired on the German garrison of the adjacent trench until the French troops, who were in close support, came forward and captured the trench and garrison. Two other tanks reached the enemy intrenchments but were unable to go farther or assist in their capture. Even the Schneider tanks which were called from the reserve to aid in the attack at this point were unable to wipe out the resistance after reaching the position and, as the infantry could not take the trenches alone, the attack was halted.

A German account of this action which obviously refers to the successful use of the Schneider tanks, states: "These tanks were able, for the first time, to show their full worth without heavy losses; their achievements were material and tactically conclusive. Many machine gun groups and many guns fell a sacrifice to their work. French artillery succeeded in obscuring the front with a smoke screen, thus interfering with German observation service, while at the same time the French fliers had acquired control of the air. Since these important conditions had been fulfilled the tanks were able to act with but slight losses."

During this action 3 Schneider and 3 St. Chamond tanks were destroyed, 2 officers and 11 men were killed and 14 officers and 49 men were wounded.

Malmaison, October 23, 1917.

Thirty-six Schneiders and 56 St. Chamond tanks supported this action which took place over the Malmaison plateau.

The 8th and 11th Schneider Groups, with 12 tanks each, supported the 21st Corps in the attack on Forêt de Bellecroix and the Montparnasse Heights; the 12th Group with 12 Schneider tanks supported the 11th Corps attacking old Fort Malmaison initially and later the line Chavignon-Many Farm; St. Chamond Group No. 31, with 28 tanks supported the 27th Division against Hill 170 and the town of Fruty; and Group No. 33 supported the 28th Division in the direction of Allemant.

Following the infantry at the start of the action, Group No. 12 later passed it and took part in the capture of the Quarries of Bohery, the crest of the Chemin des Dames and the Malmaison Fort.

The most of Group No. 8 was prevented, by very heavy artillery fire, from catching up with its infantry during the first part of the action. A few tanks of this group passed the infantry and did useful work. One of these wiped out some machine guns that were holding up an infantry company and soon thereafter it also stopped a counter attack.

St. Chamond Group No. 31 left two tanks, which became mired in the first trench, but continued with the remainder. At Gobineaux they destroyed a machine gun. Another tank of this group caused the crew of the first machine gun encountered to abandon it before the tank reached it. Later, this tank wiped out three machine guns and a field gun on the St. Guilain plateau. It then protected the infantry troops for an hour while they were consolidating the position. Other tanks of this group reached the Soissons-Laon highway.

St. Chamond Group No. 33 had very difficult terrain due to rain and many shell holes, and these tanks were unable to do much. By accident one of them got stuck in a trench near a dugout in which 15 Germans had taken refuge. The Germans surrendered to the tank crew.

The second phase started at 9:15 AM. Some of the 12th Schneider Group, which had entered the Malmaison Fort area, became stuck in the fort excavations and could not advance against Chavignon-Many Farm. Part of the 8th Schneider Group were prevented from starting by artillery fire Only a few tanks from this group aided the infantry at this time but they almost made up for the lack of support from the others. One is credited with having captured Oubliette trench, the edge of the Bois de Hoinets, and it then went around the Bousseux ravine, attacking German dugouts from the rear. This tank also protected the infantry while it consolidated the captured position. Another tank rendered great assistance near Bellecroix when it advanced about 100 yards ahead of the infantry against a quarry, capturing 10 men and a machine gun. This tank reached the objective just north of Chavignon plateau about 11 AM.

A part of Group No. 8 had been held in reserve. These were ordered forward at 9:15 AM. They overtook and passed the infantry and overcame a German battery in the Bois de Hoinets.

Schneider Group No. 11 was also ordered up from the reserve for the second phase. After passing the infantry it caused the crews of a German battery of 77 mm guns to withdraw from the edge of the Bois de Bellecroix, and wiped out several machine guns located at the same place, enabling the infantry to advance. Other tanks of this group helped capture trenches near Vaudesson village and put many machine guns out of action which, until then, had held up the infantry.

During this action 6 tanks were knocked out by artillery fire. The personnel losses, including those among the infantry sections which had accompanied the tanks were: 3 officers and 25 men killed, and 15 officers and 104 men wounded.

Adelpare Farm, April 5, 1918.

Six Schneider tanks supported two infantry regiments in the attack on Adelpare Farm and Sauvillers-Mongival. In this action little or nothing was accomplished by the tanks. They were not in the best of mechanical condition due to a long approach march and there were too few of them to cover the area assigned. Only one tank reached its objective at Adelpare Farm.

Grivesnes Park, April 7, 1918.

Six Schneider tanks were used to support one company of infantry in an attack through the woods at this place and, although the tanks reached their objective, they were forced to return because the infantry remained too far in the rear to take advantage of the opportunity presented by the advance of the tanks. Some of the tanks did not return. Captain Dutil reports that: "Those returning were found to be riddled with armor piercing ammunition which, however, had not been able to penetrate the second layer of armor, and the crews suffered only from bullet splash through the eye slits."

Bois Senecat, Du Gros-Hetre, Castel, April 8, 1918.

Three Schneider batteries aided by a heavy fog, attacked over a three kilometer front in advance of two infantry regiments. The infantry maintained close contact with the tanks throughout the action. Bois Senecat and Bois du Gros-Hetre were taken after severe fighting. The German troops here made special efforts to fire through the tank eye slits. Many machine guns were destroyed by the tanks in the woods. Two Schneider tanks cleared the way for the infantry as far as Castel a distance of 900 yards from the line of departure. Three tanks were lost in this action.

Cantigny, May 28, 1918.

Another instance which illustrates the results normally to be expected where trained tank units and trained infantry units cooperate in the preparation for and in the execution of the attack occurred at Cantigny when three batteries of Schneider tanks supported the advance of the 28th U. S. Infantry. This action started with a short but intense artillery preparation. The plan for this preparation arranged for lanes to be left free from fire so that the tanks could advance without difficulty. The tanks led the attack and, advancing on the town from three directions, entered just ahead of the closely following infantry. After the town was captured the tanks covered the reorganization of the position, after which all the tanks withdrew to their rallying point.

CHAPTER VII.

TANK COMBAT HISTORY. FRENCH (Continued).

THE RENAULT TANK.

Renault. Produced in 1916, 1917, and 1918, by Renault and Company. Total production, approximately 5000.

Crew: 2.
Armament: One 37 mm (1.46 in.) gun or one machine gun.
Armor: 0.3 in. to 0.6 in.
Maximum speed: 6 mph.
Suspension: Coil and leaf springs and pivoted bogies.
Tracks: Flat plates with single grousers, width 13 in., pitch 10 in.
General arrangement: Driver in front, gunner in center, engine and final drive in rear.
Dimensions: Length 16 ft. 5 in.; width 5 ft. 8 in.; height 7 ft. 6½ in.
Weight: 7.4 tons.
Engine: Renault, 4 cylinder, 39 HP, thermo-syphon cooled.
Horsepower per ton: 5.3.
Transmission: Sliding gear, 4 speeds forward, 1 in reverse.
Obstacle ability: Trench 6 ft. 6 in.; stream 27 in.; slope 45 degrees; vertical wall 24 in.; tree 10 in.
Fuel distance: 24 miles. **Fuel capacity:** 24 gallons.
Special features: Idlers made of laminated wood. Turret 360 degrees traverse. Front doors well protected against bullet splash by angle irons. The U. S. Six-Ton M 1917 tank was copied from this tank.

Plate 10.
Renault.
(Photo by Signal Corps, U.S.A.)

The Renault Tank. List of Actions in which Engaged

Date	Battle	Units	Tanks
May 31, 1918	Ploisy–Chazelle	3 battalions	
June 2, 1918	Faverolles–Corcy	1 battalion	
June 2, 1918	St. Paul Farm		5
June 3, 1918	Corey–Vouty		15
June 3, 1918	Faverolles		5
June 3, 1918	Pernant and Missy-aux-Bois Ravines		10
June 4, 1918	La Grill Farm		5
June 4, 1918	Montgobert		10
June 5, 1918	Chavigny Farm		5
June 6, 1918	La Grille Farm		5
June 12, 1918	Corcy–La Grille Farm		15
June 15, 1918	Coeuvres	1 company	15
June 18, 1918	Chafosse	2 sections	10
June 28, 1918	Cutry–Saint Pierre–Aigle	4 companies	60
July 15, 1918	South of the Marne	2 sections	10
July 18, 1918	Oeuilly	2 sections	10
July 18, 1918	Soissons	6 battalions	255
July 23, 1918	Marfaux	1 section	5
July 23, 1918	Connantreuil	1 company	15
July 23, 1918	Bullin Farm–Espilly	2 sections	10
July 23, 1918	Savarts	1 company	15
July 24, 1918	Bois des Dix Hommes, Bois de Reims	1 section	5
July 26, 1918	Bois de Fleury	4 sections	20
Aug. 1, 1918	Beugneux-Hill 205	2 companies	30
Aug. 8, 1918	Hangest en Santerre	2 battalions	
Aug. 10, 1918	Ressons sur Matz	1 battalion	45
Aug. 17, 1918	Roye	1 battalion	45
Aug. 20, 1918	Nompcel–Bleroncourdelle–Nouvron Vingre	1 battalion	45
Aug. 28, 1918	Crecy au Mont Crouy	7 battalions	
Sept. 12, 1918	St. Mihiel (American crews)	2 battalions	174
Sept. 12, 1918	St. Mihiel (French crews)	3 battalions	135
Sept. 14, 1918	Mennejean and Colombe Farms	3 battalions	135
Sept. 26, 1918	Foret d' Argonne (American crews)	2 battalions	141
Sept. 26, 1918	Foret d' Argonne (French crews)	4 battalions	214
Sept. 26, 1918	Champagne	7 1/3 battalions	330
Sept. 26, 1918	Hooglede-Thielt	2 companies	30
Oct. 17, 1918	Petit-Verley region	1 battalion	69
Oct. 25, 1918	South of Guise	3 battalions	
Oct. 25, 1918	Hunding Stellung	3 battalions	
Oct. 31, 1918	Advance to the Escaut	2 battalions	

Defense of the Foret de Retz, May 31 to June 18, 1918.

The successful German advance in May 1918 had arrived at the Marne and large forces were advancing in the direction of the Forêt de Retz which Marshal Foch considered the key to the Paris defensive positions. At this time, the French Tank Corps was making every effort to build large numbers of Renault tanks. It had been decided that these new light tanks should not be used in action until at least twelve battalions were available so that they could be used in large numbers. However, since every available means was required to halt the German advance, the three battalions of Renault tanks of the 501st Tank Regiment, then ready, were sent to General Mangin to support his counter attack with six infantry divisions on May 31st.

Ploissy-Chazelle, May 31, 1918.

The long move by the tanks necessary to reach their positions in the Forêt de Retz was made hurriedly by truck, tractor and under their own power. The tank machinery was not benefitted by this move. Only one company, the 305th, had reached its detrucking point at Saint Pierre-Aigle on the night of May 30th and it still had to go to Calvaire that night. A part of the 304th and 305th Companies arrived on May 31st, at 11 AM. The attack was ordered for 12 noon.

No smoke screen was available to cover the tanks, and in order to join the infantry, these 30 tanks had an approach march of about 1600 yards over open, flat ground, and in daylight, in plain view of the German artillery. The French army had been retreating and, consequently, the artillery was not well organized, hence efficient artillery support could not be counted upon during the counter attack.

The six platoons of five tanks each were divided into two groups. One group was to support the attack on Chazelle and the other to aid in the capture of Ploissy. The infantry did not advance far before they were halted by machine gun fire. Passing through the infantry the tanks put the machine guns out of action. After crossing the Soissons road they came under artillery fire which was called for by a German plane that had observed the tank advance.

Continuing, the tanks overcame the resistance on the plateau and along the sunken road between Cravançon Farm, Ploissy, and the Chazelle ravine, their objective.

The Moroccan infantry were slow in advancing even after the tanks were on the objective so the tanks had to return to lead them forward. This procedure was repeated several times and finally a few of these troops came forward to the ravine. Two German guns were then observed. One of them was knocked out by a tank 37 mm gun, but a shot from the other struck a tank, setting it on fire. At this time the tanks, having taken the objective assigned and led the infantry to the position, were withdrawing. The position was not held very long, however, only a few of the Moroccans having come forward and these troops also withdrew when the Germans counterattacked.

Faverolles-Corcy, June 2, 1918.

The 3rd Renault Battalion, with Companies 307, 308, and 309, supported the 11th Corps in this counter attack, which had for its purpose the halting of the German advance through the Villers-Cotterets forest on their way to Paris.

The Germans had crossed the Laviere at Corcy and installed many machine guns in the rye fields on the plateau. One platoon of the 308th Company supported the attack on this area, coming under the fire of the German machine guns at close range. The infantry gave good support to the tanks during this action and succeeded in taking up a position about 250 yards beyond the objective assigned. The Germans moved forward south of this point on the night of June 2-3, and held a line from Vouty to Faverolles. On the 3rd, the 309th Tank Company led an attack which drove the Germans back to the east side of the Corcy-Faverolles road, the Germans retreating without waiting for the tanks and without offering resistance.

A section of the 307th Company led an attack on the right and with their machine guns forced the leading elements of the German advance out of the Faverolles ravine.

In these actions the morale factor and the reputation the Renault tank had made resulted in a considerable withdrawal of the German troops and the saving of many French lives.

The German right flank continued to advance on the 3rd until the 1st Renault Battalion, which was in concealment near the northeast corner of the Forêt de Villers-Cotterets and watching the advance, moved out with its infantry. Passing the foot troops, the tanks attacked the Vertefeuille Farm which had just been captured by the Germans. The latter retreated at once and the farm was occupied by the French.

Pushing ahead the next day west of Longpoint, the Germans penetrated several hundred yards into the forest and captured the Grille Farm. A counter attack was started at 2 PM supported by tanks. The woods were so dense at this point that the tanks were severely handicapped and cooperation between them and the infantry was impossible. They took the farm from the Germans but the infantry could not hold it. This farm was taken the next day with the aid of the 1st Battalion tanks, and consolidated by the infantry.

The Chavigny farm was taken by the Germans, after which the position was attacked by French infantry without success. Tank assistance was then requested and an experienced platoon was detailed to support the next attack. The position was taken and with it, 50 prisoners and 10 machine guns.

Foret de Retz, June 2 to 12, 1918.

The 3rd Renault Tank Battalion was assigned to the 11th Corps to support counter attacks against the German forces which had just captured Longpoint, Corcy and Faverolles. Five tanks were used on June 2nd, in support of infantry units ordered to counterattack towards St. Paul farm. The German machine guns and infantry were well concealed in the rye fields in this vicinity and, being well trained troops, they made good use of this cover, remaining concealed until the tanks approached them when they directed their fire against the eye slits. However, cooperation between the French infantry and the tanks was good and the former was enabled to consolidate a position about 200 yards beyond its designated objective.

During the night of June 2-3, the Germans captured positions between Corey and Vouty, and in the ravines southwest of Faverolles. They had just succeeded in consolidating the latter positions when French infantry, supported by three Renault sections, advanced from an adjacent forest at 5:30 AM on the 3rd. Surprise aided this small counter attack, the Germans were forced back, and the French occupied the positions marked by the Corey-Faverolles road. Later, another section supported an infantry attack in front of Faverolles and, firing from the edge of a ravine upon the German position, forced the Germans back and aided the infantry advance to the Troesnes road.

Having been checked in the vicinity of Faverolles, the German advance was continued north of this area until the Forêt de Retz, in the vicinity of the Pernant and Missy-aux-Bois ravines, was reached. Here they captured the Vertefeuille farm and started into the forest. The First Renault Tank Battalion had just gone into position near this place and, at 6 PM on June 3rd, two sections of tanks supported an infantry battalion in a successful counter attack, the German position being outflanked on two sides. The French infantry were soon reestablished in their position at the farm. At about the same time another tank section was used on a 200 yard front to the north of this area to force back the German machine gunners who were infiltrating into the woods.

An unsuccessful counter attack occurred on the 4th when one section the 302nd Tank Company was used against La Grille farm which had been taken by the Germans about 10 o'clock that morning. This attack was ordered and executed upon short notice, without arrangements for effective liaison. The tank section commander was killed early in the fight. The tank crews could see very little due to the thick woods in which they had to maneuver and, although both the infantry and some of the tanks attacked the farm, they had to withdraw. On the same date, two other tank sections made an unsuccessful effort to force the Germans from the northeast corner of the forest near Montgobert. The tanks were compelled to remain on the paths in this part of the woods as it was not possible to operate through the trees and brush. They could see nothing of the enemy and could only search the brush with more or less ineffective fire so they were finally withdrawn without having accomplished their mission.

An unsuccessful infantry attack occurred west of the Chavigny farm on June 5th, when a part of the 136th Infantry advanced alone against the German salient at this point. After this effort failed the same troops again attacked. In the second attempt one section of tanks was assigned to their support. Paths through the woods permitted the tanks to move quickly into action on the flank of the German position. The position was captured, with ten machine guns and 52 prisoners, and the French reoccupied the edge of the woods.

On the 6th of June another attack was made on La Grille Farm, one section of tanks being assigned to support the infantry allotted for this action. However, the Germans again successfully defended their position. The tanks and about twenty infantrymen reached the farm in spite of the heavy fire but the rest of the infantry did not. After two unsuccessful efforts to get the infantry forward, the tanks withdrew and the farm remained in the hands of the Germans.

On June 12th, a counter attack was made for the purpose of stopping the German advance between Corcy and La Grille Farm. One company of Renault tanks supported this attack. The tanks were very successful here in spite of the thick woods, the abatis, the grenades, and the heavy machine gun fire from German troops which were well hidden in the underbrush. The French infantry advanced several hundred yards and held the position taken.

The following comments by Captain Dutil well describe the situation and the circumstances under which these small tank actions were fought:

These engagements by small fractions supporting such small infantry units that they were occasionally insufficient to hold the ground after it had been taken by the tanks, this prolonged waiting in the immediate vicinity of the front line, this perpetual alert which did not permit giving the machines the necessary care nor the men their indispensable rest, were certainly conditions unfavorable to the employment of tanks. These conditions had been imposed by circumstances, by the critical situation of corps and divisions which had no reserves, by the exhaustion of units in the lines which needed immediate assistance to disengage themselves from the enemy's too active pressure or to help them to regain lost positions. Men and machines had been subjected to a harsh test which they had passed with flying colors. Had they not been capable, the Foret de Retz would have been lost. On June 3rd or 4th the Germans would have been in Villers-Cotterets, perhaps in the heart of Valois.

Counter Attack on Mery-Belloy-Lataule-St. Maur, June 11, 1918.

The Germans continued to advance at various points and on June 9th they had succeeded in reaching a line Mery-Belloy-St. Maur in the Matz valley.

General Mangin, with four divisions, to each of which a groupment of tanks had been assigned, prepared to counterattack on the 11th. The 10th Groupment of St. Chamond tanks supported the attack toward Courcelles-Epayelles-Mery-Wacquemoulin. The 3rd Groupment of Schneiders supported the attack on Mery-Cuvilly; the 12th St. Chamond Groupment were used by a division attacking Belloy-Lataule and the Bois de Lataule, and the 11th St. Chamond Groupment supported the attack on St. Maur.

The 10th Groupment with its infantry, started forward about 9:45 AM under heavy artillery fire. These St. Chamond tanks aided in the capture of German positions southeast of Courcelles, silencing a large number of machine guns which were well concealed in the growing crops. They overcame various points of resistance and had to return frequently and aid their infantry to move forward.

Both the infantry and tanks went past Courcelles, encountering only a few German troops but the tanks came under severe artillery fire from near Rollot and many of them were put out of action. Later the tanks stopped a counter attack coming from the Bois de Mortemer. Sixty-eight per cent of the tanks of this groupment were lost on this date.

The 3rd Schneider Groupment tanks passed Mery and then aided their infantry in the reduction of a strongly held trench system. Many of these tanks were put out of action, in passing Hill 134, by direct fire from German artillery in that area. This fire was so severe that the advance was stopped. To the right of this hill the tanks went forward but the infantry were repeatedly held up by machine gun fire from near Belloy. The tanks stopped to allow the infantry to overtake them, but while waiting a number of them were struck and put out of action by artillery fire. Reserve tanks were ordered up and succeeded in stopping a counter attack, but soon thereafter many of them were also put out of action. During this day, every tank of the left group was knocked out and all of the rest of the groups had lost heavily, the total tank losses for the groupment being over 50 per cent.

The 12th St. Chamond Group caught up with its infantry, which was held up on the whole division front, about noon, June 11th. It passed through the infantry, and the German machine gun crews at once retreated to Belloy. This place was finally taken. It was surrounded on three sides by French infantry and tanks but both suffered severe losses from machine gun and small caliber artillery fire. This groupment lost 37 per cent of its tanks on June 11th.

The 11th St. Chamond Groupment started about 11 AM and supported the attacks on the ravines north of Neufvy, wiping out the machine guns ahead of the infantry and later aided in the attacks on Hill 110 and certain ravines and farms near St. Maur. A 77 mm gun and a great many machine guns were silenced in the attacks mentioned, and a number of prisoners taken by the tanks. Other tanks stopped two counter attacks and aided in the capture of several strongly held trenches. In one of these actions the Germans surrounded a few of the tanks but were repulsed by pistol fire delivered by the tank crews. Nineteen per cent of the tanks of this groupment were lost on June 11th. On the next two days, one of the 11th Groupment batteries aided its infantry in the capture of about 100 prisoners, and several of the tanks are credited with having rendered assistance by overcoming machine guns.

Coeuvres, June 15, 1918.

Attacking along the front between Vingre and Corcy, with special attention to the plateau north of the Forêt de Retz and with the area between the Forêt de Compiègne and the Forêt de Villers-Cotterets as the apparent objective, the Germans had, on June 12th, succeeded in capturing Coeuvres-et-Valsery and the valley drained by the Saint Pierre-Aigle brook before their advance was stopped.

On June 15th, the 303rd Renault Tank Company supported an infantry division in an attack which had for its object the occupation of Coeuvres and the area to the east of this village. Before the tanks could support this attack they had to cross the above-mentioned brook, a natural obstacle between the French line and the plateau east of Coeuvres, which could be crossed by only one bridge. After the brook was crossed the tanks would then have to climb a steep, brush-covered slope.

After the bridge was taken by the infantry at 3:15 AM, the first two platoons of tanks, arriving 15 minutes later, crossed the bridge as planned and without losses, but due to mechanical trouble or shell fire, the third platoon lost three tanks at this point. Climbing the slope, the other tanks overcame the initial resistance and, after they had deployed, advanced ahead of the infantry to the Cutry-Saint Pierre-Aigle road, forcing the Germans back and enabling the infantry to advance to and occupy their objectives with very little loss.

Chafosse, June 18, 1918.

Two platoons of Renault tanks supported an attack east of Montgobert which was directed against the German troops occupying the plateau towards Chafosse. One platoon operated along the edge of the forest south of this place but the other platoon's mission took it through the forest. This attack succeeded in spite of the difficulties encountered by the platoon in the forest. After leading the infantry to its objective, this platoon aided other infantry on the right, to which no tanks had been assigned and which had been unable to advance until the tank platoon arrived.

Cutry-Saint Pierre-Aigle, June 28, 1918.[1]

Four companies, or 60 Renault tanks, were assigned to the 153rd French Division and the 3rd Moroccan Infantry Brigade for this action, which had for its purpose the occupation of the slopes of the plateau east of the Ru de Retz and to reach the line: Fosse en Haut—intermediate points—western part of Saint Pierre-Aigle.

The artillery was given certain definite objectives and protective missions, it was not to open fire until H hour, 5:05 AM. The infantry, supported by the tanks, was to move forward at H plus 2, following the rolling barrage at the rate of 100 meters in three minutes. An examination of the field order of the 153rd Division will show that this was a well-staged attack, and definite instructions, which were well fitted to the situation, were issued to the infantry, artillery, air service, tanks, engineers, and flame throwers. The missions assigned to these branches had been well thought out and so tied together as to assure the maximum amount of teamwork.

The tank companies were directed to support the specified infantry regiments and *assure their installation* on their objectives. One platoon of each company was held as temporary reserve to be thrown in against counter attack, or, as the situation required, and to protect the infantry by fire or by advanced patrolling until the infantry weapons were in position to repel counter attacks.

The instructions pertaining to tank cooperation issued to the assaulting battalions merit special mention:[2] "The cooperation of the tanks may give the attack an irresistible impetus, but an infantry would be dishonored who would subordinate its advance to that of the tanks. These may suffer mechanical troubles, may be bellied, etc. The infantry has no cure for such difficulties and must alone obtain the results sought after; it knows, moreover, that the forced stops of the comrades of the tanks will never be of long duration." Both the infantry and the tank commanders were reminded that the tank units should, after the operation, return to their rallying points only after having been released by the infantry unit commanders under whom they were serving; that it was the function of the platoon commanders to suggest that this order be given, in case the commanding officer of the assaulting battalion delayed giving it as soon as the captured position was sufficiently organized by the emplacement of the machine guns.

The approach march was exceedingly difficult on account of having to cross a ravine with marshy banks. There were only two possible crossings and these were swept by hostile artillery. However, the tank units successfully crossed this obstacle, with slight losses, and entered the fight on time.

On account of the smoke and dust which prevented the troops from seeing more than a few paces, the liaison between the 307th Tank Company and its infantry was poor, so the tanks operated against the resistance in its path somewhat independently and soon reached their objective. One platoon left its zone, due to an error in direction given by its infantry, and operated ahead of an adjacent battalion, reducing the resistance encountered. This platoon was subjected to heavy fire, all five tank commanders being either killed or wounded, but it brought the 9th Zouaves to Les Trois-Peupliers and then moved along the road toward Saint Pierre-Aigle where

---
[1] From *La Revue d' Infanterie*.
[2] *Les Chars d' Assaut* by Captain L. Dutil, French Army.

it delivered effective fire on the House of the Three Gables and its grounds. Hostile machine gun nests at this place, which had been holding up the infantry, were wiped out. The other two platoons reached their objectives and, being unable to establish contact with their infantry, considered that their missions had been accomplished and so returned to their rallying points, During their withdrawal these tanks came under heavy artillery fire and six tanks were struck, two of which had to be abandoned. The personnel losses of this company were five killed and twenty wounded.

The 308th Tank Company used two of its platoons to support the 9th Zouaves assault battalions and kept its 3rd platoon a short distance in rear as a reserve. Dust interfered with liaison between the tanks and the infantry but did not prevent liaison between the infantry units themselves. Although the infantry could not see the signals given by the tank crews, the report on the action states that the cooperation with the tanks was complete and continuous. All resistance was reduced, many machine guns were captured and the objectives were taken. The reserve platoon was used early in the action on the left of the company zone and in cooperation with the tanks of the adjacent regiment. Although several of its tanks were struck, the damage was slight. They were repaired on the field and again took part in the action. All of this company's tanks returned to the rallying point after the action, although some had to be towed back. During this unusually successful action, the only casualty among tank troops was one man wounded.

The entire 309th Tank Company was assigned to support the south battalion of the 1st Moroccan Infantry on account of the difficult ground over which this battalion was to attack, and the probability of a counter attack. The first platoon was directed to move along the edge of the Rue d'Eau woods and dispose of the hostile machine guns which might otherwise be able to enfilade the assault waves of the Moroccans. This platoon lost two tanks (temporarily) as it came forward over unreconnoitered terrain during the approach, so it entered the action with but three tanks, and due to the time lost, it was only able to participate in the latter part of the action. It was successful in reducing the resistance in and about Cutry. The two ditched tanks and the others were brought back to the rallying point. The other two platoons were used directly in the attack and were able to complete their missions over their entire front in spite of the temporary loss of liaison with their infantry. Although some of these tanks were put out of action temporarily, both platoons are credited with having greatly assisted the progress of their infantry. The losses in personnel were: 1 killed and 9 wounded.

No records are available covering the action of the other tank company assigned to this fight.

Porte and Loges Farms, July 9, 1918.

Special preparations were made prior to this action in order to demonstrate what the tank personnel believed could be expected of tanks used on suitable terrain if proper attention was given to efforts to secure surprise, to make detailed preparations in advance, and to secure a rapid execution of the plan of the attack. The objective was somewhat limited and was marked by the above mentioned farms on the sunken road from St. Maur to Anteuil.

Four Schneider tank batteries were assigned to the assault battalions. The approach march was made at night and was masked by harassing artillery fire and, since the attack was started very early, before daylight in fact, surprise was attained. The ground over which the attack was made was under hostile observation but this handicap was quite well taken care of by a heavy smoke screen which was placed on the German observation posts, and the German batteries were subjected to a heavy gas attack.

The objectives were taken according to schedule and, having aided their infantry to capture and consolidate its position, the tanks were withdrawn one hour and ten minutes after the engagement. The cost of the action in tank casualties was small and this was attributed to the very careful preparations made before the action and the promptness with which the tanks were withdrawn after a limited exposure on the field.

Counter Attack South of the Marne, July 15 to 17, 1918.

The German offensive on July 15th had driven the French troops south of the Marne and back of the southern edge of the Bois de Conde-en-Brie. Anticipating the attack in this area, the 3rd Corps of the Sixth French Army had been prepared for the counter attack. One battalion of the 502nd Tank Regiment was to support this movement east of the Sarmelin river and at 7 PM two sections of the 314th Renault Company, which had reached Celles-les-Conde, were used. This late attack was successful, the infantry and tanks caused the Germans to withdraw into the Conde forest.

Moving forward again on the morning of the 16th, the Germans were met by the tanks and infantry fire. The action continued throughout the 16th and 17th. The tanks were used at several points between the German lines and the towns of Celles-les-Conde, Montmirail and Connigis. They are credited with having given effective assistance to the infantry of the 3rd Corps in stopping the German offensive in this area.

Oeuilly, July 18, 1918.

The tanks of the 4th Renault Tank Battalion were moved up by tractor and trailer on this date to relieve the 5th Battalion tanks which had been engaged for three days under severe terrain conditions. The 4th Battalion tanks reached Bois de la Grange-Fosse late in the afternoon and two sections were placed in support of infantry units attacking German observation posts south of Oeuilly. The attack started at 6:30 PM with no time for the tank crews to rest and little or no time for reconnaissance. The two sections were used in two attacks and, in spite of the haste with which they entered the fight, they reached their objectives in both instances and enabled the infantry to reach and consolidate them.

On July 19th one tank section supported an unsuccessful attack on the Bois de Leuvrigny. On the following day the German positions south of the Marne were subjected to attacks from different directions but little or no resistance was encountered as the German troops in this area were withdrawing across the river to the north. The French troops inflicted some losses among the last of the retreating forces as they crossed the Marne. In all, two groups of Schneider and four companies of Renault tanks were assigned to the various attacks on July 20th.

CHAPTER VIII.

TANK COMBAT HISTORY. FRENCH (Continued).

Soissons, July 18 to 26, 1918.

In this action, three groupments (123 Schneider tanks), three groupments (90 St. Chamond), and three battalions (130 Renaults) were assigned to the Tenth French Army. In addition to these, three battalions (approximately 125 Renaults) and one group (approximately 10 St. Chamond tanks) were assigned to the Sixth French Army. Thus an approximate total of 478 tanks was used. Among the troops participating in the Battle of Soissons were the 1st American Division, to which the 11th and 12th St. Chamond Groupments (of about 30 tanks each) were assigned, and the 2nd American Division, to which the 1st Schneider Groupment (48 tanks) was assigned. The two American divisions formed a part of the Tenth French Army.

Where divisions advanced with brigades abreast, tanks were allotted to the assault, support, and reserve battalions and, where divisions advanced in column of brigades, each brigade received an allotment of tanks. With this arrangement, each division had its own tank reserves. In addition, three of the Renault battalions allotted to the Third Army were held in army reserve. The higher commands were charged with providing protection for the tanks by the usual means, and assault positions were selected with a view to concealing the tanks until the last moment. Every effort was made to retain and profit by the element of surprise. No preliminary bombardment was permitted and the final movement of tanks to their assault positions was made at night.

At 4:35 AM on July 18th, French batteries opened fire and the infantry and tanks moved forward. The German troops were surprised and offered little resistance during the first part of the advance. In the zone of the 3rd Groupment (Schneider tanks) the resistance stiffened about 11 AM and from this time on the tanks sustained casualties. From about 3 PM the infantry was practically pinned to the ground a few hundred yards east of the Saconin Missy-aux-Bois ravine. Several of the St. Chamond tanks were stopped by mechanical trouble but the remainder assisted in the capture of the above mentioned ravine, and broke up a counter attack. Later they led their infantry beyond the Soissons road. One group of St. Chamond tanks assisted the 18th American Infantry Regiment as far as its final objective.

One groupment of Schneider tanks assisted the Moroccan Division in capturing Dommiers, the Ferme de la Glaux, and the clearings of Jardin and Translon. The 2nd Brigade of this division, after it had leapfrogged the 1st, in cooperation with its Schneider tanks, carried the attack beyond Chaudun. The tanks cleaned out Chazelle and Echelles ravines and later in the day supported their infantry in the second attack on Vierzy.

One Schneider group aided in the capture of Vertefeuille farm. They then helped to clear out the ravine between Chaudun and Vaux-Castile, and about 11 AM led their infantry in the first attack on Vierzy. Although this attack was successful, a counter attack drove the French back at this point and Vierzy was again taken by the 2nd Brigade, as mentioned above.

A group of St. Chamond tanks assisted in the capture of Villers-Helon and led its infantry one kilometer beyond this place. This group was replaced in the afternoon by a group that had been used in the attack on the Bois de Manloy. This attack was not carried to completion, however, as the infantry was exhausted and was driven out by a counter attack.

The three Renault battalions had followed the attack in reserve and were intended to be used in case they were needed. The first battalion of Renaults was used about 7 PM by the 2nd American Division. It supported an attack from Vaux-Castile toward Concroix and Hartennes woods for a distance of three or four kilometers. One company of Renaults was used late in the day at the Echelle ravine. In general, the tanks sooner or later caught up with and passed the infantry during the first day of the battle and led it into action until it could advance no further.

The tank support in the Sixth Army was also very effective. Two companies of Renaults supported the 2nd French Division in the capture of Marizy, Sainte Genevieve, Passy-en-Valois, Montron and Macogny.

One battalion of Renaults aided the 47th French Division in reaching and taking Dammard, Lessart Farm, Monnes, and Cointicourt. One St. Chamond group was unable to pass through its infantry on account of the speed of the infantry advance and, consequently, it was not able to render much assistance. One other company of Renaults is credited with having taken a very effective part in the capture of the Orme wood north of Courchamps. The Sixth Army advanced about five kilometers during the first day.

Of the 223 tanks of all types actually engaged in the Tenth Army on July 18th, 62 were put out of action by artillery fire and 40 more by mechanical difficulties, etc. Tank casualties were thus 102, and of the tank personnel, 25% were casualties. The tanks of the 12th Groupment (St. Chamond), assigned to the 2nd American Division, were known to be in poor mechanical condition at the start. Many breakdowns were expected and for this reason two battalions were given to this division. During the night of July 18-19, all tanks that could be were repaired and returned to duty. Tank units in the lead were reorganized where possible and reserve units moved forward to replace the units which had sustained the greatest losses.

The 3rd Schneider Groupment was able to turn out only two platoons of six tanks each. These supported the attack to the main Soissons road. Here another reorganization took place, and, one platoon being left, the attack was started again at 5:30 PM and the line was carried to the Mont de Courmelles farm. At this point only two tanks remained, the rest had been put out of action by artillery. The 4th Schneider Groupment, with eight tanks, supported the Moroccan Division in the attack on Charantigny. Two of the tanks were knocked out but two of the others entered the village and cleaned it out while the other four fired on the ravine north of the village.

The 11th St. Chamond Groupment placed 14 tanks in the field on July 19th with the 2nd Division but, due to the fact that the tanks and infantry were not well coordinated, only a small advance was made.

One company of Renaults operated with Moroccan troops in the attack on Echelle ravine, but this attack failed.

The attack by the American Division in the direction of Tigny and Hartennes was delayed. A false start revealed to the Germans the presence of the tanks. Having been located, one company lost five of its eight tanks at the assault position, and another company, which advanced against Tigny, came under direct artillery fire and lost seven of its eight tanks. Eight tanks led American infantry to Tigny and other tanks were followed by American infantry, in spite of severe losses, to a point about one kilometer beyond Taux. These gains, however, were not held.

During the afternoon a group of St. Chamond tanks assisted in the capture of Blanzy. This company preceded the infantry and surrounded the village, losing eight of their nine tanks in this action. Ten Renault tanks also aided in this attack. Other tanks aided the attacks on Saint Remy-Blanzy and on Parcy-Tigny.

The advance of the Sixth Army was assisted by the tanks used the previous day, to which had been added two St. Chamond groups. The Neuilly-Saint Front was taken by the 2nd French Division and three companies of Renault tanks.

A St. Chamond group did good work in breaking the resistance ahead of the infantry in the advance on Rassy, and this group also later assisted in a second attack on Menuet woods. The first attack had occurred in the morning when French infantry, having waited in vain for the arrival of these tanks, had made an attack but had been unable to advance.

A company of Renaults aided in the capture of the Monnes-Neuilly-Saint Front railway line and another Renault company was used to advantage in the Rassy attack. Continuing its attack of July 18th, another Renault company led its infantry from Orme Signal to Sommelans.

Thus it is seen that, in the actions on July 19th, many of the tank units which were in action on the previous day were still able to give assistance, although, as on the 18th, they suffered many casualties. Of 195 tanks, 50 were destroyed by artillery and 22% of the personnel were casualties.

As far as the tank units were concerned, the remainder of the Soissons battle, from the 20th to the 26th, was a series of small actions executed by small numbers of tanks. Repair work was carried on in all units night and day and almost continual reorganization was necessary to get tanks together for participation where needed. The assistance rendered by the tanks from July 20th on gradually became weaker as the personnel and vehicles became fewer. The German antitank measures were becoming better organized and more efficient.

The 3rd Group had only three Schneider tanks in operating condition for an attack upon a well fortified strong point on July 20th. They were inadequate to deal with the many machine guns which had stopped the infantry. These tanks were sacrificed without having accomplished their mission, all being put out of action. On July 21st two platoons of three tanks each were brought forward and the attack on the same strong point was continued. Three of the six tanks were knocked out without having accomplished their mission.

On July 20th an attack was attempted, with tank assistance, in the sector of the 20th Corps. Ten Renaults of the 2nd Battalion took part in the attack of a battalion of infantry on the trenches located on the plateau between Charantigny and Villemontoire. This resulted in very little gain. Six of the tanks were hit, three of them being destroyed.

On July 21st an attack was made against Villemontoire and Tigny with three infantry regiments, together with the 1st and 2nd Renault Battalions and four platoons of the 4th Schneider Battalion. The advance was started without preliminary bombardment and at first gave promise of being successful. The Schneider tanks entered Tigny, but the infantry was reduced in personnel and could not follow. A counter attack forced the French back to their line of departure. Another attempt was then made and again the advance progressed well until the infantry was pinned to the ground by machine gun fire, and the attack had to be abandoned, leaving a number of the tanks in the German lines.

In the zone of the 30th Corps, five separate attempts to take the village of Le Plessier-Huleu occurred on July 20th and 21st, with different infantry organizations and tank units. One group of St. Chamond tanks and two platoons of Renaults made three attacks on the 20th. Two attacks were made on the 21st. In the final attempt, the tanks succeeded in approaching the village from a flank and destroyed the machine gun nests there. The village was then occupied. Before withdrawing, the tanks broke up a counter attack but, as they withdrew, a German gun appeared at the edge of woods close to the tanks and five of the six tanks were put out of action by it. The French were driven out of the village and it was necessary to take it again on July 22nd. Three St. Chamond tanks took part in this attack but the infantry did not follow them. However, the village was partly evacuated by the Germans, and the French occupied it in the evening.

Units which had been held in reserve were used with the Sixth Army on July 20th, 21st, and 22nd and worth-while advances were made. One Renault company aided in taking Vareille farm and the Grisolles road; another company pushed forward with its infantry to the main Chateau Thierry-Soissons road and the Chatelet wood; still another assisted in the attack on Maubry and the Latilly wood. Two St. Chamond platoons of the 41st Company had been brought up and gave material assistance at Menuet wood.

The exploitation of the initial success of July 18th was over, but still one other attempt was made by the Tenth Army on July 23rd, in which one groupment of Schneiders, and one of St. Chamond tanks, 37 in all, were allotted to the 20th Corps, and 6 St. Chamond and about 62 Renault tanks to the 30th Corps. The main objective was the Orme de Grand-Rozoy, where the Germans were in well prepared positions and defended them with efficient artillery. Some success was had at the start, but the number of tanks was inadequate to affect the outcome materially. Of the 69 used, 48 were knocked out. In one case a battery of German guns opened fire within 50 yards of a tank platoon and entirely destroyed it. The remaining tanks and their infantry withdrew. The losses in tank personnel amounted to 27%. The remaining tanks of the Tenth Army were withdrawn and placed in reserve on the evening of the 23rd, after having lost 102 of the 223 tanks engaged and 25% of the tank personnel since July 18th.

The tank units of the Sixth Army continued, as far as they were able, to accompany the advance until July 26th. Two Renault companies aided in the attack on Cheneviere farm on July 23rd, and on the 25th and 26th another Renault company attacked Tournelle wood. These companies had been 48 hours on the battlefield of which 30 had been spent in actual combat. By this time the Sixth Army tanks were unable to render further appreciable assistance, so they were withdrawn. The tank units with this army had lost 25% of their officers, 6% of their men and 58 tanks.

Tanks gave a great impetus to the Soissons advance at the start and the formation of the tanks in depth permitted them to move as far and as fast as the infantry could follow. It became apparent that fresh infantry, to take prompt advantage of opportunities created by the tanks and to hold the ground gained by them, was one requisite to success. This action also proved that small numbers of tanks could not overcome great numbers of field guns and machine guns and, when used in small numbers, they only served to draw upon themselves, and upon the infantry units, hostile artillery concentrations. Finally the Soissons battle showed that the principal enemy of slow tanks at that time was the hostile artillery. This fact the French division commanders often overlooked. They considered their artillery insufficient in quantity and did not want to allot a portion of it for what they considered to be the benefit of the tanks alone. "The employment of smoke shells was little honored by many artillerymen, and the batteries, in their endeavor to have the largest possible supply of explosive shells on hand, had only a small amount, if any, of smoke shells at those moments when they were needed."[1]

---
[1] *Les Chars d'Assaut*, Captain L. Dutil.

Marfaux, Connantreuil, Bullin Farm, Espilly, Savarts Farm, July 23, 1918.

After the German retreat as a result of the action initiated by the French on July 18th, several attacks were started to the south and east of the Soissons area and a few of the tanks were used in some of these actions.

At Marfaux one section of Renault tanks supported the 186th British Infantry and aided in taking that place, although most of the tanks were lost in the action. The strong point at Connantreuil was attacked on this date by a group of chasseur battalions, supported by a company of Renault tanks. The Germans had mounted a great many machine guns in the park and woods about the chateau of Connantreuil and had cut down trees for use as tank obstacles. Working together the tanks and infantry finally captured the woods and park but could not take the second objective, Bois des Hommes, on that date. Another attack was made on the 23rd by a Scotch Infantry organization to which two sections of Renaults were assigned. The objectives, Bullin Farm and Espilly, were captured after a great many machine guns, located in the woods through which the troops advanced, were wiped out.

The attack on Savarts Farm, which was made by an infantry regiment and one company of Renault tanks, did not succeed. Lack of preparation, insufficient liaison, and well concealed machine guns in the very thick woods about the objective were among the causes of this failure.

Bois des Dix Hommes and Bois de Reims, July 24, 1918.

One section of Renaults supported a successful advance into the Bois des Dix Hommes on this date but the attack on the Bois de Reims failed. The reasons given for this failure are that the distance covered was too great and that the advance was made too rapidly.

Bois de Fleury, July 26, 1918.

Four Renault tank sections were assigned to support the 9th Division in its mission of straightening out the line west of the Bois de Fleury. This attack failed due to the successful German counter attacks which repulsed the French infantry. The tanks then covered the retreat of the infantry.

Beugneux and Hill 205, August 1, 1918.

A general attack started on this date in which the 30th Corps participated. The mission of this corps was to capture Hill 205 which commanded Grand-Rozoy, a village that had been occupied by the French on July 30th, and Beugneux. Two Renault companies attacked in line, protected somewhat by mist and smoke. The Germans defended this position with a great many machine guns which were located in the Beugneux woods. These were wiped out by the tanks and the woods and hill were occupied by the French.

The plan now called for a passage of lines by the 27th Division supported by the 330th Renault Tank Company. An unfortunate delay prevented the tanks from leaving with or ahead of the infantry and when they did start the mist had cleared away. These tanks soon encountered direct fire from two 77 mm guns. Ten of the tanks were struck and set on fire. This misfortune held up the tank advance and the attack made no further progress.

Hangest-en-Santerre, August 8 to 10, 1918.

The French First Army advanced on August 8th, as a part of a general Allied offensive, on the right of the 4th British Army which was operating between the Ancre and the Amiens-Roye road. Two battalions of Renault tanks had been allotted to the First Army and assigned to the divisions on the front southeast of Thennes.

Following a brief artillery preparation the infantry and tanks advanced, surprising the Germans and forcing them back toward Fresnoy-en-Santerre. The attack was continued on August 9th when Fresnoy was captured. On the 10th, Renault tanks took the ridge between Hangest-en-Santerre and Fresnoy, overcoming very strong resistance in the form of machine guns and antitank guns, and the French line was extended to Hangest.

Ressons-sur-Matz, August 10, 1918.

The Third French Army advanced on the right of the First Army. The 10th Battalion of the 104th Renault Tank Regiment supported the center of this attack with Ressons-sur-Matz as its objective. The tanks led this attack and are credited with having aided in the capture of La Taule Park, Machet Mill, Ressons Station, Ressons, and La Neuville.

Tillaloy, August 16, 1918.

One groupment of Schneider tanks, consisting of three groups, with a total of 32 tanks, was assigned to the French Reserve Army for the attack on this date. Owing to the poor mechanical condition of these tanks it was decided to withhold them from the attack until after the first objective had been taken, as they would not be effective fighting machines in the trench system to be crossed.

Only eight batteries participated in the attack on August 16th, north of Tillaloy. The tanks wiped out a number of machine guns but as the infantry did not follow they were withdrawn. Adjacent troops, which were to attack with the troops supported by these tanks, did not advance on August 16th. They did attack on the 17th, however. Eight tanks of one group became stuck in the trenches, as no crossing points had been prepared for them, and they accomplished nothing. The tanks of another group passed the trench system and reached the vicinity of Laucourt where they were engaged for an hour and a half but as the infantry did not advance, the tanks withdrew. As they went to the rear, the Germans followed, trying to surround the tanks. The tanks turned to face the enemy occasionally and were also defended by their infantry. The Germans regained the position from which they had been driven.

Roye, August 17, 1918.

The 11th Battalion of Renault tanks supported the 31st Corps in its attack on Roye. The left division was, on August 16th, held up by a very strong trench system on the line Goyencourt-Camp de Cesar-St. Mard, and the tank battalion was assigned to this division to capture the trenches and the railway back of them. The attack made little headway on the 17th as many tanks became stuck. The infantry advanced about 700 yards on this date. On the following day the Bois Fendu and the Grange Farm were captured and occupied after severe fighting. Other points were captured but not held. Ten tanks were destroyed in this operation.

Nouvron-Vingre, August 20, 1918.

One battalion of Renaults, one group of Schneiders, and 30 St. Chamond tanks were used by the Tenth French Army in this action. The 7th Corps, to which the Renaults were attached, was to attack the plateau north of Nampcel; the 18th Corps, with the St. Chamonds, was to capture Lombray and La Fresne; and the Schneiders were to support the attack on Nouvron-Vingre by the 11th Division.

About half of the Schneider tanks were stopped by German defensive measures and rendered little or no assistance. The remaining Schneiders led their infantry over the plateau from Nouvron to Bieuxy, encountering heavy artillery fire toward the end of this advance which knocked out all of these tanks.

The Renaults aided a Senegalese battalion to cross a ravine where machine guns had stopped it. They then placed their fire on gullies leading to the Nampcel plateau thus assisting the infantry to move on, and later helped capture Bleroncourdelle at the far side of the plateau. In this successful action six tanks were knocked out, eight broke down and 30% of the crews were killed or wounded.

The St. Chamond tanks did not start well, many of them being held up by mechanical trouble. Three of them went with their infantry and later ten of the others succeeded in getting forward to their infantry and assisted in the capture of Camelin, La Fresne, Besme and La Jonquiere and Javelle farms.

Lombray, Le Fresne, Nampcel, Nouvron-Vingre, August 20, 1918.

On the above date, tanks joined the attack which had been started by the 10th French Army on August 18th, against the main position on the north slope of the Nampcel-Morsain ravine. Tanks had been available since the start of the action but were not used in the beginning due to the fact that the German trenches to be taken were too wide for the tanks to cross in action.

The 5th Renault Tank Battalion, the Schneider Group 11, and 30 St. Chamond tanks were used in the action. The 7th Corps, on the right, with the Renault battalion, was to attack the plateau north of Nampcel; the 18th Corps, supported by the St. Chamond tanks, was to capture Lombray and Le Fresne; while the Schneider tanks were to support the 11th Division in its attack on Nouvron-Vingre.

During the action, two Renault companies aided the 48th Division in crossing the Nampcel ravine by breaking up the resistance located in Nampcel which had stopped the 3rd Senegalese Battalion. These tanks also aided in the advance on the plateau and in the capture of Blerancourdelle. Shell fire accounted for six Renault tanks and eight were ditched. Thirty per cent of the personnel was killed or wounded. The St. Chamond tanks were reduced in number by many breakdowns. They were barely able to keep up with, but could not advance ahead of, their infantry. Three of them reached Lombray with the infantry and two others gave some assistance in the attack. One of the two batteries of Group 11 (Schneiders) had difficulty in the trench area and could not catch up with its infantry. The other battery gave excellent assistance. It advanced about one mile east of Bieuxy and continued in action until all of its tanks had been struck by shells and stopped.

On the night of August 20-21, the St. Chamond tanks were reorganized and a group of ten was assigned to support the attack on Camelin and Le Fresne. Only one of these successfully crossed the wooded defile en route to Camelin but this tank moved around this village and aided in its capture. By evening four more tanks had been unditched and had crossed the defile so they were used in the attack on Beseme.

On August 22nd, seven tanks aided in wiping out the resistance coming from the line, Bourguignon-Besme. One of these supported its infantry in attacking La Jonquiere and three others reached the line, Javelle Farm-St. Paul, forcing the Germans to retire from this line. From here on until the Ailette canal was reached no more resistance was encountered.

Crecy-au-Mont—Crouy, August 28, 1918.

Seven Renault battalions were used by the French Tenth Army in this five-day action. The French line in this area, Manicamp-Ailette Canal-Guery-Pont St. Mard-Juvigny-Cuffies, marked the beginning of a severe struggle, as the Germans were determined to hold their positions opposite the above mentioned line. Three battalions were allotted to the 30th Corps and used initially as follows: one battalion supporting the attack upon Crecy-au-Mont and Limonval Farm, one company in the attack on Le Banc-de-Pierre and the plateau of Mont de Leuilly, two companies supporting the 32nd U. S. Division against the plateau north of Juvigny and Fontaine St. Remy. Three battalions were allotted to the 1st French Corps in the southern sector. One battalion was kept in army reserve.

Two unsuccessful attacks were made against the defenses north of Cuffies on August 28th. Four tank sections were used in the morning attack but twelve of these twenty tanks were ditched or destroyed, and the attack failed. Two sections supported the second attack in the afternoon but it also was stopped without having accomplished its mission. On the 29th, four companies led an attack between Crecy-au-Mont and Juvigny against determined resistance consisting of great masses of machine guns which had been organized in great depth. The tanks successfully crossed the trench systems and wiped out a large number of machine guns but were unable to reach all of them. The remaining guns stopped and held the infantry from advancing except in the center of the attack where a limited progress was made in the direction of Terny. One section was used on August 30th against the Bois du Couronne. This section captured its objective and turned it over to the infantry.

The advance was renewed all along the line on August 31st, following a four hour bombardment. Antitank measures were very efficient and German observation was good on this date. The 6th Tank Battalion supported a Moroccan Division and the 66th French Division. The 32nd U. S. Division was supported by the 4th Tank Battalion. The tanks on the left came under direct observation from north of the Ailette and 24 tanks were knocked out. The remaining tanks continued in action and assisted the infantry in the capture of the assigned objectives. The third German line was penetrated near Terny-Sorny by the 32nd U. S. Division and its tanks.

In the 1st Corps zone, the tanks assisted the 59th Division to advance about one mile to the Bois de Faucon and to the south of Leury. One tank section assisted in the capture of Crouy during a very heavy bombardment by the German artillery. The French infantry was stopped at this village at the close of the day's fighting. A well prepared attack on September 1st, when two tank sections supported two infantry battalions, advanced the line about one kilometer or to the National road near Leury. The advance was again started by both the 30th and the 1st Corps on September 2nd. The tanks started before H-hour in order to stay close to the rolling barrage which started at the end of the short artillery preparation. In their intitial advance, the tanks were covered by smoke screens. They did good work and accounted for many machine guns. The infantry and tanks were able to advance the line about one mile during the day although twelve tanks were put out of action in this advance by the well organized antitank measures encountered. Two tank companies supported two divisions farther south to the National road, and two other companies aided the 69th Division, coming from Crouy, to attack the heights known as the Vergny plateau. These tanks avoided a frontal attack on this plateau by climbing up through the ravines. They were subjected to a heavy gas bombardment but, by this approach they were able to destroy a number of machine guns. One section attacked Bucy-le-Long where 80 prisoners were captured and this aided the 5th Division to cross the Aisne. After the 69th Division had taken its position upon the plateau, a counter attack drove them back a short distance but they remained upon the plateau.

On September 3rd, minor operations were carried out by several tank sections and small infantry units. Perriere Farm was captured and attacks were made upon the Bois de Terny and upon Bucy-le-Long. At the latter village the tanks engaged in street fighting and pushed through to the far edge but the infantry was not able to follow, so only the western portion was occupied.

Mennejean and Colombe Farms, September 14, 1918.

A combined attack by the 1st and 20th Corps was planned for the purpose of capturing the Mennejean and Colombe plateaux upon which the above mentioned farms were located. The 8th and 9th Renault Battalions supported the 1st Corps against the Mennejean farm. The 7th Renault Battalion was assigned to support the 20th Corps in its attack on the Colombe farm.

The ground over which the attack was made unsuitable for tank operations, due to continued rain, and certain ravines near the Colombe plateau, which were to have been captured prior to the action, were still occupied by the enemy. So this attack was made more difficult than was at first expected and the infantry, worn out by previous actions, did not reach the plateau. This attack was aided by the division on the right, which had reached its first objective at the west edge of Vailly, and which was then ordered to make a flanking attack on the Colombe Farm plateau, supported by the 320th Renault Tank Company. This division succeeded in changing the direction of its attack and moved north. The tanks cleaned up many machine guns and assisted the infantry to secure a position on the plateau. These troops repulsed a strong counter attack during the night, and continued the advance on Setpember 15th. One section was used on the 16th in this area, but the concentration of artillery fire was so severe that four of the five tanks were quickly put out of action.

In the Mennejean area, an advance of about one mile was made. The tanks reached Mennejean farm but the infantry troops were held back by fire from the Colombe sector, where the attack on September 14th had not been successful, so the gains made by the tanks in the Mennejean sector could not be held. The tanks cleared out some of the trenches north of Sancy, but many tanks were stopped by artillery fire and the infantry did not continue the attack.

Champagne, September 26 to October 3, 1918.

This series of actions was carried out at the same time that the American forces were engaged in the Argonne offensive. On account of the condition of the terrain, tanks were not expected to participate in the first part of the attack of the Fourth French Army in the Champagne area; they were to get forward to a position where they could be used in the second phase of the action, and pioneer crews were detailed to assist them in getting forward.

The 16th Battalion, with 45 Renault tanks, the 10th Company, with 15 Renault tanks, and the 15th Group, with 12 Schneider tanks, were assigned to the 2nd Corps. The 2nd and 3rd Battalions, with 45 Renault tanks each, and the 4th and 9th Groups, with 10 Schneider tanks each, were assigned to the 21st Corps. To the 9th Corps were assigned the 10th and 11th Battalions, with 45 Renault tanks each. It was also planned that the Army should have the 17th and 18th Renault Battalions and two St. Chamond groups in reserve. One supply and repair section was attached to the tanks allotted to each corps.

An artillery preparation was used in this action, starting at 11 PM on September 25th. From 5:25 AM (zero hour) on the 26th, the infantry attack progressed rapidly, and the pioneers, about 2800 men detailed from engineer and infantry regiments, assisted the tanks over trench and terrain difficulties, found and destroyed, or aided the tanks to avoid, the mines and tank traps provided for them, and in some cases constructed roads for the tanks. This efficient cooperation enabled the tanks to join the troops to which they were assigned for the second phase on September 27th.

Although the use of tanks was not contemplated in the first phase on September 26th, some of them were called into action during this phase by infantry units of the 11th Corps. Early in the day one Renault platoon successfully attacked machine guns which were holding up the advance in one locality, and about 5 PM another Renault platoon wiped out the resistance which had been preventing the infantry from leaving the Mecklenburg and Schwerin trenches.

Several platoons were engaged on September 27th. Two Renault companies aided in the attack on La Dormoise in the morning. Two platoons were used in an attack against the Bois de la Tourtelle although the terrain in this area contained many serious obstacles for tanks in the form of steep ravines. German artillery fire and smoke blinded the tank crews at this point and many tanks upset, seven out of eight being abandoned in the ravines. Two Renault platoons drove the Germans from Oiseaux ravine and later aided in the capture of the "Battoir." This point was recaptured by the Germans, but they were driven out by another tank platoon and fresh infantry. These tanks then advanced to Croix Muzart, but were not followed by the infantry.

In the zone of the 21st Corps, the infantry was stopped by fire coming from Wurzburg trench and the Manre Tunnel until this resistance was overcome by Renault tanks from the 3rd Battalion. The Germans were forced back and the infantry occupied La Pince Heights and the Bois de Bouc. In this zone several counter attacks were stopped during the day by both Renault and Schneider tanks. In the zone of the 11th Corps, Renault tanks of the 10th Battalion encountered stiff resistance while assisting in the capture of the Rhenans, Hohenzollern, Eperon, and Karlsruhe trenches. One platoon of Renaults aided in the capture of Salzbourg and Stuttgart trenches and a company of Renaults cooperated with its infantry in the capture of the Bois de Grand and then assisted in taking the final objectives.

On September 28th the various corps advanced individually as rapidly as possible in their own zones, some being more successful than others. In the 2nd Corps one platoon supported an unsuccessful attack east of Manre. Here the infantry was held up by machine gun fire. A tank company attacked the machine guns and made five different efforts to take the infantry forward but the infantry did not follow. The tanks then passed through Manre but they were soon all hit or failed mechanically and the infantry did not occupy the town. Renault and Schneider tanks advanced against the Croix Muzart and Neckar trenches, forcing the Germans back, but the infantry did not occupy these trenches and, when the tanks returned, the German troops had reoccupied them.

In the zone of the 21st Corps the tanks encountered many field guns and antitank rifles, and several tanks were lost. The infantry was becoming exhausted and only small advances were made. There was some progress in the afternoon when two platoons of Schneiders led an attack on the Bois de l' Araignée. Three Renault companies led a division attack on the Prussian and Essen trenches. Aided by smoke shells which blinded the German observation posts this attack was successful, but these objectives could not be held. A counter attack drove the French back. But they attacked again, the infantry and tanks were again successful, and the position was retained.

Similar actions occurred in the zone of the 2nd Corps. The first attack by infantry and tanks was successful, a counter attack drove them back to the Somme-Py Heights, and a second attack by the French was also successful. Farther to the west two Renault companies led the infantry forward to the railroad which crossed their zone but, due to the excellent use made by the Germans of the numerous OP's and artillery, few tanks could get past the railroad. A few crossed it but could find no place to cross the river Py. Three tanks passed Somme-Py but were quickly put out of action.

No important results were achieved on September 29th, the infantry divisions having been relieved, and the tank units, which remained in the line with the new divisions, were able to supply only a few tanks. The reserve tank units were employed and several different platoons supported successful attacks at various localities and broke up numerous counter attacks. On the other hand, the tank platoon used against the Sainte Marie A-Py station was knocked out by an antitank gun, and other tanks were hit while seeking a crossing over the Py river. The infantry made no progress at this point. All tank units were ordered withdrawn for reorganization and reconditioning on September 29th, but one battalion did not receive the order until the 30th, when it was preparing platoons, which had already been reorganized four times, for use in support of an attack on Pine wood. This attack was successful but many tanks were lost, one company having only two tanks left when the objective was taken. The personnel losses during the past three days had been heavy and many tanks had been put out of action but the reorganization was rapidly carried out and, by October 1st, 27 platoons were formed. The 18th Battalion arrived in the line, making 36 platoons or 180 tanks available for action in the surprise attack prepared for October 3rd.

On October 3rd, the 21st Corps carried out the main attack on Mont Blanc-Medeah Farm-Orfeuil Ridge, and the 11th Corps covered this movement on the left by attacking Notre Dame-des-Champs. Twenty-nine platoons were assigned to the three assault divisions, the 2nd U. S. Division was given the 2nd and 3rd Battalions; the 43rd French Division, the 16th Battalion; and, to the 167th French Division, the 18th Battalion of tanks was assigned.

The Germans made effective use of their artillery along the Py and this artillery fire, together with the difficulties of getting across the river itself, delayed the 11th Corps. The 2nd Division made good progress and the tanks joined the assault troops about 7 AM. Two platoons supported an attack near Medeah farm during which they took many prisoners. A little later another platoon stopped a counter attack and one company wiped out a number of machine guns and three antitank guns. One tank company reached Medeah farm in advance and delivered it to the infantry, one tank company passed the farm but was not followed by the infantry, and one company captured Aure trench and went on to the Medeah-Orfeuil railway.

The tanks with the 43rd Division reached the heights east of Orfeuil but machine guns in this town stopped the infantry. Another attack was arranged and one company of tanks went into the town where they were engaged for two hours. Although they cleared the Germans out of the town, the infantry was exhausted and could not occupy it. However, the objectives set for the attack had been gained.

CHAPTER IX.

TANK COMBAT HISTORY. FRENCH (Continued).

Flanders Offensive.
Hooglede-Thielt, September 26 to October 20, 1918.

Two Renault companies from the 501st Regiment reached the front occupied by the Belgian Army in time to take part in the advance following the capture of the Foret d' Houthulst. Only one of the tank companies supported the advance from this forest in the direction Hooglede and Reygerie. A few days later, as the advance neared these towns, both companies were used against the well organized German defensive positions found in this area. The tanks were unable to overcome all of the machine guns in Hooglede and Reygerie as these guns were mounted in concealed positions in specially prepared concrete emplacement in the ground, in houses, belfries and chimneys of mills. The infantry was held up at these places. Later, due to the success of the Second British Army south of this area, the Sixth German Army withdrew from Armentières and vicinity, relieving the pressure in front of the Sixth French Army and the Belgian Army.

In the second phase of the Flanders advance, starting on October 14th, the movement toward Thielt and Gand was renewed, supported by three Renault companies and one groupment of St. Chamond tanks. The terrain in this area was not suitable for tanks. The marshy ground was badly cut up by numerous ditches and recent rain storms had not improved the situation. The Germans made good use of smoke, blinding the tanks and hindering their advance. The Hooglede Hill machine gun batteries held up the infantry for several hours until late in the day, when they were silenced by the tanks.

The St. Chamond group used between Staden and Roulers became stuck in the marshy ground. Renault tanks flanked Hooglede and caused the Germans to retreat. The St. Chamonds reached Gitsberg with the infantry. A Renault company, with its infantry, captured the village of Geite-St. Joseph.

On October 15th, a Renault company aided its infantry in the capture of the Roulers-Throuroult road and another Renault company participated in the reduction of determined resistance at Gitsberg and its railway station, forcing the Germans to the east. One platoon of Renaults supported the attack on Beveren in the afternoon. This action continued on the 16th, when two tank platoons were used. The Germans were now in retreat and these tanks led their infantry beyond Ardoye.

On October 17th two Renault platoons were directed to flank Zeswege. The Germans retreated and the tanks entered the town, capturing many machine guns. Two Renault tanks went on to Hille and, after this place was taken, the cavalry moved forward. Two Renault companies led their infantry toward Coolscamp, silenced the machine guns which had stopped the infantry, and aided in an advance of about 8.5 miles.

The advance continued with little resistance until it neared the town of Thielt where the infantry was again held up by machine guns. One platoon of Renault tanks silenced these guns. In this action an antitank gun knocked out two of the tanks, but the remaining three supported the advance for about a mile from this point, where it was stopped in front of Thielt. On the north of this advance, one platoon of Renaults, on October 18th, aided its infantry in the direction of Beer, one of the tanks continuing until a stream was reached which could not be crossed. A heavy fog interfered with the operations in this area.

The advance between Beer and Thielt was stopped by machine guns at certain farms east of Hooilhoek, until eight Renaults were sent against these farms. Fire from the tanks soon decided the issue and the German garrison surrendered. The attack on Thielt was resumed on the 19th, following an artillery bombardment. Two platoons of Renaults participated in this attack, which did not succeed on this date, but the Germans withdrew that night.

Fifty per cent of the tank personnel and equipment had been lost since September 26th. The remaining tanks were withdrawn to a rest area south of Roulers on October 20th, to be reorganized for future actions.

Seboncourt, Petit-Verly Region, October 17 to 20, 1918.[1]

The 19th Tank Battalion with 69 Renault tanks, was used in this action in cooperation with two division of the 15th French Corps. This battalion, a part of the 507th Tank Regiment, was engaged in a regimental movement by rail which would place the entire regiment under the orders of the First French Army when, on the 15th at about 6 PM, the battalion commander was advised that his tanks would detrain on the 16th at the Saint Quentin station and support the 15th Corps in an attack to be made on October 17th near Seboncourt, which was more than 20 kilometers from the detraining point. Eighteen Pierce Arrow 5-ton trucks were to be ready at the Saint Quentin station to transport the tanks by successive trips to their destination in rear of the front lines. These trucks did not arrive at the time specified but did arrive at 5 PM and two companies of tanks were transported without serious difficulty although the trucks were not equipped to carry the 7½ ton tank. On account of the delay in their arrival, the number of trucks was increased to 30.

Due to this truck delay and to the arrangement for using only two companies on October 17th, the third company was later transported by rail to Escaille which was within two kilometers of the front lines. It was found necessary to secure permission from the Fourth British Army to use a good road, which crossed its zone, in order to move the truck-carried tanks forward. The granting of this request permitted the delivery of some of the tanks to be made in time for the action.

The number available for the attack was small in comparison with the front to be covered, so the missions of the various divisions and the terrain over which they were to attack were studied in order that the tanks could be utilized to the best advantage. Two tank companies were assigned to the 66th Division on account of its mission, which was to flank the Andigny forest by the south; and one company to the 123rd Division, to aid the 66th Division in carrying out the main effort of the Corps. The 126th Division, being unable to use tanks to advantage, was not provided with any. Only one company of the tanks assigned to the 66th Division was placed in position at the start of the action, due to the narrow front of this division at that time. The general plan called for the expansion of the front assigned to the division, so, although the second tank company was placed temporarily in corps reserve, it was moved forward in rear of the first company and was ready to take its place on the line at the appointed time.

On the morning of October 17th a heavy fog appeared and gave considerable protection to the tanks and infantry. No preliminary artillery fire was permitted. The front line divisions and the tanks started forward at 5:30 AM behind a rolling barrage. Two platoons of the 355th Tank Company, one following the other, were in the zone of the right assault battalion of the 411th Infantry at the start of the action. The leading platoon moved

[1] From an article by Lt. Col. Clayeux published in *La Revue d' Infanterie.*

out against a group of machine guns located about 600 meters from the French line of departure and silenced them, enabling the infantry to occupy the crest upon which the guns were located. The fog made this difficult. One of the tanks became separated from the platoon, one broke down and another was struck by shell fire. The company commander, with the aid of his command tank, was keeping in close touch with his platoons and, having observed the situation in which the leading platoon was placed, directed another platoon commander, whose tanks had been echeloned behind the leading platoon to move forward and take the place of the leading platoon, which was ordered to rally under cover of the nearby woods.

The second platoon was used to clean up Grougis Mill and the area west of same, enabling the right infantry battalion to occupy this area, but machine guns, located about 600 meters north of the mill stopped the advance on the left. At this time, the third platoon, which had been delayed in reaching the front lines, arrived. Securing the consent of the commanding officer of the local infantry unit, the tank company commander immediately put this platoon in on the left of the second platoon against the machine guns mentioned, enabling the infantry to occupy the village of Grougis. In the meantime, the first platoon had been put into shape and had joined the other two platoons ready for action. Thus the entire 355th Tank Company was prepared to continue the fight, but the infantry did not again attack on that day. The echelonment in depth of the tank platoons and companies and the presence of the tank company commander on the field were factors in the excellent work performed by the tanks in this action.

The 356th Tank Company used two platoons to support the two assault battalions in the sector assigned to the 66th Division on October 17th. Their attack advanced without difficulty toward Petit-Verly. The tanks wiped out the machine guns which barred the way of the infantry to the line: Hill 132—crest south of this hill—Marchavenne Farm. They then moved against Petit-Verly, forcing the Germans from the south side of this village and enabling the French infantry to occupy a part of the village. The tanks rallied behind the infantry while it was consolidating the position and were thus in position to and did break up a counter attack.

At this time the 66th Division had a foothold in the village but its right flank was exposed to fire due to the 123rd Division being held up near Grougis. The commander of the 66th Division decided to discontinue the attack because his right flank was exposed and, believing that the units in Petit-Verly were too far advanced, he ordered them back from that point to avoid being captured. The 3rd platoon of this tank company had arrived in the area after the attack started and was held in reserve during October 17th.

The method used by the two first platoons to protect the infantry while they were consolidating their position at the south edge of Petit-Verly was not in accordance with the French practice. However, if the tanks had remained in front of the infantry, they would have been in view of the enemy gunners and might have been destroyed, and they might not have had as good liaison with the infantry as they did have from the rear of these troops. By keeping his tanks in rear and in a covered position, the tank commander protected them and they were in a position to move promptly against any counter attack. This officer returned to the front lines, after having placed his command in the rear, and was one of the first to observe the approaching counter attack, which he took prompt and effective measures to stop.

In general, the 15th Corps had reached its first intermediate objective on October 17th. The 356th and 357th Tank Companies were sent to their rallying points. The wheeled echelons of these companies, which left Saint Quentin about 5 PM, reached the rallying points that night with supplies. The repair and supply unit had not yet been detrained, so the field train section of the Army was directed to establish an advanced gasoline depot at Seboncourt for the three companies of the 19th Battalion.

The plan for October 18th provided that the 123rd Division, which had become exhausted by the action on the 17th, was to be replaced by the 46th Division after the first objective was taken. In the attack of the 18th one platoon of the 355th Tank Company was to support the left battalion of the 123rd Division in the attacks on Grand-Thiolet and Petit-Thiolet. The guides appointed to bring the platoon forward became confused in the fog and the tanks did not arrive in time to accompany the infantry. The attack, which took place without them, failed. The attack was resumed with two tank platoons at 1 PM, in accordance with an order from the corps commander, and the villages mentioned were occupied. The 123rd Division reached its first objective too late for the 46th Division to pass it and attack the second objective, so the tank platoons were released.

In the 66th Division area, the attack was to be made by three groups of chasseurs, in the order: 8th, 7th, and 9th Groups. The 8th Group was to lead the attack on Petit-Verly at 5:30 AM, supported by the 357th Tank Company. The 7th Group and the 356th Tank Company were to pass the leading troops at 8 AM and take the first and second objectives while the 8th Group organized the position at Petit-Verly. After the second objective was taken, the front of the 66th Division was to be enlarged and the 9th Group, supported by the 357th Tank Company, was to advance to the left of the 7th Group. In preparation for these maneuvers, the tank officers were directed to get in touch with the officers of the three groups of chasseurs, in advance of the action, in order that they might cooperate with each other. As it turned out, this was a wise provision, because the 357th Company became lost in the fog during the approach march. The battalion commander, who was present and in close touch with the tank movements, knew that the 356th Company had arrived at the head of the 7th Group, so, in order that the tanks might play their part in the attack, he at once ordered the 356th Company to support the 8th Group, and personally led this company to its point of departure. The 357th Company was found and ordered to support the 7th Group which, according to the original plan, was not to enter the action until 8 AM. This arrangement could not have been made at the last moment if the battalion commander had not been present and if the officer commanding the 356th Tank Company had not been familiar with the work to be done by the leading group.

Two platoons of the 356th Company moved out in advance of the 8th Group against Petit-Verly. One platoon cleared a portion of the village south of the church and turned it over to the chasseurs. Two of the tanks were lost in this action and the platoon leader was seriously wounded, but the remainder of the platoon continued in action. The other platoon cleared the north side of the village. Later, the platoon commander noticed that the infantry could not leave the village on account of German guns mounted in houses south of Mennevret which was outside of his zone of action. These guns had also stopped the advance of the troops in that zone. This platoon commander moved promptly into the neighboring zone against the machine guns there located, enabling the battalion in this zone to advance and occupy Mennevert. He then returned to his own zone, dispersing German reserves preparing to counter attack Petit-Verly. When nearing the Petit-Verly cemetery, his tank was struck and set on fire and his driver was killed. He would have been captured or killed had it not been for the prompt action of a gunner in another tank, who kept the enemy under fire with his one-pounder while the platoon commander was rescued.

Petit-Verly was occupied about 7:30 AM and the tanks rallied back of the village. Later one platoon was ordered out to repel a counter attack against the village. This platoon had only three tanks; it did not succeed in stopping the attack and the Germans entered the village. The 7th Group was to pass the 8th Group at this time, but, as the 123rd Division had not yet captured Hill 180 which commanded the area on the right of the 66th Division, the passage of lines was postponed until 1 PM, the time set for the 123rd Division to attack.

The 357th Tank Company led the 7th Group to the attack and the three platoons disposed of enemy troops which had come forward south of Petit-Verly. The strong point formed by Saniere farm and adjacent woods was reduced by the use of all three platoons of the 357th Tank Company, the flank platoons enveloping the area by the north and south while the center platoon made a direct attack. This maneuver succeeded and the 7th Group occupied its first objective at 1:45 PM, having thus, with tank assistance, made a bound of over 2 kilometers in 45 minutes. While rallying his tanks in the depression west of the farm, after the troops had occupied the captured position, the tank company commander was killed by a shell.

On the 19th of October, the 19th Tank Battalion sent one platoon of tanks with each of the three groups of chasseurs as they advanced toward the canal leading from the Sambre to the Oise. The tanks advanced with the first echelons, ready to support them when needed. However, the Germans had crossed the canal on the previous night and the 66th Division did not have to fight in order to take its objective, so there was no need for the tanks and they returned to Petit-Verly.

When the 19th Battalion left this place it took along 62 of the 69 tanks brought into action on October 17th. Of the seven tanks left on the field, five had been put out of action by hostile artillery and two by mechanical difficulties. Lieutenant Colonel Clayeux attributes the small losses in tanks to the very foggy weather and to the efficient action of the French artillery, which supported the attacks and which was well supplied with smoke shells. Smoke was used to good advantage in blinding hostile observation during the attacks on Petit-Verly.

The German antitank rifle was used during these actions and a number of them were captured by the French. It was found that with direct impact, the bullet from this rifle would penetrate the Renault tank armor.

Offensive of the First French Army
South of Guise, October 25 to 30, 1918.

The 501st Tank Regiment (Renault) was attached to the First Army for this advance, which was a continuation of the advance starting October 17th, during which Petit-Verly, Petit-Thiolet, and the Derni-Lune farm were captured. The advance was continued toward Guise and, on October 25th, while one company of tanks was in action over the Heine-Selve plateau, German mines were encountered. These blew up two tanks and eight more were knocked out by various antitank agencies, but the objective (a village) was taken and many prisoners captured. Courjumelles and Signal d' Origny were captured on the 26th along with about 500 more prisoners.

On October 30th, the three tank battalions supported the 20th Corps in the advance south of Guise on Andigny and Flavigny. The tanks led the infantry under cover of a smoke screen until shortly after the Valenciennes-Marle road was cleared, when an adverse wind destroyed the smoke screen. The German machine gunners quickly stopped the infantry and prevented them from holding the ground captured by the tanks. Although the tanks continued in action until several of them were knocked out, no further advances could be made.

Attack on the Hundung Stellung, October 25 to 30, 1918.

The 502nd Tank Regiment (Renault) supported the Fifth French Army which was advancing between the Aisne river and Sissonne. In this advance, the second great defensive line (which paralleled the Hindenburg line) known as the Hundung Stellung, was encountered. Only two Renault battalions were available at the beginning of the action. These tanks had been carried on trailers and towed by tractors from Reims to Nizy le Comte. In this action the tanks were to follow the infantry, moving forward when needed at each objective. Airplanes were used to drown the noise made by the tanks as they came up to their assault positions on the night before the attack.

A strong antitank defense was anticipated by the French, so provisions were made to neutralize the German observation posts with smoke, and groups of artillery were designated to deal with antitank guns. They found that the Germans had prepared a well organized defensive system. Large numbers of small minenwerfer were placed in the brush and grass; 77 mm guns were emplaced in camouflaged positions and dug in practically to the level of the ground, commanding the ravines; mine fields were placed along roads and near villages; and machine guns were held underground until the French artillery preparation had passed. The advance was checked by these machine guns until the tanks advanced and silenced them. St. Quentin le Petit was flanked by one tank platoon and taken after two tanks had been destroyed by mines. The tanks in the center of the 3rd Division zone lost heavily but aided the infantry to occupy a part of its objective. On the right of this division the tanks reached their second objective only, as all but one were knocked out by one 77 mm gun. Antitank guns caused many losses among the tanks of the right division but these tanks silenced many machine guns and thus assisted the infantry advance.

The 2nd Tank Battalion reached the lines on October 27th and, on the 29th, two of its companies supported the 151st Division in its advance, which started at 11 AM. These tanks had orders to advance as far as possible before night. They aided their infantry in taking the Recouvrance and Conde objectives and then stopped because the troops on their flanks had not kept up with the advance.

The Hundung Stellung had been broken. 51 tanks were lost in this advance.

Advance to the Escaut, October 31 to November 2, 1918.

The Flanders Army Group made a combined effort, starting October 31st, to force the German troops across the Escaut river. In this last instance of tank employment by the French, very little resistance was encountered by the tanks or infantry except rear guard actions by the retreating Germans. On October 31st, the 7th and 8th Battalions accompanied the infantry of the 30th and 34th Corps to the Deynze-Audenarde road. On November 1st, the 7th Battalion passed through Nazareth and reached Eecke, while the 8th Battalion advanced to Synghem after a march of 25 miles.

On November 2nd some of the tanks from these battalions supported a reconnaissance to Seevergem without gaining contact with the German forces. An operation was scheduled for November 11th to force the German army back across the river, but the armistice had been signed and the war was over.

Renault Tanks in Morocco, 1925[1]

The 337th Tank Company, which was sent to Morocco on July 16th, 1920, and used in drills and maneuvers, was returned to France in March, 1922. Tanks were not used again in Morocco until June, 1925, when the 1st Moroccan Tank Battalion debarked at Casablanca. This battalion was to consist of a battalion headquarters and staff, three companies and one tank supply section, but only two of the companies, one from the 504th and one from the 511th Tank Regiment, were shipped at this time. The third company, from the 61st Tank Regiment, was shipped later.

The Moroccan Tank Battalion Headquarters consisted of 3 officers and 14 men, equipped with 4 vehicles. A tank company consisted of 4 officers and 106 men. It was equipped with 13 tanks, 10 (later 15) tank carrying trucks and about 12 auxiliary vehicles. The company was organized into a command section, which was equipped with means for communication; 3 combat platoons (each platoon having one 37 mm gun tank and two machine gun tanks); and a train. The train had a replacement section of 3 tanks; a transportation section of 15 tank carrying trucks; a supply section of a shop, two trucks, and a maintenance crew of 11 men. The supply section carried the supply of spare parts.

A variety of missions were performed by the tanks in Morocco in addition to the principal one of supporting the infantry in action. Among the unusual missions carried out were: flank guards, patrols, raids, mobile blockhouses between the infantry and the enemy, and the evacuation of the dead. Another feature in the employment of these tanks was the unusual distances travelled by them in the accomplishment of their missions, both under their own power and when carried by trucks.

The 504th Company went into service on June 13th, making the journey from Fez to Ain Aicha, a distance of 50 miles over a rough trail, by trucks, arriving on the 15th. This company took part in the evacuation of Tacunat Post from June 20th to 23rd. It was next used with the advance guard of a column en route to Tissa on the 24th, and on the 25th it returned to Fez, a distance of 34 miles.

The 511th Company left Fez on June 27th by truck, en route to Taza, a distance of 78 miles. The tanks were unloaded several times to cross streams by fording or to cross over bridges, some of which had to be strengthened in order to carry one vehicle but would not even then carry both the tank and the truck. This trip was made in three stages. The tanks saw no action on this trip and, on July 4th, were ordered back to Fez. The return trip was made in two stages.

This company was sent as a patrol to cover the right flank of the column on August 17th. The column was preparing a point of departure at Gros Rocher Post which it would leave en route to occupy the Maila Post, evacuated on July 12th. This flank guard duty involved travel on very rough terrain. During this march, one platoon was used to support a reconnaissance on the 20th by the Spahis which was completed without incident. On July 25th a platoon supported two Tunisian battalions in an attack on DJ. Semiet. Being untrained in working with tanks, these troops did not take advantage of the opportunities created by them. During the action, the battalions changed direction to the right but failed to notify the tank platoon in front of them. The tanks were called back by orders from the tank company commander who had observed the unexpected change in

---
[1] From an article in *Revue Militaire Francaise*.

direction. The tanks of this company took part in the capture of Semiet on July 26th, and a platoon was used as a patrol in the Maila valley in company with the colonial armored cars on July 29th. In the seven days fighting about Taza, the tanks were used in direct support of the infantry only a few times although they were given several special missions such as flank guards, patrols, etc.

The 504th Company was sent into the Ouezzan region on account of the threat made by the presence of Adb-el-Krim's regular troops at Chechaouen. It made the trip from Ouezzan to the Bab el Moroudj saddle on September 5th, the last 12 miles of the trip, over very rough terrain, being made by the tanks under their own power and at night, arriving at 4 AM. The tanks then proceeded to Issoul Bas Post, leaving this point at 7:45 AM, to aid the infantry to occupy the plateau. After this mission had been accomplished, a platoon returned to Issoul Bas Post in order to aid in its defense. The other two platoons took positions ahead of the infantry to protect it while organizing its positions on the plateau. These tanks had traveled 18 hours under their own power since the previous evening. The exceptional performance of both trucks and tanks in the Moroccan campaigns was undoubtedly due to the fact that the tank companies were provided with special maintenance crews and facilities for this important work.

The platoons defending Issoul Bas Post maintained the same positions from September 7th to 10th. The platoons protecting the infantry were kept out as mobile blockhouses. These were fired on by the Riff artillery but sustained no losses. The 504th Company took part in the attack on Fort de Bab Haouceine on September 11th and 12th, and, during this action, a tank was sent to the rear to help an officer save a machine gun, the crew of which had been killed in a rear guard action. Another special mission given to a platoon was the evacuation of the dead and wounded. While these were unusual missions for tank units, it should be remembered that these French troops were fighting under unusual conditions. It is not improbable that in the future many unusual missions will be found for tanks in open warfare such as existed in Morocco. The company moved down on Mjara on September 15th and, during this move, it was found that the tank trucks, which had been used to supply the column while the tanks were engaged in fighting, were in very poor condition.

The 61st Tank Company had joined the battalion about the end of August, reaching Taza August 21st on tractor-drawn trailers. It was sent on the 23rd to Bab Moroudj, north of Taza. This was a very difficult trip and the tanks had to make the last 19 miles under their own power. On the 27th of August, the company entered the action on the southwest slopes of Amessef at an altitude of about 4183 feet. The orders given the tanks were not clear, there was no liaison with the infantry, and several mistakes were made by the tanks in this action. They fired on their own infantry units which reached the objective ahead of them; one platoon went beyond its objective without waiting for the infantry and the enemy came in behind the platoon, making it necessary for the replacement section tanks to be used; and another platoon ran past its objective while keeping the retreating enemy under fire.

The 61st Company was returned to Taza on September 3rd and then sent at once toward Es-Sebt to join another column near Amellil which was then engaged in action. Due to the slow speed of the tractors, the action was practically over when the tanks arrived. The Company returned to Taza on the 8th and was ordered to take part in an expedition to supply the Moulay and Kelaa Posts north of Taza. One of the platoons had a march of 15 miles in order to approach the posts from the north. This distance was the extreme radius of action with the fuel carried. After five hours marching, the platoon was recalled.

This company was used again on September 15th to relieve and supply Tiffilassenne and, during this day, the tanks covered 19 miles under their own power. Towing sleds behind the tanks, the company supplied the Dahar Post on the 20th. The only real difficulty encountered during the trip was caused by the breaking of the towing chains. These had to be repaired under fire from the Riffs. However, the mission was accomplished without loss, whereas previous supply expeditions, when tanks were not available, had resulted in many casualties.

The 504th and the 61st Companies were then assigned to the 19th Corps, which was operating in the mountains north of Kiffane. The 504th Company was sent on September 25th over the trails toward Caid Medboh. The rains made bogs of the trails and the tanks had to tow the 61st Company tractors, which had been loaned for the trip, as well as the artillery trucks, in bad going. Arriving at Kiffane on September 29th, the company aided in the attack against Djebel Kerkour on the 30th. The objectives in this action were high rocky peaks and, although the tanks were at times in dangerous positions on sharp edged rocks, the operation was well executed and was successful.

Following this action the company supported the infantry in the valleys north of Kerkour, being relieved on the 8th of October. While returning to Taza it was found necessary to camp one night away from the other troops, so the tanks were formed in a circle with the crews in the center. On this trip the tanks made 50 miles under their own power over very poor trails. On September 30th, the 61st Company, while operating with the right brigade, east of Kerkour, came under the fire of the French long range artillery at Kiffane which was firing on Kerkour. The infantry and the tanks moved back, down the valley, and the tanks successfully utilized a prearranged signal for increasing the range.

On October 1st, the 61st Company was sent on a reconnaissance to the village of Ouizert where a number of inhabitants had indicated a desire to talk with the French troops. Upon approaching the village a 37 mm gun tank was used to take the intelligence officer forward. This tank was fired upon from the village. The company then destroyed several houses with gun fire whereupon a white flag was raised. The rebels came forward, interviewed the intelligence officer and, as a sign of submission, cut the throat of a ram in his presence. They were then taken in charge by the tank crews and brought into the French lines. The company was used in mopping-up operations on October 3rd, 4th and 5th in the valleys about this region.

The 2nd Tank Battalion, which left Marseilles for Casablanca on July 18, 1925, and which was made up of the 506th, 507th and 508th Tank Companies, was sent to the Quessan region as soon as it debarked. The tanks of the 508th Company were equipped with the Kegresse-Hinstin traction system with rubber tracks.

The 506th Company was engaged at Azjen beginning on August 2nd, and was used to flank the east and west sides of an olive grove and to mop up caves located on both sides of Azjen where the enemy was concentrated. One platoon was used on the latter mission. The tanks were guided by an Arab lieutenant of Scouts to the entrance of the caves from which point fire was opened at short range, wiping out the resistance. Owing to the location of some of the caves, it took several hours to overcome the resistance. Some of the tracks were broken and were repaired by the crews within 25 yards of the enemy, under cover of the fire from tank weapons. The tanks were withdrawn after dark, and this time the Moroccans did not follow.

This company was used again on August 10th to support a column en route to Zitouna and Amezzou. The Moroccans had reoccupied Azjen so it was decided to take this place on the way. Upon arrival at Azjen, the tanks were unloaded and an operation similar to that carried out on August 2nd was started. The Moroccans did not wait for the attack in this case, however, so the tanks were again loaded on the trucks and the column proceeded on its way. Zitouna and Amezzou were taken without difficulty and the tanks were returned to Ouezzan on the 13th. The trails had been made very difficult by heavy rains and the tanks had to be unloaded and used to tow the trucks for about three miles. The tank company had to provide its own security at night on this trip.

The 506th Company was used on several expeditions up to the 11th of September but was not molested by the Moroccans and did not have to fight in order to occupy the positions taken.

The 507th Company took part in the operations against the heights of Mostitief on the 11th of September, starting from Teroual on each side of the Teroual-Ain Ben Aissa trail. The tanks moved in column on the trail, and the infantry, which was in the lead, advanced rapidly until stopped by the fire from nearby ravines. The tanks moved forward, firing on the resistance, causing the withdrawal of the Moroccans, and allowing the infantry to occupy the first objective at Si Allal-Zrari. The advance started for the second objective at Mostitief but again the infantry was stopped by fire from this place and from the third objective at the Portuguese ruins. One tank platoon was sent against Mostitief followed by the infantry, and this point was captured. Two platoons were used to flank the Portuguese ruins whereupon the Moroccans withdrew. The entire tank company then started for the fourth objective at Haddarine, which it captured without resistance over an hour ahead of the arrival of the infantry. This company led the advance to a depth of nine miles, without losses to the French infantry. The success of the operations up to this point was attributed to the fact that the attacks were well-planned and the tanks were given definite orders and well-marked objectives.

The infantry changed the direction of the attack to the east on September 12th but the tanks were not notified until 5 AM on that day and loss of contact resulted. The tank personnel then had to search for proper routes from the heights, upon which they were located, down to the valley. This hurried movement without proper reconnaissance resulted in the tanks becoming lost and separated from the infantry by about three miles. They then had to return to the starting point, which was reached after seven hours operation of the tanks under their own power.

The two companies participated in an attack on September 16th against the heights of Bibane. This point was reached without resistance as the Moroccans would not oppose the advance of the tanks.

The last important action in which these tanks are reported to have been used occurred on September 26th, when an expedition moved out to supply and reinforce the Bou Ganous Post. This post was garrisoned by about 40 men, and located eight kilometers north of Ouezzan. Its communications with the other posts had been cut. Previous supply operations had been costly in casualties, as the column was under fire from all sides during most of the trip. The Moroccans had dug trenches blocking the approach to the post since the last expedition, and they had received reinforcements so it was recognized that the relief of the post in this instance would involve severe fighting. With reference to the terrain, the commander of the 2nd Battalion states: "To these difficulties provided by the enemy were added those of the terrain. From Ouezzan to Bou Ganous the trail was dominated on the east by the heights of Hamar, the village of Harrara and Hill 505. From the Ouezzan to the post (the trail) crossed numerous ravines, real death-traps, and it was dominated on the north by a rolling terrain whose deep cuts favored infiltration. Under these conditions every column marching from Ouezzan to Bou Ganous had to 'parade' up and back in a true corridor of fire. The command resolved to clean out this region and to assure its security by the establishment of a large post consisting of several company strong points."

Preparations for this operation were carefully made. Reconnaissance from an observation balloon and from the ground by field glasses gave a good idea of the terrain and helped to locate the tank objectives. Tank and infantry officers made their plans together for the attack. The 506th Company supported the right column, and the 507th, the left. The tanks started before dawn and about 30 minutes ahead of the infantry. This was done to get as close to the Moroccans as possible before they knew the tanks were coming. The report of the battalion commander states: "In their advance the tanks emptied the enemy trenches of their occupants at all points. The latter fled from the battlefield or took refuge in deep shelters hollowed in the sides of the ravines, particularly between Hamar and Hill 505 and in the ravine south of Lalla Chakria, where the very deep caverns opened opposite the Bou Ganous post, * * * these caverns were inaccessible to the tanks, but (were) however, easily swept from a distance. The tanks took up positions everywhere they were able to in front of the mouths of the rebels' hiding places, holding them there all day. They facilitated their cleaning out by the infantry * * * *"

The tanks were used on September 27th to oppose the infiltration of the Moroccans while the post was being supplied and the new organization effected. The French troops were not molested so the tanks were sent back to Ouezzan.

During some of the actions, the Moroccans used a new method of attacking tanks. Having no antitank guns they had to resort to other means. Four men, each equipped with a short iron bar, lay in holes or behind rocks until a tank came near them. They then placed the bars between the tracks and the drive sprockets, locking these parts and stopping the tank.

The 508th Company, equipped with Renault tanks using the Kegresse-Hinstin rubber tracks, supported the Spahis. These troops were to maintain contact with the 3rd Moroccan Division and to establish liaison with the Spanish forces advancing from Melilla. The tank company was transported from Casablanca to Taza and, on October 3rd, it was used on the right of the French command, supporting a battalion of the Legion. The Moroccans were on a hill which commanded the plain over which the cavalry was operating.

The tanks were followed by a company of the Legion and their right was covered by a Spahi squadron. During the attack, the foot troops exposed themselves needlessly by following closely behind the tanks but, after a three-hour action, the position was taken and consolidated.

The purpose of this action was to deny the use of this hill to the enemy during the passage of the French forces; it was not to be held permanently. Therefore, at the proper time, it was necessary to disengage the troops in possession of the hill from the enemy forces. The withdrawal was not well coordinated, the cavalry and the Legion company ignoring the tanks in this movement. One tank broke down on rough terrain. One of the rubber tracks came off and, while the tank crew were trying to replace the track, they were fired on by enemy infantry and cavalry. Other tanks then went into action and prevented the capture of the disabled tank. It was towed until this became impossible on account of the rocks projecting from the ground which caused the tracks to be twisted off. Finally four of the eleven tanks being used were out of action. They were not repaired until the following day. The records available do not state whether or not any of these tanks were captured by the enemy.

The conclusions drawn from this experimental action were: that the replacement of rubber tracks took too much time, three hours being necessary for a trained crew to effect the replacement; that the tanks with this type of track should only be used on good ground as the rubber was too fragile for rocky ground; that the tracks were rapidly torn to pieces by such ground; and that the rubber tracks caused the tank to slide or skid easily.

Renault Tanks in Syria, 1920.

In July 1920, a strong French mixed brigade with some artillery and a platoon of Renault tanks opened up the road to Damascus, which was being blocked by Arabians and Turks under the command of a former Turkish officer. A well organized defense in two lines was broken, with small losses, by the use of tanks in a flank attack.

CHAPTER X.

TANK COMBAT HISTORY. GERMAN.

Information concerning German tank actions has been obtained with the consent of E. S. Mittler and Son, publishers, from *Die Deutschen Kampfwagen im Weltkrieg* by Lieutenant Volkheim.

THE A 7 V.

(Allgemeine Kriegsdepartement 7 Abteilung Verkehrswesen.)

Produced in 1918 by Daimler Motor Co. Contracts let for 100, of which only 20 were built as tanks, the remainder were, when built, to be cargo carriers.

Crew: 18.
Armament: One 57 mm (2.24 in.) rapid fire gun forward and six machine guns at sides and rear.
Armor: 0.59 in to 1.18 in.
Maximum speed: 8 mph.
Suspension: Coil springs and bogies.
Tracks: 2 curved grousers to each plate, width 20 in., pitch 10 in., ground length of track only about 16½ ft. Upper part of track carried by rollers.
General arrangement: 57 mm gun in front, engines in center, driver and tank commander above engines, final drive in rear.
Dimensions: Length 24 ft.; width 10 ft.; height 11 ft. 2 in.
Weight: 33 tons.
Engine: Two, Mercedes-Daimler, 4 cylinder, sleeve valve, 150 HP each, forced water cooling.
Horsepower per ton: 9.1.
Transmission: One sliding gear transmission for each engine. 3 speeds forward and 3 reverse.
Obstacle ability: Trench 6 ft.; slope 30 degrees; vertical wall 18 in.
Fuel distance: 50 miles. **Fuel capacity:** 132 gallons.
Special features: Forward rollers (idlers) mounted on pivoted arms. Fuel tanks, transmission and all machinery under floor. Engines and radiators encased and cooled by two fans for each engine. Radiator at each end of each engine. Engines side by side. Cooling air drawn from interior of tank and forced downward and out through floor. Apron of armor hinged at forward end of tank between tracks. Poor trench crossing and cross-country ability due to overhanging hull. Armor is proof against 37 mm gun fire but admits bullet splash. Magneto ignition used with a starting hand magneto. Starting the engines difficult in winter and the following means were used: Priming pump, Anlasz electric starting motor, Bosch atomizer, hand crank for 3 men, acetylene. A dynamo provided interior and exterior lights. Temperature inside the tank reached about 140 degrees F., and engines at times gave trouble through overheating. Either track could be put in forward or reverse separately to facilitate short turning.

The A 7 V Tank. List of Actions in which Engaged

DATE	BATTLE	NUMBER EMPLOYED	
		Units	Tanks
March 21, 1918	St. Quentin	1 section	5
April 24, 1918	Villers Bretonneaux	3 sections	13
June 1, 1918	Soissons	1 section	5
June 1, 1918	Reims	1 section	5
June 9, 1918	On the Matz	2 sections	10
Sept. 1918	Fermicourt	1 section	5
Oct. 11, 1918	North of Cambrai	1 section	5

In addition to the above, German tanks are reported to have been used in small numbers in minor actions upon a few other occasions.

Captured British Tanks. List of Actions in which Engaged

DATE	BATTLE	NUMBER EMPLOYED	
		Units	Tanks
March 31, 1918	St. Quentin	1 section	5
June 1, 1918	Soissons	2 sections	10
Sept. 1918	Fermicourt	1 section	5

Captured British tanks are believed to have been used in very small numbers upon a few other occasions.

96   THE FIGHTING TANKS SINCE 1916

St. Quentin, March 21, 1918.

Germany had two tank detachments of five tanks each in readiness on March 21st. Section No. 1 consisted of five German A 7 V tanks and Section No. 2 of five captured British tanks. The British tanks, evidently some of those captured at Cambrai, had been taken back to Charleroi, Bel-

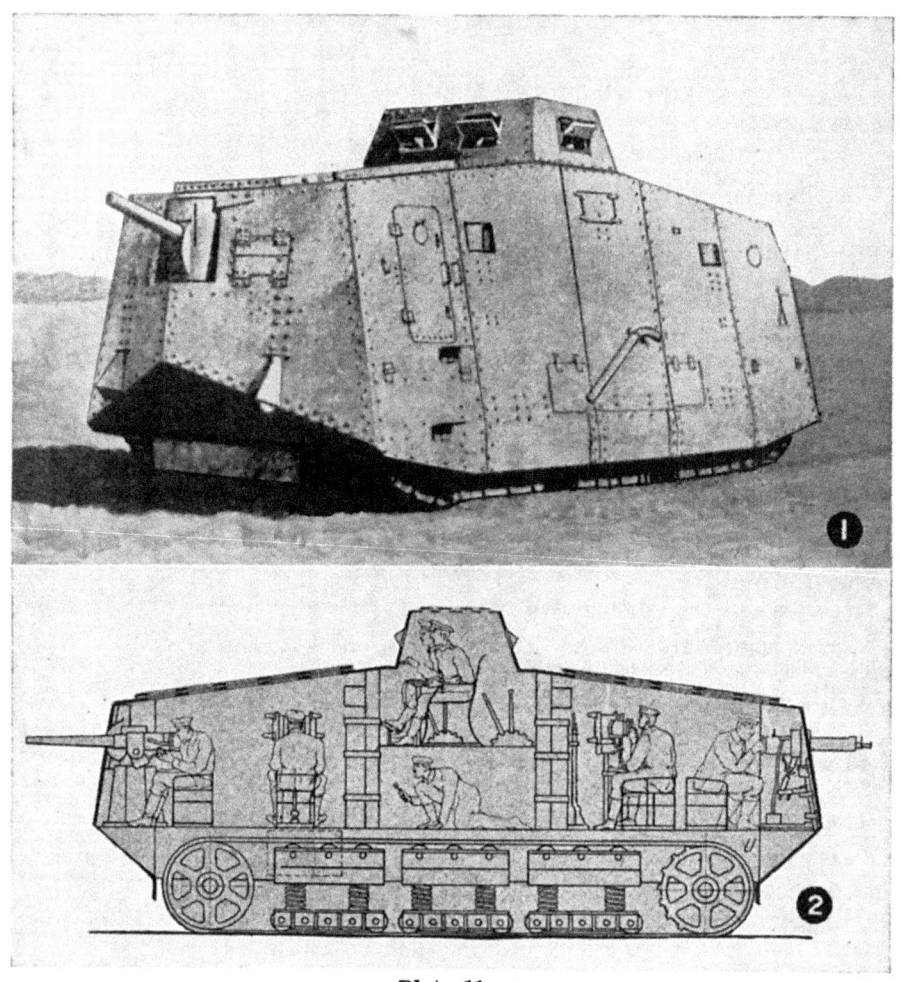

Plate 11.
1. **A 7 V.**   2. **A 7 V.**
(From *Tanks* by R. Krüger.)

gium, the Headquarters of the German tank troops, and there reconditioned. Captured Russian 57 mm guns were installed in the captured tanks.

With the exception of the gunners, none of the tank crews had had previous experience at the front. Most of the tank commanders had not had much experience except in the training camps.

The battle mission given to the tanks, to be carried out in cooperation with an infantry assault battalion which had received some rear area training with tanks prior to the action, was to attack enemy trenches and machine gun nests passed by the first assault waves of the general attack.

In this initial action, a heavy fog at first concealed the tanks but, as soon as they were discovered, they were subjected to severe machine gun fire. The machine gun nests were wiped out by shrapnel and the British trench and strong point garrisons were forced to surrender.

Of the five German tanks, two were credited with having fully accomplished their missions, the other three being put out of action just before or during the action by mechanical trouble. Two of the British tanks were damaged by artillery fire, two were stopped by mechanical trouble, but the other one, although it lost its way in the fog, appeared in the sector of another division and is credited with having given assistance to these troops. Seven of the tanks used had to be repaired before they were ready to fight again. Little experience was gained by the Germans in introducing their tanks in this small action and, by using them prematurely, they notified the Allies of the existance of German tank units.

Villers-Bretonneux, April 24, 1918.

Fourteen heavy tanks were assigned for this action. They were formed into three groups and assigned: three tanks to the 228th Infantry Division, six to the Fourth Guard Infantry Division, and five to the 77th Reserve Infantry Division. The first two groups were to attack Villers-Bretonneux in conjunction with the infantry and the last group was to attack Cachy.

Among the instructions given to one group commander the following items appeared: "No. 3, the commander's tank will be the guide; the other two will follow at a distance of 200 meters in echelon to the right and rear. If, during the combat, the infantry should request a tank, their request is to be granted in any case. Six men of the 207th Infantry will be assigned to each tank as patrols."

Two motor trucks, loaded with fuel, ammunition, intrenching tools, etc., were assigned to each group. These vehicles were to follow their groups, by bounds.

The terrain was very favorable for the use of tanks. There were few obstacles and the fields over which the attack was to take place, were dry. There was a heavy fog at the start and a heavy bombardment was carried out during the approach march to prevent the British from hearing the noise made by the tanks.

Engine trouble in one of the tanks reduced the number participating to 13. Group No. 1 left its starting point, just in rear of its infantry front line, at 6:50 A.M. crossed the German front line at 7 A.M. and the British front line shortly thereafter. The British troops and especially those in well concealed machine gun nests which, due to the heavy fog, the Germans had not discovered, put up a good fight, bringing all available weapons to bear upon the tanks. After a short but sharp engagement, the British infantry and machine gunners surrendered to the tank personnel and were turned over to the German infantry.

The commander's tank of Group No. 1 advanced under heavy infantry and artillery fire, to within 100 yards of Villers-Bretonneux, when it was discovered that the infantry was not following. It turned back to regain contact and soon wiped out four more machine gun nests which had been firing on the tank from the rear. Rejoining the infantry, this tank moved to the eastern edge of the town under heavy machine gun fire and overcame several machine gun nests at this point. The tank and its infantry then entered the town.

The other two tanks of this group cleaned out strongly intrenched machine gun nests which were holding up the German troops and, after reaching the town where they again supported the infantry attack, they joined tank No. 3, according to plan, near the tile factory. This factory had been made into a large machine gun nest. The three tanks, using their heavy guns, shot it to pieces. Six British officers and 160 men surrendered. After the

German infantry had arrived at the tile factory, tanks No. 1 and No. 2 moved against an airdrome, also heavily armed with machine guns, and destroyed it. After wiping out several machine guns in houses, where more prisoners were taken, they reached their objectives at about noon, and, having reported their departure to the infantry commander, returned to their starting point.

Group No. 2 crossed the German front line a little after 7 AM and attacked a strong point along the railroad embankment from the front, flank and rear, silencing its guns and permitting the infantry to advance. One tank cleaned out a trench nearby and captured 15 prisoners. Two of these tanks moved past the railroad station and one of them fired upon approaching British reinforcements. The other tank was having trouble with its gun recoil mechanism but managed to silence several strong points, and the two tanks, by opening fire on the Bois d' Aquennes and the British reserves west of it, aided the German infantry to enter these woods.

Tank No. 3 cleaned out the British first line, caused several casualties, and took 30 prisoners. It then captured a switch trench with 40 prisoners and moved toward a fortified farm. It reached the farm after a breakdown, and silenced the machine guns located there. The mechanical trouble continued, but before the tank stopped the crew was able to break down strong resistance south of the railroad station, capturing one officer and 174 men. Finally, the carburetor jets became stopped up and the tank could not be moved, so the crew went forward without it. Later, the commander returned to the tank, changed the jets and made another attempt to move the tank. He succeeded in getting it started but soon ran it into a large shell hole which had just been dug by a shell that exploded in front of the tank. As the tank entered the hole it turned over. It was therefore temporarily abandoned but later brought back to a safe position.

Tank No. 4 reached the British front line trenches at 7:10 AM, cleaned them out and attacked a fortified farm south of the town, where it cleared the way for the infantry. Joining tanks No. 1 and No. 2, the three vehicles moved against Bois d' Aquennes and stopped a British counter attack. Tank No. 5 became lost on account of the fog. It came under heavy machine gun fire and the driver was wounded. When he was hit, he lost control of the tank. The engine stopped and the tracks were held fast by the brakes, which jammed. The commander used some of his men as an infantry detachment until the tank was repaired, when, with the men remaining, he moved the tank toward the Bois d' Aquennes, cleaning out a few machine guns which were in the trenches crossed by the tank.

Tank No. 6 advanced at the proper time but its infantry did not follow. The tank came under heavy fire but went on until it was about 20 yards from the British line, when both engines stopped due to overheating. The driver had been wounded and the substitute driver was not with the tank. After the engines cooled off the commander brought the tank back to the German lines.

Group No. 3 lost a tank soon after the action started. This tank advanced with its infantry's first wave, successfully attacked several machine gun nests and portions of the trenches, but soon thereafter it struck a hole and turned over on its side. According to the account of this action, the British troops had started to lay down their arms and the tank commander had ordered his crew out of the tank to support the infantry troops on foot, when the British took up their arms and shot most of the tank crew. One member of the crew succeeded in getting back to the German lines and one was captured by the British. The captured man gave information to the British concerning the German tank troops. The German infantry retreated at this point and the tank was blown up by a German officer since it could not be brought back. Apparently this officer did not make a good job of it for the tank was later captured by the British in fair condition.

Tank No. 2 moved toward Cachy and attacked several machine gun nests including one which had held up the infantry advance for over an hour. This tank then advanced to a point about 700 yards from Cachy, firing on the British position at the village. At this point British tanks appeared and the first, and much discussed, tank-against-tank action occurred. The German account states that one of the German tanks was stopped by artillery fire and another one was forced to retreat in the initial encounter. As the second tank was moving back, it was put out of action by a direct hit from the right. Another shell struck the oil tank. However, the commander finally succeeded in saving the tank and moved it back, a little over a mile, to the German lines.

The British counter attack won back part of the ground captured by the German advance and this caused a change in the plans for using tank No. 3. It was intended that this tank should support the attack on Gentelles, but, since this attack failed, the tank was sent against Cachy. There it fired upon the eastern edge of the village. Later, however, since the German infantry did not plan to storm Cachy, the tank was released, whereupon it returned to the assembly point.

Tank No. 4 was also used in the attack on Cachy. It succeeded in cleaning out several machine gun nests and got into position where it could enfilade a 200-yard trench, thus causing some casualties and driving the remainder of the garrison back. Toward noon the tank commander noticed that the German infantry were retreating from the direction of Cachy. He turned his tank in that direction, stopped the retreat, and moved his tank toward the village. When within about 900 yards of Cachy, he came upon a number of British tanks which were approaching from the German right flank. Shortly afterward other British tanks made a frontal attack. The British tanks opened fire with their machine guns and the German tank replied with its heavy gun. The second shot struck a British tank and set it on fire. Soon thereafter this gun struck another British tank, The crew of one of the British tanks evacuated their tank and were shot down. The other British tanks left the field, being followed by machine gun fire to within 200 yards of Cachy. During this action the German cannon failed after the second British tank was struck, so, had the British known it, they were on even terms as regards type of weapons. The German infantry again moved against Cachy but, as they did not enter the town, this tank was released and returned to its assembly point after having been in action eight hours.

These detachments entered the action with 22 officers and 403 men. Of this force, one officer and eight men were killed, three officers and 50 men were wounded, and one man was captured. Twelve of the 13 tanks were brought back to the German lines.

An account of the use of German tanks,[1] written by an English officer who commanded a front line company which was attacked by these tanks on April 24th, states that unusually accurate machine gun fire was being received on his support trench and that orders were given for his men to keep their heads down. When this fire ceased he stood up to observe the sources of the fire and saw an enormous and terrifying iron pill-box with automatic weapons bearing down upon him. He got down in the trench and the tank passed over him. The tracks of the tank were within three feet of his face as he lay in the trench. After it had crossed he stood up and fired his pistol at the water jacket of the rear machine gun. Being warned by his men, he looked around quickly and saw a large German crash

---
[1] From an account published in the British *Army Quarterly*.

into the trench, his bayonet sticking into the parados. Several other Germans ran toward the trench but they were all shot down by the garrison. Next, another German tank appeared, moving along and shooting the men in the front trench, crushing them, or firing into them if they tried to leave it. In this advance, the tanks were aided by German light automatic gunners who followed the tanks. In addition to these light guns, the German foot troops carried flame throwers which they used on the trench garrison. However the flames only reached to the parapet, so that men were not severely burned. They were scorched, however, and had to throw off their equipment. Having cleared up the first line trench, the tanks went on to the second trench, and now a third German tank appeared followed by German infantry. These troops bayonetted the remaining members of the first trench garrison. When the third tank started for the second trench, the officer and the garrison of the second trench retreated. All but five of this group were shot down before a nearby railway cut was reached. The first tank approached the cut firing on the group at this point as they ran down the railway. These shots went over their heads, however, as the machine gun in the tank could not be depressed enough to strike them. Removing his collar and tie for easier breathing, the officer reporting this action, a member of this group of five, outran the German infantry. He organized a counter attack later with men from various regiments. He was wounded during this affair and, while on his way back to the first aid post, met a tank company commander to whom he related the attack by the German tanks. This officer at once ordered British tanks forward to attack the German tanks.

Soissons, June 1, 1918.

The 1st Tank Detachment, with five A 7 V tanks and the 13th and the 14th Tank Detachments, with 5 captured British tanks each, were assigned to this action. The plans included a provision for an assault detachment to accompany each tank. This detachment was detailed for the special purpose of taking immediate advantage of the opportunities created by the tanks. The tanks were to start in time to cross the hostile front line at 4:40 AM. All front assault waves were to wait for the tanks.

The 1st Detachment reached the starting point at 4 AM. It had to leave tanks Nos. 4 and 5 at this point on account of mechanical trouble. From here on the advance was slowed down as the tanks had to pass through a forest and crush many trees en route. Tanks No. 1 and No. 2 became ditched in a tank trap in the French front line and could not move forward. These tanks were in plain view, and the enemy artillery became very active, so the detachment commander ordered the tanks to turn back. Both tanks were moved back out of the trap and one of them succeeded in getting back to its starting point, but, due to mechanical trouble, the other one could not be kept in motion. Hostile artillery and machine gun fire being concentrated upon the latter, the order was given to take out the machine guns and most valuable parts of the tank machinery. Soon thereafter a shell struck the command turret and destroyed the interior of the tank.

Tank No. 3 did not get far from the starting point before engine trouble developed. By the time it was put in order and the tank started, the others were returning, so this tank was taken with them back to the assembly point. The tank detachment casualties were two officers and four men wounded.

The 13th Tank Detachment lost two tanks while advancing to the starting point. Tank No. 2 fell into a ditch and tank No. 3 broke a differential gear, so both were abandoned. The approach march of the other tanks of this detachment was slowed up by a truck lying across the road, by a fuel and ammunition dump, and by troop columns using the road.

After crossing the hostile front line trench, tank No. 5 used its gun and machine guns effectively against visible targets until the engine, which had become overheated during the approach march, became ineffective and could not pull the tank out of a hole. By the time the engine cooled off sufficiently to move the tank, the infantry had retreated to its front line, leaving the tank between the two lines. Eventually the tank was started back to the German line but the engine failed again and the tank had to be evacuated and demolished, under infantry and machine gun fire.

The 13th Tank Detachment had been made up of three of its regular tanks and two others, one from the 11th and one from the 12th Tank Detachment. The tank from the 12th Detachment did not come under fire until it approached Fort Pompelle, where it was fired upon by artillery directed from an airplane. It was evacuated temporarily but the crew later returned to it and continued forward. At the request of the infantry, the tank commander took his crew and three salvaged machine guns into action as an assault detachment to aid in repelling a counter attack which started at 10:45 AM. The German infantry was retreating and this assault detachment joined the retreat, abandoning the tank. The tank from the 11th Detachment had lost some time during the approach march in trying to pull tank No. 2 out of the ditch. However, it crossed the hostile trench under artillery fire and penetrated the enemy front line about 5:30 AM. A loss of several minutes occurred when the foot brakes jammed and burned. At about 5:37 AM, this tank crossed one of the trenches, from which sufficient machine gun fire was coming to hold up the infantry. It overcame this resistance and enabled the German infantry to proceed. The tank then went toward Fort Pompelle, cleaning out several machine gun nests en route. For some distance, the tank had no infantry support. It was subjected to intense machine gun fire until it came close enough to break up this resistance. The tank machine gun fire turned back advancing enemy infantry units which suffered severe casualties as they withdrew. Soon thereafter hostile artillery fire was brought to bear on the tank while it was working along the enemy main line, wiping out machine guns.

Fort Pompelle had fallen before the arrival of this tank, so, upon being informed that the German infantry at the Aisne-Marne canal were in need of his assistance, the tank commander moved his tank toward the canal, coming under violent machine gun and artillery fire and, when within 150 yards of the canal, the tank was put out of action by a shell which broke the right track. The tank guns were removed and brought back under fire. Soon thereafter the tank was attacked by an airplane but the plane was driven off by two machine guns. The tank casualties were one NCO killed, one officer and four men wounded.

The 14th Tank Detachment preceded its infantry at 4:40 AM as scheduled, and, after crossing three lines of trenches, turned to the west, attacking and cleaning out the machine guns. Some of these tanks moved along the British main line, mopping up. However, due partly to insufficient infantry support, some of the well-concealed machine gun nests escaped detection. One of the tanks met strong resistance at the 3rd line and was here also exposed to heavy artillery fire. It started back but became ditched, standing up on end in a large trench. As it could not be moved the guns were removed and it was blown up.

Tank No. 2 drove along the main line and twice crossed the trenches in this line, mopping up and clearing out such machine gun nests as were found. This tank returned to its starting point without being harmed. Tank No. 3 also completed its mission, but engine trouble developed as it was returning to the German lines. It had to be abandoned and later was completely destroyed by enemy artillery fire.

Tank No. 5 silenced the machine guns in its path and turned towards Fort Pompelle, using its H.E. shells at first, and then, as it came nearer, using shrapnel. The infantry stormed the fort and the tank then started on its return. On the way back, this tank was put out of action by a direct hit. The machine guns on the side of the tank away from the enemy, together with other valuable parts, were removed. With the heavy gun and the machine guns toward the enemy, the crew, from within the tank, continued the fight against the nearest hostile machine guns. Finally the tank received two direct hits and caught fire. It was then evacuated. The casualties in this detachment were one NCO killed, one officer and one private wounded.

Reims, June 1, 1918.

The 2nd Tank Detachment, equipped with five German A 7 V tanks was placed under the orders of the 242nd Infantry Division for this action. The attack was to be made at 7 PM and each tank was assigned definite missions, including the mopping up of definitely located machine gun nests in its zone.

Tank No. 3 was delayed for an hour at the start due to trouble with its fuel lines, and tank No. 1 was held up for a while with overheated roller housings. The last mentioned tank moved on at 9:30 PM, after lubricating the rollers and repairing some slight damage to its water pump. The other three tanks left their starting point, just in rear of the German line, at 7 PM. The two leading tanks crushed wire and other obstacles that blocked the road over which the tanks had to pass. All three tanks came under enemy artillery fire at close range from the left flank at about 8:20 PM. The leading tank returned the fire with its cannon and machine guns until it was struck by a shell from a trench battery which suddenly opened fire with gas shells. The tank ceased firing and had to be evacuated. This battery was one of those not located by the Germans prior to the action. At another point where the tank personnel were told to expect one machine gun nest, a battery of four guns was found.

Tank No. 5 was stopped by a direct hit while moving along a road and, since wide trenches flanked the road, the other tanks could not get around the disabled tank, so they turned back. A counter attack forced the German infantry back to their lines and tank No. 5 was captured. The losses in tank personnel were three men killed, two wounded, and two officers and one man gassed.

On the Matz, between Montididier and Noyon, June 9, 1918.

The 1st and the 3rd Tank Detachments were assigned to this action. Tanks No. 1 and No. 2 of the 3rd Tank Detachment were allotted to the 227th Infantry Division and tanks No. 3, No. 4, and No. 5, to the 19th Infantry Division. Only three of the tanks belonging to the 1st Tank Detachment were fit for service, and these were assigned to support the 3rd Infantry Division.

The terrain over which the 3rd Tank Detachment was to attack was a rolling hilly country, planted with tall grass and grain which had been beaten down by German artillery fire. Many trees and much underbrush remained and, during this action, a very heavy fog covered the field.

Tank No. 3 moved out to Orvillers, overcoming the machine gun nests south of that town. It then took part in the attack on the second line. At the entrance to the village of Cuvilly it broke down a strong barricade where very strong resistance was encountered and, with the infantry troops which came forward in response to a bugle call, it entered the village and forced the enemy troops to withdraw. After passing through the village, this tank came under heavy artillery fire and, as the infantry did not follow and further advance was considered impossible, the tank returned to its assembly point, north of Cuvilly.

Tank No. 5 advanced in the direction of Biermont-Hagrand wood where it mopped up machine gun nests. After this task had been accomplished, it supported the infantry attack on the 2nd line southeast of Cuvilly. It then entered Sechelles wood to regain contact with its infantry and, while in this woods, it captured one officer and 25 men. Discovering that the engine was knocking badly and the fan belt broken, a halt was made to effect repairs, after which the tank was returned to its assembly point.

Tank No. 1 became ditched in a deep trench south of Conchy-les-Pots and was hauled out by Tank No. 2. It moved up via Viermont-Hagrand wood-Sechelles wood without encountering resistance. The 147th Infantry requested assistance in an attack on Ressons wood, so the tank started on this mission, accompanied by a company of 20 men who fought with the tank. The tank and this small group soon lost contact with the remainder of the infantry while moving toward Bellicourt farm and, coming under direct fire from hostile artillery, they turned back.

Tank No. 2, after pulling the other tank from the trench, continued on its way until its tracks were caught in a heavy wire entanglement and the tank was stopped. Considerable time was lost before the tank could get started again and it did not catch up with its infantry until it arrived at the Sechelles woods. It entered the action at this point but had to stop when the water supply pipe of one of the engines broke. By the time this was repaired, the need for the tank had passed, so the tank returned to its assembly point.

Tank No. 5 became stuck twice in shell holes while advancing to the attack but managed to get out under its own power. It crushed a barricade in Orvillers and went on to the other side of and beyond the town. At the far side of the town the tank entered a smoke and gas cloud. Gas masks were used but these decreased the already poor vision, and the tank in consequence fell into a pond and went out of action. It was rescued at 4 PM with the aid of a tractor and some French prisoners, but the engines were found to be defective so it was returned to the assembly point. When the tank went out of action, the commander and crew went forward as an assault detachment with the infantry. The 3rd Tank Detachment lost one man killed in action.

While the 1st Tank Detachment of three tanks was en route to its starting point, one of the tanks was put out of action by artillery fire which also wounded the tank commander and some of the crew. After tanks No. 4 and No. 5 arrived at the starting point, they were subjected to heavy artillery fire, and two officers, two NCOs, and one private were killed, and one officer and several men were wounded. The two tanks being still fit for action, at 3:50 AM, they left the starting point and arrived at a point within 50 yards of the infantry line at 4:20 AM, ready for the attack. The first mission was to aid in the attack on Hill 110, where a strong point was located. This attack was made in the heavy fog which covered the ground at the time. After a sharp fight, the guns on this hill were silenced, and the tanks then started for the village of Mortemer. Notwithstanding the fact that liaison agents had been sent ahead to investigate the condition of the ground, a very large hole in the path of the tank had not been discovered and, owing to the fog, the driver could not see it, so the tank fell into it. The infantry had not followed, and since the tank could not be moved, the crew removed some of the machine guns and set them up at the edge of the hole. The infantry reached the tank about an hour after the accident.

Tank No. 4, in the meantime, started for Mortemer but, before this place was reached, the transmission was damaged, the engine gave trouble, and several of the rollers broke, so this tank was also out of action. All three of the tanks in this detachment were eventually saved and returned to Charleroi. The losses in the detachment were two officers and four men killed, and two officers and eight men wounded.

Fermicourt, September, 1918.

An unsuccessful attack was made with A 7 V tanks toward the end of September in the Cambrai-Bapaume region.

Orders were received at 11 PM, for the attack which was to start at 5:35 A.M. An approach march of 20 kilometers had to be made and only three of the five tanks used succeeded in entering the action. The German infantry was evidently not notified in time of the inclusion of tanks in the attack and two of the tanks were disabled by fire from their own troops.

North of Cambrai, October 11, 1918.

One section of A 7 V tanks and one section of captured British tanks remained as a last reserve on the morning of October 11 when the British advanced northeast of Cambrai. The German forces withdrew and, at the time the tanks were called into action, the British cavalry patrols were reported to be in rear of the German position.

The two tank sections were ordered to support the 371st Infantry in a counter attack. The captured tanks were unable to make much progress, due partly to the condition of the terrain, and were all immobilized within a short time. The A 7 V tanks operated over more suitable ground and, aided by a light fog, they advanced rapidly. The infantry and artillery were retreating when the tanks entered the action.

Some of these tanks worked their way forward until they were in rear of the British lines. One went about five miles back of the lines and came unexpectedly upon reserves which were soon scattered by the fire from the tank. A British battery, going into action about 250 yards from the tank, was dispersed by a shell from the tank's heavy gun. This tank was later disabled by artillery fire and half of its crew were severely wounded, including the tank commander. However, the energy with which the tanks entered into the action is reported to have aided materially in stopping the British advance at that point and to have enabled fresh German infantry to reorganize and occupy the former German line.

The German Tank Situation.

An interesting comment upon the German tank situation is contained in the following extract from a statement by Captain Wegner, of the German army:

The idea of an armored and armed vehicle on caterpillar tracks, to be used against the enemy, was offered to the German War Department by a German firm in December, 1913. A torpedo-shaped body carried on two tracks was indicated on a rough sketch. Undoubtedly a tank could have been developed therefrom under the direction of the Army, with the assistance of technical engineers; then Germany would have had the lead over the allied powers in this respect as she had in so many others. However, the military necessity was not recognized. The German Army was built and organized for short heavy blows, but not for a four year trench war.

In 1915 a German firm built a gun-carrying vehicle with caterpillar tracks and demonstrated it before a military commission. The firm wanted so much for it and, as it had to be improved to be considered practical, the matter was dropped. The question of the need (for such a vehicle) was not considered acute and, in view of the increasing scarcity of horses and oats, it was thought to be more important to procure caterpillar tractors for munition transport.

Hence, nothing was accomplished in regard to tanks up to harvest time, 1916. Therefore, the German War Department was in no way prepared to meet an entirely new situation when the English first attacked with tanks on September 15, 1916.

The 27th Infantry Division writes of the April 11, 1917, attack:
The moral impression of tanks on the infantry is very great. Also the fire effect of the tank guns and tank machine guns must not be underrated. The 124th Infantry Regiment suffered severe losses from this fire on April 11th. But if the infantry was supplied with plenty of "K" ammunition and trench cannon, they would have weapons in hand herewith to put an end to tank attacks.

On November 20, 1917, the tank attack took place at Cambrai. By concentrating all his tanks, our adversary achieved some success. If the English Army leader had taken full advantage of his opportunity, it would not have been possible to limit his break-through in that region. The attack was a complete surprise. It came without artillery preparation, on a "quiet front," well fortified, but weakly garrisoned. Without tanks it would have been impossible to overrun the excellently built and invulnerable positions in the sector Havrincourt-Cannelieu. The enemy lost many tanks, mainly through our motorized artillery (1. Kraftw. Flanks). The 2nd Army alone, counted 64 tanks, 26 male and 38 female, that fell into our hands through artillery fire and breakdown.

After the lost ground had been recaptured through our counterattack, the captured tanks were towed off and sent to Charleroi, where a large tank plant was erected. (Later, Charleroi became headquarters for the German tank troops.) Here the tanks were repaired and restored to battle condition. Unfortunately it was impossible to protect the tanks against theft and wanton destruction at the hands of our own troops; and at that time prize money was given in order to obtain and protect the valuable and unreplacable parts and instruments. Transmissions were blown up and motors broken in order to obtain an insignificant piece of brass. Owing to the need for man-power, the Armée Ober Kammand was disinclined to furnish guard for the tanks. In refitting the captured tanks, the 57 mm guns, captured from the Russians, came in very handy.

Expert engineers were sent to Charleroi to study the captured tanks for ideas that could be applied to our own tank construction. Copying the English tank, with some improvements, was proposed. The amount of raw material, metal and reclaimed labor necessary was so great, that, in view of the obtainable delivery number and delivery time, it was decided that the proposition would not pay. The motor question was particularly difficult, as all that could be turned out were needed by the flying troops and by the artillery for tractors, and could not be spared for anything else. The O.H.L. was furthermore forced to curtail all new orders to a minimum on account of the reserve situation of the infantry. Hence, this tank building scheme was, after long conferences, abandoned.

In the meantime, after many difficulties in its production had been overcome, the German A7V was so far advanced that it could be employed at St. Quentin on March 21, 1918. On January 6, 1918, the first tank section with 5 A7V (we organized sections of five tanks each) was sent to the Western Front. They were supplied with the most necessary transportation, repair shops, etc., and were to be trained for action with a shock battalion. According to all experience on our and the enemy's side, the closest cooperation between the infantry and the tanks is imperative. Therefore, it was necessary for the infantry, particularly the shock troops, to learn how to work together with the tanks. The time from January to March was further utilized in improving the tanks and installing signal apparatus.

Through its neglect thereof, the German War Department is to blame for our failure in tank building. In connection herewith, the following should be emphasized:

The high level in mechanical engineering that we had attained in Germany would undoubtedly have enabled us to develop the tank to perfection provided that at the beginning of the war, we had been familiar with the caterpillar traction system. However, it would have been necessary on the part of the O.H.L. to express loudly a demand for tanks, probably as early as the latter part of 1914 when position warfare had set in on all fronts. In the fall of 1916 it was already too late to start building such a complicated weapon and reach mass production. In order to have put as many as one thousand tanks in the field in March, 1918, it would have been necessary to have had the tanks delivered by New Year, 1918. The late start in 1917 would also have confined us to the then known model; that is, we had three months at our disposal in which to produce a useful trial tank, a time that was not sufficient to produce even the most essential drawings, much less to build a new trial tank and to correct errors that would have developed. A well-known firm estimated that the time necessary for the production of the drawings alone, would be from eight to nine months.

Our attacks, even without tanks, were successful, but tanks would have saved us blood. Tanks, to insure success, must be used in large numbers and this in turn, demands mass production. But with the abnormal shortage of raw material and labor, this could not be done without curtailing production of other important raw materials. Hence, the tank was a field in which we were forced, with heavy hearts but through necessity, to let the enemy take the lead.

The German Tank Corps had altogether four detachments or sections of five A 7 V tanks each, and five sections of 5 captured British tanks. Additional sections were authorized to be formed as soon as more captured British tanks became available.

CHAPTER XI.

TANK COMBAT HISTORY. AMERICAN.

St. Mihiel, September 12, 1918.

In this action, which was carried out by the American First Army, it was planned to reduce the St. Mihiel salient by two simultaneous attacks; one from the south, with the 1st Corps on the right and the 4th Corps on the left, and one from the west by the 5th Corps. The 505th French Tank Regiment, with three battalions of 45 Renault tanks each and two groups of a total of 36 French St. Chamond tanks, was in support of the 1st Corps. The 304th American Tank Brigade, with two battalions of Renault tanks and two groups of a total of 24 French Schneider tanks, supported the 4th Corps (Included in these two Schneider groups were 8 Renault tanks used mostly as command tanks.) No tanks were available for the 5th Corps. The 304th Brigade was to hold 41 Renault tanks in a reserve which was to follow the 344th (formerly 326th) Tank Battalion as flank protection for the 4th Corps, and to be used later as an exploitation force. The 345th (formerly 327th) Battalion was to follow the infantry until the D'Houblons trench had been passed, as lack of cover prevented these tanks from using a starting line closer to the infantry line. They were then to pass through the infantry and lead the attack. The French Schneider tanks were to support a part of the 42nd Division in the left half of the sector. The 344th Battalion was to take the trenches near Richecourt and then proceed to Nonsard. It being apparent, toward the last part of the approach march, that the last company of the 345th American Battalion would be late, that company was shifted to the reserve.

The attack started as planned and proceeded without difficulty until the tanks encountered the first German trenches. Here a great many had trouble in crossing. A large number became stuck and some time, as well as the expenditure of a large amount of fuel, was required to get them out of the trenches. One platoon of the 344th Battalion crushed the wire ahead of the infantry and attacked the Harem salient, but was unable to cross the trenches encountered. The remaining tanks passed between Richecourt and the left bank of the Rupt de Mad toward La Hayville and then moved against the machine guns in the Bois de Rate. Twenty-five tanks reached Nonsard in advance of the infantry, silencing the machine guns and 77 mm guns along the eastern edge of the Bois Quart de Reserve and Bois de Rate. Owing to the great number of difficult trenches in the path of the tanks, they used up their fuel much faster than was expected and were practically out of gas by 3:00 PM. Additional fuel was taken up to them on sleds towed by tanks.

The 345th Battalion also had difficulty in crossing the wet and crumbling trenches. It came under heavy shell fire while crossing the D'Houblons trench and two tanks were put out of action at this point by direct hits. However, five tanks entered Essey ahead of the infantry and ten prisoners were captured by them in this village. After inspecting the bridge across the Le Madine river, three tanks crossed and moved against Pannes, leading the infantry to the town. Thirty prisoners were captured here by the tank personnel. When the infantry halted on its objective, the tanks withdrew and later preceded other infantry to the woods northwest of Beney. The

Plate 12.

Brigadier General S. D. Rockenbach, Chief of Tank Corps, U. S. Army

On December 23, 1917, General Pershing assigned Colonel Samuel D. Rockenbach, (Cavalry) Quartermaster Corps, to duty as Chief of Tank Service, American Expeditionary Force. On January 26, 1918, the Tank Corps of the U. S. Army was created and Colonel Rockenbach was assigned as its chief. Returning to the United States in July, 1919, General Rockenbach continued as Chief of the Tank Corps, U. S. Army until the Corps, by Act of Congress, ceased to exist on June 30, 1920.

tanks drove out machine gun crews here, then overran a battery of 77 mm guns in the village of Beney, and later rallied in Pannes. During the advance on this date, the artillery gave material assistance with H.E. and smoke shells along the edges of woods as well as on the ridges east and west of Maizerais, where the Schneider tanks operated. The Schneiders were unable to precede the infantry but followed the first wave as far as Maizerais where they halted in accordance with instructions. Two Schneiders were put out of action by mines and, near Maizerais, one was struck by a 150 mm shell which put it out of action and resulted in 15 casualties. Of the 174 tanks that entered the action with the 304th Brigade, 22 were ditched and out of action all that day, and 21 were out of action due to mechanical trouble.

Of the Renault and St. Chamond tanks supporting the 1st Corps, 16 Renaults and two St. Chamond tanks reached their objective, the Bois d'Heiche. Four St. Chamond tanks remained at the starting point on account of mechanical trouble and two were mired at that point. Of eight tanks belonging to one group which attacked, six were stopped by the mud at the German trenches. There were 159 Renault tanks of the 505th Regiment and 18 St. Chamond tanks available for use on the 13th.

No tank action of importance was carried out on the 13th of September. The principal reason for this was the fact that the tank units were unable to secure fuel due to a road jam at Flirey which prevented the gas trucks from getting through. The roads were so badly congested that it took these trucks 32 hours to move nine miles. Seven of the reserve tanks were supplied with gas drained from other tanks and these were sent to cooperate with the 1st Division in its advance on Vigneulles. These tanks had considerable difficulty with fan belts. They encountered no resistance from the Germans. They passed through Vigneulles and went on to Hatton-Chatel, which they reached ahead of the infantry at about 5:30 PM. At 2:00 PM, the gas for the rest of the 344th Battalion arrived. Fifty of these tanks arrived at Vigneulles at midnight. Thirty of the 345th Battalion tanks and 22 French tanks moved up to St. Benoit.

On September 14th, the 344th Battalion moved through St. Maurice to Woel in an effort to regain contact with the 1st Division. When within about two kilometers of Woel, information was received that the Germans had just been driven out of Woel and that the town was held by twenty French soldiers. Instructions were requested from the brigade headquarters, and an officer's patrol was sent into the woods to the south in an attempt to gain touch with the IV Corps. While the reconnaissance by this patrol was in progress, gas trucks arrived with a supply of fuel. These trucks had been attacked on the way to this position by German airplanes using bombs and machine guns, but no serious damage had been done. The tanks were concealed in hedges between Avillers and Woel at this time. Two German planes flew low over the tank position, and an American plane, which attacked the German planes, was shot down by them west of Woel.

No information having been received regarding the 1st Division by noon, a patrol of three tanks and five dismounted men was sent through Woel and two kilometers down the road toward Benoit. Having failed to locate the Germans, the platoon commander returned to Woel and started toward Jonville. As the platoon neared Jonville, it met a German battalion marching in close column toward Woel. The tanks and the dismounted men opened fire, scattering the battalion, and followed it back to Jonville. Here the tanks came under fire from a battery of 77 mm guns. The platoon attempted to withdraw but one of the tanks failed mechanically so a message was sent to the battalion commander requesting assistance. Before the five tanks which were sent up arrived on the scene, another tank failed and the platoon commander hooked both tanks to his and towed them back to the battalion area.

Believing that a counter attack was imminent, the battalion commander decided to remain in position and surprise the Germans when they came over. No counter attack was made, however. About 6:00 PM another low flying German plane passed over the tank position, and, believing his tanks had been located, the battalion commander ordered them to another spot. Accordingly they moved about 500 yards to the south and had just left the first position when a heavy artillery concentration was put down on it by the German artillery. Both American battalions and the French tanks were ordered back that night to the Bois de Thiaucourt and they were shelled en route to that place.

Meuse-Argonne, September 26 to November 1, 1918.

In this action the 304th American Tank Brigade with its two battalions of Renault tanks, and two groups of Schneider tanks, the latter operated by French tank troops, supported the 1st Corps on the left of the 1st American Army front. One of the American battalions was to follow in the support echelon 1500 meters in rear of the front line. The 505th French Tank Regiment (Renaults) plus one battalion (a total of four battalions) and two groups of St. Chamond tanks supported the 5th Corps. No tanks were assigned to the 3rd Corps, on the right. The tanks supporting the right division of the 5th Corps were ordered to assist the advance of the 3rd Corps whenever practicable. The two groups of Schneider tanks with the 1st Corps were to be held in the reserve echelon until the 1st Corps objective at Very was reached and were then to support the attack from that point on.

The Malancourt, Cheppy and Montfaucon woods and the Forges river were considered too great an obstacle for tanks and, the country back of the German front line for a distance of about five kilometers being covered with shell craters, it was decided that the tanks could be counted upon only for supporting the attack north of the Gercourt-Cuisy-Very-Baulny line.

The resistance encountered on September 26th, especially along the eastern edge of the Argonne Forest and near Cheppy and Varennes, was such that, contrary to the plan, most of the tanks entered the action the first day. The chief resistance encountered was from machine guns and artillery. Many machine guns were put out of action. The tanks reached Varennes at 9:30 AM, four hours ahead of the infantry.

The French heavy tanks had difficulty in crossing the German trenches. At 1:00 PM they had arrived within 400 meters of Cuisy wood when the infantry of an adjacent division was stopped by intense machine gun fire from an area about 200 yards ahead of the tanks. Two Renault companies attacked the machine guns, cleared Cuisy wood, and delivered it over to the infantry. At 6 PM the assistance of the tanks was again sought and one section of Renaults was used against the woods bordering the Malancourt-Montfaucon road. The Germans retreated upon the approach of the tanks. Another section was engaged at 7 PM and led the infantry to the defenses south of Montfaucon.

Upon the request of the infantry, a French tank detachment entered Cheppy, where it encountered severe resistance from German antitank guns. The machine guns which were at the edge of the town were not observed by the tank crews and, while the tanks were engaged in the town, these guns held up the infantry following the tanks until another French battalion intervened. After the village was taken the tanks supported the infantry to a point beyond Very.

On September 27th, 11 tanks from one of the American battalions led an attack along the edge of the Argonne Forest northwest of Varennes, silencing many machine guns and capturing a number of machine gunners, who were turned over to the infantry. Repeated requests for assistance came from the infantry on the east bank of the Aire but no concerted plan of attack could be arranged at this time. Two platoons supported an attack on the plateau north of Very.

One of the French Renault companies supported an attack intended to surround Montfaucon. This was accomplished, and the infantry entered the village. At 5:30 PM, French Renaults cleaned up the ground in advance of the infantry between Montfaucon and Septsarges, the southern edge of Beuges wood, and on the ridge north of Nantillois. The infantry followed very poorly and, although Nantillois was evacuated by the Germans, the infantry remained 1500 meters to the south.

Another battalion of French Renaults supported an attack by an infantry regiment against the German forces intrenched on the Charpentry-Romagne road. These tanks quickly reached their objective, and the infantry followed. Two of the tanks reached Charpentry, but the infantry organized its position on the ridge to the southeast of that village.

On September 28th, in compliance with requests for tank assistance which had come in on the 27th, the tanks of the American battalions were assigned as follows: 15 to the 28th Division, 5 to the 91st Division, 36 to the 35th Division, and 27 to the reserve. Six French Renaults were also assigned to the 35th Division. Throughout the action on this date, well organized tank defense was met, consisting of antitank rifles, direct firing artillery and frequently artillery barrages, the latter being brought down with little delay. In some sectors, infantry support was poor. The tanks took Apremont five times before the infantry entered the town and consolidated the position.

Two companies of French Renaults were offered to an infantry regiment and were used to support an attack on Beuge wood. No Germans were found in this wood. These tanks then cleared the edges of Woods 268 and 250, reaching Madeleine wood. The infantry did not follow on account of artillery fire. Another company supported the attack on the woods south of Cunel, destroying many machine guns in this action. However, these tanks then came under heavy artillery fire. The infantry stopped and sent orders to the tanks to retire. While complying with this order two tanks received direct hits.

A composite battalion consisting of Renaults and Schneiders aided in the attack on Charpentry-Baulny, and these tanks led the infantry beyond the objective to Chaudron farm.

On September 29th, after the repair and salvage company had worked all night, 55 of the American operated tanks were ready for action. The tanks on the left bank of the Aire remained at Apremont with a view to meeting an expected counter attack. The tanks on the right bank of this river moved to Baulny as a reserve for the 35th Division. These tanks were used late in the afternoon to aid in checking a counter attack. They were ordered to hold the line Baulny-Eclisfontaine and this was done. Tank patrols were established to allow the infantry to organize on this line, the tanks being withdrawn after dark.

Three sections of the French-operated Renaults went into action with an infantry regiment between Andon and Exermont against the village of Cierges. Heavy machine gun fire stopped the infantry, which then retreated. One tank section moved toward the village and did not return. The other two sections withdrew. One company of French Renaults was used to aid in an attack on Madeleine farm and the Bois des Ognons. The infantry followed as far as the wood only. Due to a bombardment during the night, the infantry withdrew 1500 meters, and seven tanks, five of which had failed mechanically, were left in German territory. The remaining Schneider tanks were withdrawn from the front on this date.

On September 30th, the tanks were withdrawn to reserve positions in the 1st Corps area, as were two battalions of the French Renaults on October 3rd. On the evening of September 30th, however, the 28th Division requested aid in a local attack west of Apremont. This attack was to start at 6:00 AM October 1st. Eight tanks were in position for this attack when the Germans attacked at 5:30 AM. The tanks moved forward and, meeting the German assault troops, broke up their attack, causing many casualties among the Germans.

According to a report rendered by the Chief of the U. S. Tank Corps on October 3rd, the two American tank battalions had lost 53% of their officers, 25% of their men and 70 tanks up to that time.

Three companies supported the advance of the 1st and 28th Divisions on October 4th. Artillery fire encountered on this advance was very accurate and severe. Strong resistance was encountered near Hill 240 by the 1st Division and along the Argonne Forest by the 28th. Casualties in personnel and tanks were heavy. The mutual support between infantry and tanks on this date was excellent.

Fifteen St. Chamond tanks, on the same date (October 4th), operated in the 3rd Corps sector toward Ognons wood. The tanks cleared the edge of this wood and signalled for the infantry. The infantry, which was located in Wilpré Springs Ravine, had not moved forward by 11 AM, at which time the tanks were subjected to heavy artillery fire, so they withdrew.

French Renault tanks from the 15th Battalion were used in the zone of the 5th Corps to clear the east banks of the Andon river, the western edge of Cunel wood, and in the attack on Cunel and Romagne. The infantry supported the attack effectively, assisted in taking Namelle trench, and organized on the east-west road. Owing to severe losses, they remained on this line. Many of the tanks mired in the banks of the ravines that intersected this ground.

On October 5th, 30 of the American-operated tanks were ready for action, but these were not called for. Many of the mechanical difficulties were caused by the long runs the tanks had to make in going from tank parks to assault positions. These places were usually between 10 and 20 kilometers apart. The wear occasioned in traveling such distances resulted in a great many minor mechanical difficulties which at times retarded the advance.

The 3rd Corps resumed the attack north of Wilpré Springs. Four French tanks were used in advance of the infantry. It was arranged that the tanks were to precede at 100 meters, returning to aid the infantry if it did not follow. The attack was launched at 10:20 AM and at 10:35, when the infantry failed to follow, the only tank then in running order retired. One groupment of French St. Chamond tanks was released on this date.

The 32nd Division resumed the attack, supported by two sections of the 17th French Battalion (Renaults). The infantry supported the attack effectively and the tanks cleared Gesnes, the woods east of that point, Hill 235, the Marine woods, and Chêne-Sec woods.

No tank actions of importance occurred for several days due to the lack of serviceable tanks. The salvage units were all engaged in salvaging and repairing tanks in order to prepare for future needs. It was about this time that General Pershing offered "anything in the A. E. F. for 500 additional tanks," but tanks were not to be had at any price. The American tank production program had not succeeded in completing any of the 1200 light tanks which, according to the War Department estimate, were to have been ready by July, 1918. The lack of these tanks was, no doubt, responsible for many infantry casualties in the St. Mihiel and Argonne operations.

Only eight tanks, belonging to the American tank units, were available west of the Aire river to fill a request from the 28th Division on October 7th. En route to join, one tank struck a mine and was disabled. The others reported and were used on patrol work that day. Several mine fields were encountered in the zone of the 1st Corps. The troops were usually able to avoid the mines as the Germans had forgotten to remove placards marked "Achtung! Minenfeld!"

Following the withdrawal of the French tank units, a report was made by the French officer in command of these units. His conclusions are of interest, and valuable lessons can be drawn from them. It was his opinion that the greater part of the American troops engaged knew nothing about fighting with, or supporting tanks, and his report shows many instances where the infantry not only failed to follow the tanks and take advantage of the opportunities created by them, but one or two instances where the tanks in attack were abandoned by the infantry. In other cases, where the infantry had received some training in acting with tanks, and where they supported the tanks effectively, the combination of infantry and tanks was sufficient to advance the attack and progress was made. (Cooperation between the French tank units and the American divisions was, no doubt, somewhat hindered by the lack of a common language.)

This report shows that the St. Chamond tanks were handicapped by damage to their tracks, by derailments, by the breakage of the caps of connecting rods on forward bogies and of track pins. The Schneiders were so thoroughly used up that they required a general overhaul. The power of the engines was so reduced that the tanks were unable to climb the slightest obstacle. The Renault tanks were handicapped by inefficient cooling systems. Many fan belt pins and fan belts broke. Engines overheated frequently, thus causing pistons to seize in the cylinders and the connecting rods to break from the pistons, which in turn caused the rods to damage the crankcases. Reduced oil pressure, probably caused by high engine temperatures, also gave trouble. When the engine stopped, it was frequently impossible to start it again from the inside of the tank, and it was dangerous, if not impossible, to start it from the outside if the tank was under fire. It was found that the tanks produced by a certain manufacturer were very satisfactory for the first 100 hours of operation but failed rapidly from that time on, while the tanks produced by another manufacturer gave satisfaction continuously. Gasoline supply on the battlefield was found to be difficult. The congestion on the roads prevented the effective use of motor vehicles.

Of the 214 French-operated Renault tanks, 16 St. Chamond and 22 Schneider tanks were fit for action on September 25th, just prior to the starting of the Argonne action on the 26th; 21 Renaults, 2 St. Chamond and 2 Schneiders had been hit by artillery; 1 Renault and 1 Schneider had been damaged by mines; 55 Renaults, 3 St. Chamonds, and 6 Schneiders had been left on the field in our lines; and 22 Renaults had been left on the field in the German lines. The losses in killed, wounded and missing among the French tank personnel were 18 officers and 149 men. Sixteen engagements had been fought by the four battalions of Renaults, two groups of Schneiders and two groups of St. Chamond tanks during this period; 125,570 gallons of gasoline had been consumed; and 1058 belts of machine gun ammunition, 1152 rounds of 75 mm, and 7878 rounds of 37 mm ammunition had been fired.

On October 8th, 26 of the American-operated Renault tanks were fit for action; 5 of which were in good condition and the other 21 could be expected to stand one day of operation. The divisions to which they were assigned could not use them on this date due to the unsuitability of the terrain. The commanding officer of these tank units reported that supply and communications had worked perfectly during the Argonne offensive. Forty-eight tanks had received minor repairs and were reported ready for action on the 11th of October. Complying with a request for five tanks for the 82nd Division, a total of 23 were sent by the tank brigade commander, owing to the distance to be traveled and the poor condition of the tanks. Only three of these reached the division area.

Orders were received to withdraw all but 24 tanks, these to be formed into a provisional company. This company was directed to support an attack

Plate 13.
U. S. Tank Corps Battles.

between Landres-et-St. George and St. George on October 15th. Due to the great distance to be traveled and the speed required in order to reach the attacking troops in time for the action, only ten tanks arrived, and they just reached their positions in time to go over in advance of the infantry. As they moved over the German trenches, they encountered a formation preparing for a counter attack. After dispersing the German troops, the tanks withdrew because they were not supported by the infantry.

On November 1st, 15 tanks supported the 2nd Division in an attack against Landres-et-St. George and St. George. One platoon was directed against St. George and the wire in front of that town, and the other two platoons against Landres-et-St. George and the wire protecting it. This division, composed of infantry and marines, gave the tanks excellent cooperation, and the corps commander and local commanders were well pleased with the assistance rendered by the tanks. Three tanks flanked and put out of action a German battery of four 77 mm guns. This was the last action in which the American tank units participated. These two tank battalions fought in 18 engagements during the Argonne offensive; their personnel casualties in killed and wounded were 171. Of the 141 tanks employed by them, 18 were demolished by enemy fire. 174 tanks were repaired during these actions. This was 123% of those engaged. Only one tank was unaccounted for when the tank units were withdrawn.

The following letter from General Pershing indicates his opinion of the tank units of the U. S. Tank Corps:

AMERICAN EXPEDITIONARY FORCES
OFFICE OF THE COMMANDER-IN-CHIEF
February 20, 1919.

Brig. Gen. S. D. Rockenbach,
Chief of Tank Corps,
A. E. F.

My dear General Rockenbach:—

Now that active operations have ceased and many of your personnel are returning home for an early separation from the service, I desire to express to you, and through you to the officers and enlisted men of the Tank Corps, my appreciation of the work that the Corps has accomplished.

From the beginning its history has been a consistent up-hill fight for accomplishment against almost insurmountable difficulties in the way of obtaining tanks for training or for fighting. Due to untiring efforts, a certain limited number were finally obtained from our Allies, the Corps was recruited from the pick of the personnel of all arms of the service, tank schools were started on a practical basis in France and England, and by the middle of summer the Corps took the field with several battalions. Its history in active operation, though short, is a bright and glorious one. In both the American offensives at St. Mihiel and Meuse-Argonne of the First American Army, it was of material assistance in the advance. In the breach of the Hinderburg Line with the British near Le Catelet it also won glory. The high percentage of casualties among officers and men tells the tale of splendid morale and gallantry in action of your personnel and their unselfish devotion to duty.

It gives me great pleasure to thank all officers and enlisted men of the Tank Corps and, in the name of their comrades of the American Expeditionary Forces, to convey our appreciation and admiration of their splendid work and gallant record.

Sincerely yours,
JOHN J. PERSHING.

An account of the actions in which the 301st American Tank Battalion participated may be found in that part of this work devoted to British tank history. This battalion was equipped with Mark V and Mark V Star tanks. By agreement between Great Britain and the United States, these tanks were employed only on the British front.

CHAPTER XII.

TANKS OF ALL COUNTRIES.

BRITISH.

The tanks that fought in the World War having been illustrated and described in the preceding chapters, the primary purpose of this part of the text is to illustrate and describe the more important of the experimental tanks of the War period and the tanks, both experimental and adopted, that have appeared since the War. The illustrations and data pertaining to these tanks are arranged primarily by country, as can be seen from the table of contents in the front of the book. Within the section pertaining to each country, the arrangement is, in general, chronological. The chronological arrangement is, however, departed from somewhat in order to show in direct succession the similar tanks of a series.

The same scheme of arrangement is followed, where applicable, as to the illustrations of other vehicles later in the book.

Little Willie.

Produced in 1915 by William Foster and Company, Ltd. Total production, 1.

Suspension: Rigid.
Tracks: Flat steel plates with single grousers.
General arrangement: Driver in front, engine in center, etc.
Dimensions: Length 12 ft. exclusive of tail.
Engine: Daimler, 6 cylinder, sleeve valve, 105 HP, forced water cooling.
Transmission: Sliding gear, 2 speeds forward, 1 reverse.
Special features: Two 4 ft. 6 in. tail wheels for steering, controlled hydraulically. Lips on track links engaged loosely with angle irons on hull. Turret intended but not completed. Competed with Big Willie for acceptance as first British tank and was rejected.

Big Willie.

(Also known as "Mother"). Produced in 1915 by William Foster and Company, Ltd. Total production, 1.

This was the pilot model of the Mark I. It was built of boiler plate and wood. It competed with the Little Willie and became the forerunner of the British war-time tank program. This tank was later fitted with an improved Daimler engine and a Daimler combination electric and two-speed transmission for each track. Hundreds of these transmissions were ordered but later cancelled as the system failed to stand up in use.

Flying Elephant.

Produced in 1916 by William Foster and Company, Ltd. Total production, 1.

Armament: One cannon and six machine guns.
Armor: 2 in. to 3 in. (maximum).
Suspension: Rigid.
Tracks: Outside tracks 2 ft. wide, inside tracks 1 ft. 9 in. wide.
General arrangement: Engines forward of center; final drive at rear, crew at front, rear and sides.
Dimensions: Length 26 ft. 9½ in.; width 9 ft. 10 in.; height 10 ft.
Weight: 100 tons.
Engines: Two, Daimler, 6 cylinder, 120 HP each, forced water cooling.
Horsepower per ton: 2.4.
Transmission: Sliding gear, 4 speeds forward, 1 reverse; one transmission for each engine. Final drive by worm gears.

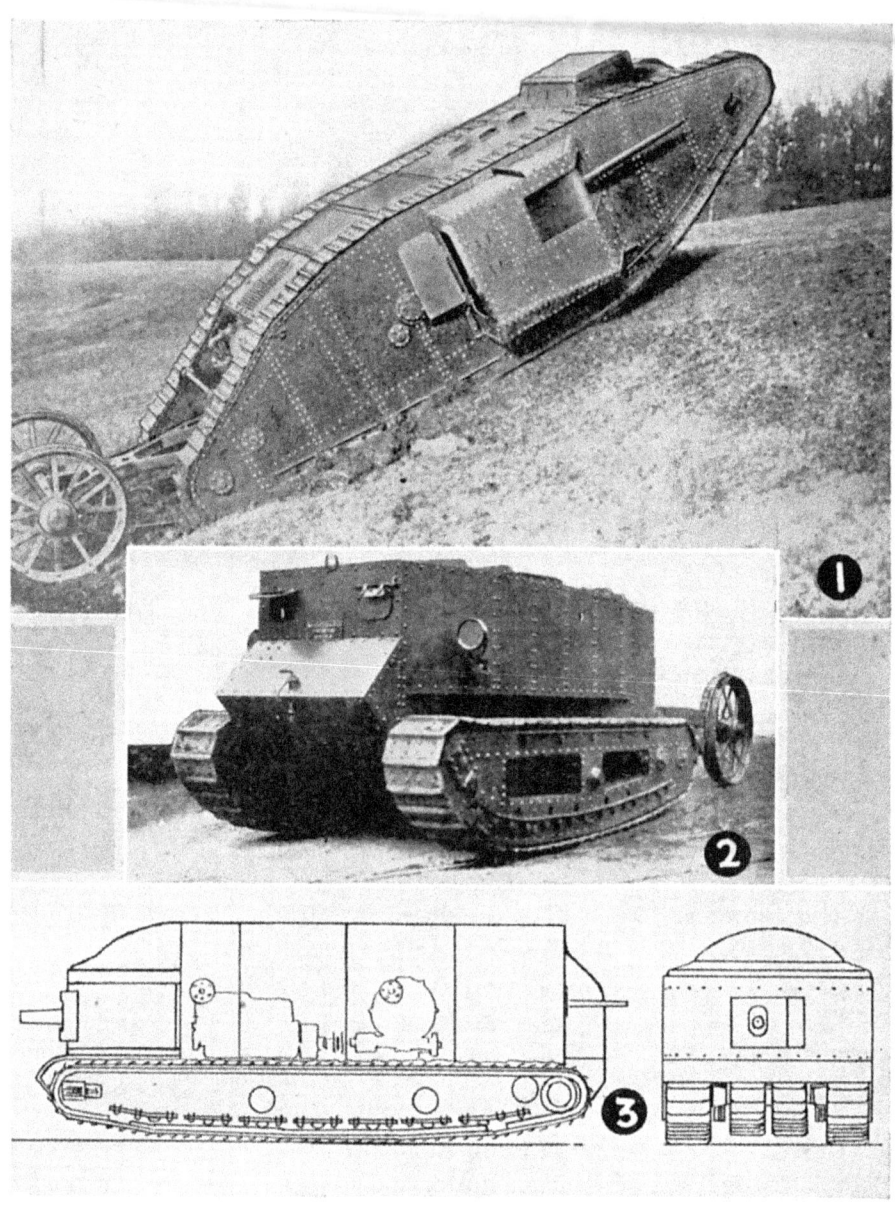

Plate 14.
1. Big Willie (Mother.)
(Photo by Wm. Foster and Co., Ltd.)
2. Little Willie.
(Photo by Wm. Foster and Co., Ltd.)
3. Flying Elephant.
(From a drawing by Wm. Foster and Co., Ltd.)

Special features: The gun had 80 degrees of traverse. Special small flywheels were used to enable the engines to be mounted close together. The tank had auxiliary tracks under the belly, coupled to the power train by a dog clutch. These tracks had 6 inches clearance. The tank was practically completed and then broken up.

Modified Medium A.

Produced in 1917 and 1918.

The modification illustrated utilized leaf springs in the suspension. An additional modification included the installation of a Rolls-Royce 175 HP engine and a Mark V planetary transmission. The hull was raised about 12 inches. These modifications made a speed of 30 miles per hour possible and thus was initiated the trend toward high speed that is found in modern tanks. Another Medium A was equipped with a tailpiece similar to the French Renault and still another with a tailpiece like that on the original Mark I.

Medium B.

Produced in 1918 by Metropolitan Carriage Wagon and Finance Co., Ltd. Total production, 45.

Crew: 5.
Armament: 4 machine guns, mountable in stationary turret or side doors.
Armor: 0.24 in. to 0.55 in.
Maximum speed: 8 mph.
Suspension: Rigid.
Tracks: Flat plates with single grousers, 22½ in. wide, pitch 10 in.
General arrangement: Driver in front; gunners in center; engine and final drive in rear.
Dimensions: Length 22 ft. 9 in.; width 8 ft. 10 in.; height 8 ft.
Weight: 20 tons.
Engine: Ricardo, 4 cylinder, 100 HP, forced water cooling.
Horsepower per ton: 5.
Obstacle ability: Trench 9 ft. 9 in.; stream 31 in.; slope 35 degrees; vertical wall 39 in., tree 11 in.
Fuel distance: 62 miles. **Fuel capacity:** 103 gallons.
Special features: Fitted with smoke emitting device. Cramped quarters. Engine inaccessible.

Medium C (Hornet).

Produced in 1919 by William Foster and Company, Ltd. Total production, 36.

Crew: 3.
Armament: 3 machine guns, which could be shifted to different positions.
Armor: 0.236 in. to 0.59 in.
Maximum speed: 8 mph.
Suspension: Rigid.
Tracks: Flat plates 19¾ in. wide.
General arrangement: Crew in front; engine and final drive in rear.
Dimensions: Length 26 ft.; width 11 ft. 2 in.; height 9 ft. 6 in.
Weight: 22.4 tons.
Engine: Ricardo, 6 cylinder, 150 HP, forced water cooling.
Horsepower per ton: 6.7.
Obstacle ability: Trench 11 ft. 6 in.; stream 31 in.; slope 35 degrees; vertical wall 4 ft. 3 in.; tree 18 in.
Fuel distance: 75 miles. **Fuel capacity:** 180 gallons.
Special features: All gun mounts in stationary turret. Engine smaller than that in Mark V with same horsepower. Considered best medium tank of the war. Horizontal louvers on the sides of the engine compartment were added later. A male tank carrying a 6-pounder and 3 machine guns was designed but never built.

Medium D.

Produced in 1919 by Experimental Establishment (Governmental). Total production, 1.

Crew: 4.
Armament: Not installed.
Armor: 0.31 in. to 0.39 in.
Maximum speed: 27 mph.
Suspension: Cable.
Tracks: Pivoted sprung plates.
General arrangement: Driver in front, gunners in turret, engine and final drive in rear.
Dimensions: Length 30 ft.; width 9 ft. 2 in.

Plate 15.
1. **Modified Medium A.**
   (Photo by F. Mitchell.)
2. **Medium B.**
   (Photo by F. Mitchell.)
3. **Medium C (Hornet.)**
   (Photo by Wm. Foster and Co., Ltd.)
4. **Medium D.**
   (Photo by F. Mitchell.)
5. **Mark IV, Lengthened Tail.**
   (Photo by Wm. Foster and Co., Ltd.)
6. **Mark V, Experimental.**
   (Photo from British Imperial War Museum.)

Weight: 15.1 tons.
Engine: Siddeley-Puma, 240 HP, forced water cooling.
Horsepower per ton: 15.9.
Transmission: Planetary, 3 speeds forward and 1 reverse.
Special features: Hydraulic controls. Cable for suspension slipped from pulleys. Track slipped from rollers. No provision for side thrust. Track higher in rear than in front. Driver was tank commander. Variations of this tank were built, most of which were amphibious with water speeds of 3 mph.

Mark IV, With Lengthened Tail.

Modified in 1918 by William Foster and Company, Ltd. Total production, several.

The purpose of the lengthened tail was to increase the trench crossing ability of a standard tank without making extensive alterations. The rear extensions at first lacked rigidity. They were later connected by platforms upon which 6-inch trench mortars were experimentally installed. The rear extensions were spoken of as "tadpole tails."

Other Mark IV Modifications.

It was generally realized that the transmission in the heavy tank was unsatisfactory. Mark IV tanks were fitted with each of the following transmissions and tested in competition: Williams-Janney Hydraulic, British Westinghouse Electric, Wilkins Multiple Clutch Gear, and Wilson Epicyclic (planetary). One tank was sent to France where a Crochat-Colardeau Electric was installed. A Hele-Shaw Hydraulic was also tested in England. The Wilson transmission was finally adopted as standard for later tanks.

Mark V, Experimental.

Produced in 1916 by William Foster and Company, Ltd. Total production, 1 wooden pilot model.

The general appearance was the same as the Mark IV but this model had a rear cab and modified machine gun mounts. It was rejected in favor of the second Mark V type. The cross shaft connecting the main sprocket pinions between the rear horns was omitted in this tank.

Mark V, With Sprung Tracks.

Modified in 1918 by John Fowler and Company, Ltd. Total production, 1.

Maximum speed: 12 mph.
Suspension: Cable suspension with cable running alternately over bogie rollers and under track frame rollers.
Tracks: Narrow steel blocks laterally flexible.
Special features: This was a standard Mark V tank converted to cable suspension. One end of cable anchored and the other end connected to a heavy spring.

Another Mark V modification was fitted with the Lanchester constant mesh gear system operated by either hands or feet. It was efficient but expensive. The tank had a cumbersome cooling system. The fuel tanks were again placed inside.

Mark V, Double Star (R. E. Bridging Tank).

Produced in 1918. Total production, 15.

Crew: 8.
Armament: Male: Two 6-pounder (57 mm-2.24 in.) guns and five machine guns. Female: Seven machine guns.
Armor: 0.236 in. to 0.59 in.
Maximum speed: 5 mph.
Suspension: Rigid.
Tracks: Flat steel plates, width 26½ in., single grousers.
General arrangement: Driver in front; engine in center.
Dimensions: Length 39 ft. 1 in.; width (male) 12 ft. 11 in., (female) 10 ft. 9 in.; height 8 ft. 7 in.
Weight: Male 39.2 tons; female 38 tons.
Engine: Ricardo, 6 cylinder, 225 HP, forced water cooling.
Horsepower per ton: Male 5.7; female 5.9.
Transmission: Planetary.
Obstacle ability: Trench 14 ft. 9 in.; stream 3 ft. 10 in.; slope 40 degrees; vertical wall 4 ft. 10 in.; tree 23 in.

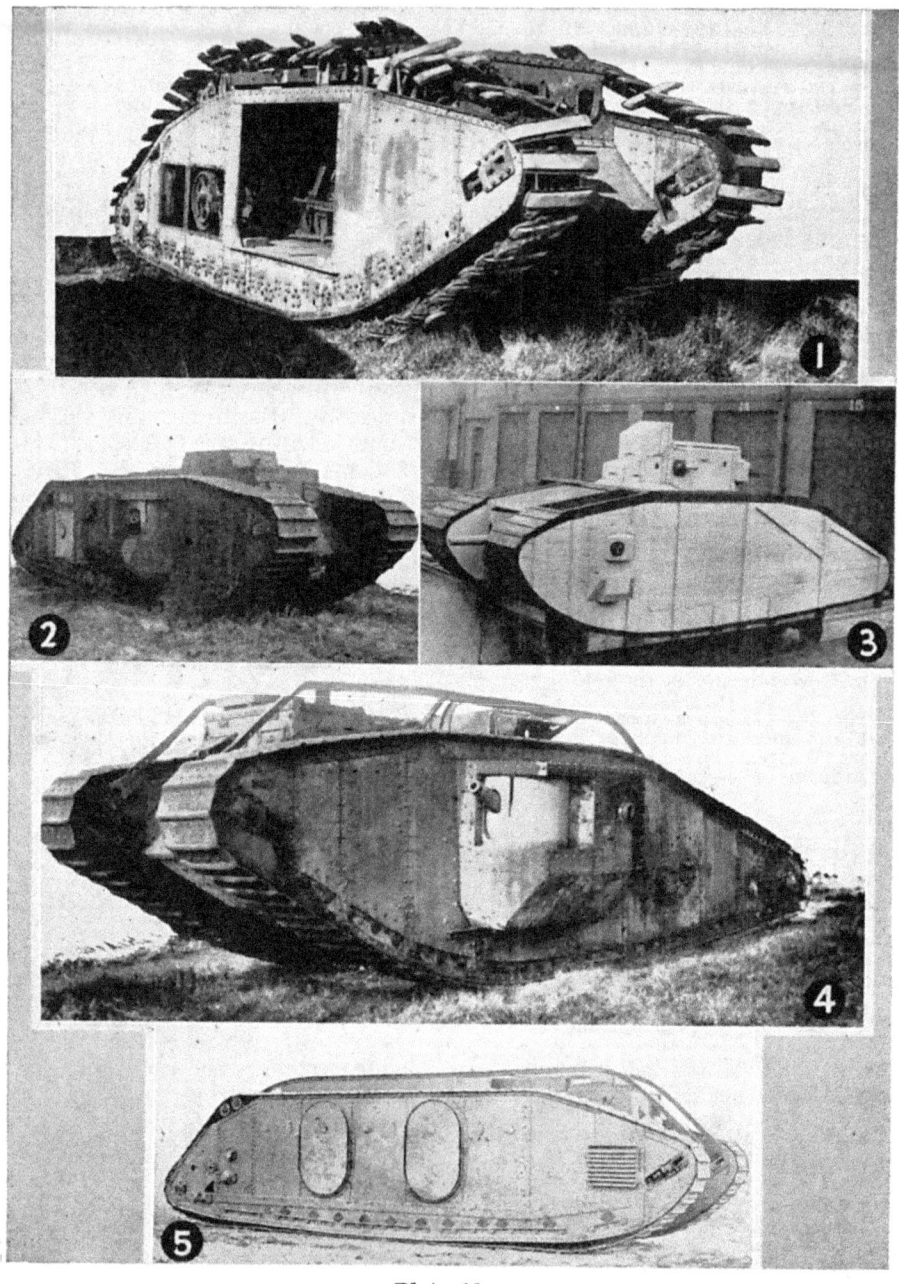

Plate 16.
1. Mark V, Sprung Tracks.
(Photo by F. Mitchell.)
2. Mark V, Double Star.
(Photo by F. Mitchell.)
3. Mark VI.
(Photo from British Imperial War Museum.)
4. Mark VII.
(Photo by F. Mitchell.)
5. Mark IX. Infantry Carrier.
(Photo by Wm. Foster and Co., Ltd.)

Fuel distance: 50 miles. **Fuel capacity:** 238 gallons.
Special features: These tanks were later modified for special purposes as follows: to carry and lay girder bridges (could span with these 24 ft. 7 in.); to carry and operate a heavy roller for destroying mines; to salvage tanks and carry pioneer supplies.

A Mark V Three Star (also called Mark X) was designed but never built.

Mark VI.

Produced in 1917 by William Foster and Company, Ltd. Total production, 1 wooden pilot model.

Armament: One 6-pounder (57 mm-2.24 in.) and four machine guns.
Suspension: Rigid.
Tracks: 36 in. wide.
General arrangement: 6 pounder gunner in front; driver and machine gunners in center; engine in rear on right side.
Engine: Ricardo, 6 cylinder, 150 HP, forced water cooling was to have been used.
Transmission: Planetary.
Special features:. Wooden model completed. Design rejected.

Mark VII.

Produced in 1918. Total production, 1.

Crew: 8.
Armament: Two 6-pounder (57 mm-2.24 in.) guns and five machine guns.
Armor: 0.2 in. to 0.47 in.
Maximum speed: 4.5 mph.
Suspension: Rigid.
Tracks: Flat steel plates with single grousers, about 26 in. wide.
General arrangement: Driver in front; engine and final drive in rear.
Dimensions: Length 29 ft. 6 in.; width 12 ft. 11 in.; height 8 ft. 8 in.
Weight: 37 tons.
Engine: Ricardo, 6 cylinder, 150 HP, forced water cooling.
Horsepower per ton: 4.1.
Transmission: Williams-Janney hydraulic.
Obstacle ability: Trench 12 ft.
Fuel distance: 50 miles. **Fuel capacity:** 170 gallons.
Special features: Elaborate engine and interior cooling systems. An oil cooling radiator and an electric self starter were used on this tank.

British Mark VIII.

Produced in 1918-1919.

Seven Mark VIII tanks, differing somewhat from the U. S. Mark VIII, were built in England. The first one had, initially, a Rolls Royce engine. Later all such tanks were fitted with two 6 cylinder 150 HP Ricardo engines arranged as one V-type engine. A Mark VIII Star was designed but never built. It was to be 44 feet long, 47 tons in weight, and able to cross an 18-foot trench.

Mark IX, (Infantry Carrier).

Produced in 1918 by William Foster and Company. Total production, 3.

Crew: 4.
Armament: Two machine guns.
Armor: 0.236 in. to 0.39 in.
Maximum speed: 3.5 mph.
Supension: Rigid.
Tracks: 20½ in. wide.
General arrangement: Driver in front, engine near front behind driver, final drive in rear.
Dimensions: Length 31 ft. 11 in.; width 8 ft.; height 8 ft. 8 in.
Weight: 30.2 tons.
Engine: Ricardo, 6 cylinder, 150 HP, forced water cooling, entrance louvers for cooling air forward and on right side.
Horsepower per ton: 5.
Transmission: Planetary, 4 speeds forward, 4 reverse.
Obstacle ability: Trench 13 ft. 8 in.; stream 3 ft. 10 in.; slope 35 degrees; vertical wall 3 ft. 7 in.; tree 22 in.
Fuel capacity: 99 gallons.
Special features: Intended to carry fifty men in addition to crew. Large oval doors on sides for entrance and exit. Poor ventilation. Ten tons of supplies

could be carried in lieu of personnel. Controls were awkward and hard to handle. Two upper track rollers on each side carried the track at the change in rear slope.

Carden Loyd One Man Tank, Mark III.

Produced in 1926 by Carden Loyd Tractor Co. Total production, 1.

Crew: 1.
Armament: One machine gun.
Armor: 0.24 in. to 0.35 in.
Maximum speed: 24 mph.
Suspension: Rubber tired rollers, no springs.
Tracks: Malleable iron plates, pitch 1.75 in.
General arrangement: Final drive in front; driver above engine.
Dimensions: Length 10 ft. 5 in.; width 4 ft. 6 in.; height 4 ft. 10 in.
Weight: 1¾ tons.
Engine: Ford, Model T, 4 cylinder, 22.5 HP.
Horsepower per ton: 12.8.
Transmission: Modified Ford planetary, 2 speeds forward, 1 reverse.
Obstacle ability: Trench 3 ft. 10 in.; stream 1 ft. 4 in.; slope 40 degrees; vertical wall 1 ft. 4 in.; tree 4 in.
Special features: 200 degrees turret traverse. Difficult for one man to drive and shoot. Driver's position hot and uncomfortable. Rough riding. Convertible from wheels to tracks and tracks to wheels. Two side wheels driven by chains from sprockets and one small tail wheel carry the vehicle when on wheels. Entire assembly raised or lowered by hand screws. The change from wheels to tracks or vice versa was made from within the vehicle, but the mechanism was neither rugged nor durable. Driver unprotected from rear. Later a two-man variety of this tank was developed which resembled the first model but was lower.

Carden Loyd Mark V.

Produced in 1926-27 by Vickers Armstrongs, Ltd. Total production, 50.

Crew: 2.
Armament: One machine gun.
Armor: 0.24 in. to 0.35 in.
Maximum speed: 21.5 mph on tracks; 31 mph on wheels.
Suspension: Rubber tired rollers and leaf spring bogies when on tracks; when on wheels, one rubber tired rigidly mounted wheel on each side near front and trailer wheel at rear.
Tracks: Malleable iron, pitch 1¾ in.
General arrangement: Driver and gunner in front with engine between, driver on left.
Dimensions: Length 9 ft. 11 in.; width 6 ft. 6 in.; height 3 ft. 4 in.
Weight: 1.25 tons.
Engine: Ford Model T, 4 cylinder, 22.5 HP, thermo syphon cooled.
Horsepower per ton: 18.
Transmission: Modified Ford planetary.
Obstacle ability: Trench 4 ft.; stream 1 ft. 4 in.; slope 40 degrees; vertical wall 18 in.; tree 6 in.
Fuel capacity: 8 gallons.
Special features: Wheel and track device operated by engine driven crank arms. Fuel tanks and cooling system in rear.

Carden Loyd Mark VI (Machine Gun Carrier).

Produced in 1929 by Vickers Armstrongs, Ltd. Total production, at least 250.

Crew: 2.
Armament: One cal. .303 or one cal. .5 machine gun.
Armor: 0.24 in. to 0.35 in.
Maximum speed: 28 mph.
Suspension: Leaf springs and rubber tired wheels.
Tracks: 106 drop forged steel open links; a double row of lugs guides the track.
General arrangement: Final drive in front; gunner on right; engine in center; driver on left; fuel tank and cooling system in rear.
Dimensions: Length 8 ft. 1 in.; width 5 ft. 7 in.; height 4 ft. (without armored head cover 3 ft. 4 in.).
Weight: 1.69 tons.
Engine: Ford Model T, 4 cylinder, 22.5 HP, water cooled.
Horsepower per ton: 13.4.
Transmission: Ford planetary with additional low gear.
Obstacle ability: Trench 4 ft.; stream 2 ft. 2 in.; slope 45 degrees; vertical wall 1 ft. 4 in.
Fuel capacity: 8 gallons.

Plate 17.

1. Carden Loyd Mark V.
   (From *Taschenbuch der Tanks* by F. Heigl.)
2. Carden Loyd One Man Tank, Mark III.
   (Wide World Photos.)
3. Carden Loyd Mark VI. With head covers.
   (Photo by Vickers Armstrongs, Ltd.)
4. Carden Loyd Mark VII.
   (From *London Daily Mirror*.)

Special features: Some models fitted with tool and ammunition boxes over fenders. Engine enclosed in asbestos lined box. This type of vehicle has several varieties having slight modifications. The tank of this type that is used in the British Army is without the head covers shown in the illustrations.

Carden Loyd Mark VII.

Produced in 1930 by Vickers Armstrongs, Ltd. Total production, several.

Crew: 2.
Armament: One machine gun.
Maximum armor: 0.5 in.
Maximum speed: 31 mph.
Suspension: Leaf springs and rubber tired wheels.
Tracks: Similar to Carden Loyd Mark VI.
General arrangement: Driver in front; gunner in rear over engine.
Dimensions: Length 10 ft. 6 in.; width 6 ft. 2 in.; height 5 ft. 6 in.
Weight: 5.6 tons.
Engine: 56 HP, air cooled
Horsepower per ton: 10.
Obstacle ability: Slope 45 degrees.
Special features: Turret slightly conical. Water cooled machine gun in armored mount and packet. High temperatures develop inside tank. Limited in obstacle ability. Driver's vision protected by triplex glass. This type of tank has been referred to by the British as an "Infighter," and also as the "Infantry tank."

(Carden Loyd) Light Tank, Mark IA.

Produced in 1931 by Vickers Armstrongs, Ltd. Total production, 6.

Crew: 2.
Armament: One machine gun.
Armor: 0.5 to 0.7 in.
Maximum speed: 30 mph.
Suspension: Springs and rubber tired wheels.
Tracks: Similar to Carden Loyd Mark VI.

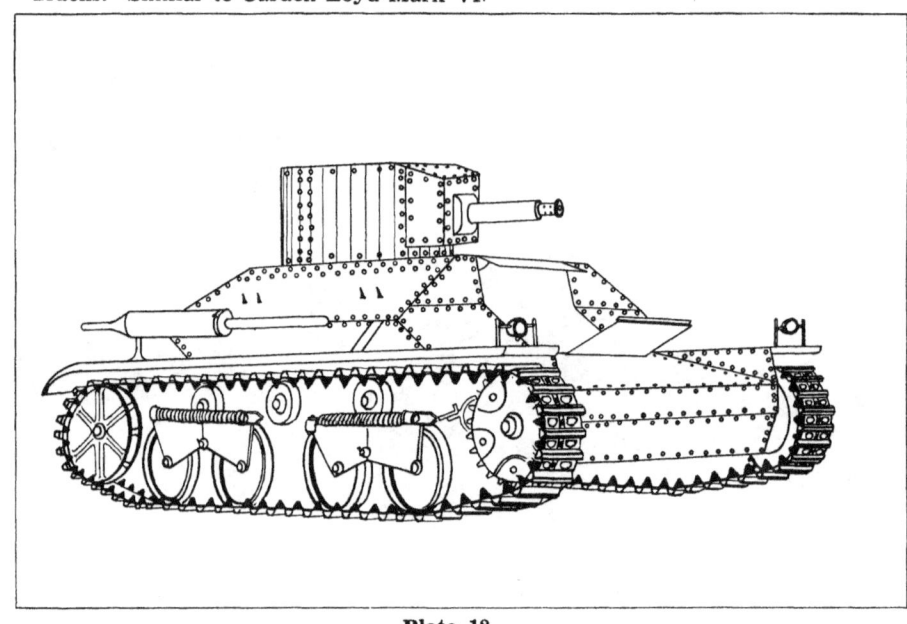

Plate 18.
(Carden Loyd) Light Tank Mark, IA.

General arrangement: Driver in front; gunner in rear over engine.
Dimensions: Length 13 ft.; width 6 ft. 2 in.; height 5 ft. 6 in.
Weight: 5.04 tons.
Engine: Meadows. 6 cylinder, 60 HP.
Horsepower per ton: 11.9.
Obstacle ability: Trench 5 ft.; stream 2 ft. 6 in.; slope 45 degrees.
Special features: This model is a development of the Carden-Loyd Mark VI, and was built for service in India. The machine gun mount projects to secure greater interior space.

Plate 19.
1. **Vickers Carden Loyd Patrol Tank. First type, leaf springs.**
   (From *Royal Tank Corps Journal*.)
2. **Vickers Carden Loyd Patrol Tank. Second type, coil springs.**
   (Photo by Vickers Armstrongs, Ltd.)

Vickers Carden Loyd Patrol Tank.

Produced in 1932 by Vickers Armstrongs, Ltd.

Crew: 2.
Armament: One machine gun.
Armor: 0.276 in. to 0.433 in.
Maximum speed: 30 mph.
Suspension: Carden Loyd type; rubber tired rollers; early model with leaf springs; later model with coil springs.
Tracks: Carden Loyd type.
General arrangement: Final drive in front; engine at left, driver at right, gunner in rotating turret in rear.
Dimensions: Length 8 ft. 6 in.; width 5 ft. 9 in.; height 5 ft. 5 in.
Weight: 2.2 tons.
Engine: 40 HP.
Horsepower per ton: 18.2.
Transmission: Sliding gear, 4 speeds forward, 1 reverse.
Obstacle ability: Stream 26 in.; slope 25 degrees.
Fuel capacity: 12 gallons.

Vickers Mark I.

Produced in 1922-1923 by Vickers Armstrongs, Ltd. Total production, 100.

Crew: 5.
Armament: One 3-pounder (47 mm-1.85 in.) gun and two Vickers machine guns permanently mounted. Four extra ports for two Hotchkiss machine guns.
Armor: 0.31 in. to 0.59 in.
Maximum speed: 16 mph.
Suspension: Rubber tired rollers, bogies, pistons, and double coil springs.
Tracks: Flat plates with double depressions, width 13¾ in.
General arrangement: Driver right front, engine left front, gunners center and rear.
Dimensions: Length 17 ft. 6 in., width 9 ft., height 8 ft. 10½ in.
Weight: 11¾ tons.
Engine: Armstrong-Siddeley, V-type, 8 cylinder, 90 HP, air cooled.
Horsepower per ton: 7.6.
Transmission: 4 speed sliding gear and 2 planetary, providing 8 forward speeds and 2 reverse speeds.
Obstacle ability: Trench 6 ft.; stream 30 in.; slope 45 degrees; vertical wall 3 ft.
Fuel capacity: 108 gallons.
Special features: Fairly steady gun platform but considerable bump and vibration on rough ground. Underpowered. Crew controlled by laryngaphone. Many tanks now equipped with radio. Track rollers unprotected. Interior of turret cooled by special fan. Double asbestos and steel partition between driver and engine. The 3-pounder gun is fired by pressing forward on elevating hand wheel.
Two female Vickers Mark I tanks, designed for use in India, were produced in 1926. Four machine guns in the turret constituted the armament. Special efforts were made to cool this type by the use of extensive asbestos linings and fans.

Vickers Mark I A.

Produced in 1923 by Vickers Armstrongs, Ltd. Total production, 50.

Weight: 13.4 tons.
Horsepower per ton: 6.7.
Special features: Armor protection for rollers. Some models had a bevel in rear of turret for mounting antiaircraft machine gun. Cooling fans in crew and driver's compartments. Engine protected by asbestos. The driver's cowl lifts up instead of swinging to both sides. Otherwise the same as Vickers Mark I.

Vickers Mark II and Mark IIA.

Mark II. Produced in 1923-1929 by Vickers Armstrongs, Ltd. Total production, 100.

Length: 17 ft. 5 in.
Height: 8 ft. 11 in.
Weight: 13.4 tons.
Special features: This tank is essentially the same as the Vickers Mark IA. Fitted with antiaircraft gun on rear of turret. Has both mechanical and electric starters. Equipped with radio. Driving is difficult, considerable fire hazard, deficient arcs of fire for machine guns, poor ventilation, unsteady gun platform.

Mark IIA. Produced in 1930 by Vickers Armstrongs, Ltd. Total production, 20.
Height: 10 ft. 2 in.
Special features: Bevel removed from rear of turret. A small rotating observation turret with beveled sides is located on the rear center of the main turret. The left ventilator is protected by a long armored box open at the top. The 3-pounder gun and one machine gun are in one mount. The machine guns (all Vickers) eject the empty cartridge cases to the outside of the tank. Otherwise substantially the same as the Mark II.

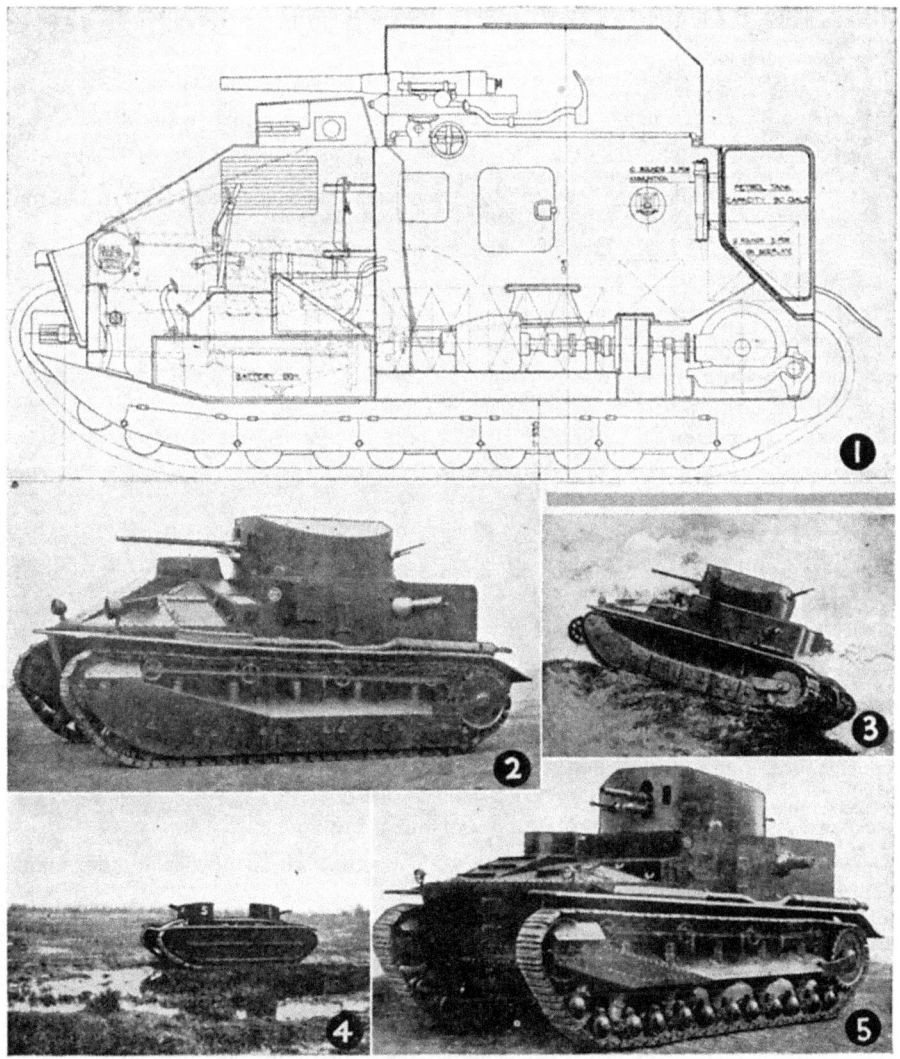

Plate 20.
1. **Vickers Mark II.**
2. **Vickers Mark II.**
   (Photo by Vickers Armstrongs, Ltd.)
3. **Vickers Wheel and Track.**
4. **Light Dragon Machine Gun Carrier.**
   (From *Royal Tank Corps Journal*.)
5. **Vickers Mark I.**
   (From *Royal Tank Corps Journal*.)

Vickers Wheel and Track.

Produced in 1926 by Vickers Armstrongs, Ltd. Total production, 1.

Crew: 5.
Armament: One 3-pounder (47 mm-1.85 in.) gun and 3 machine guns of which 1 was for antiaircraft fire.
Armor: 0.26 in. to 0.59 in.
Maximum speed: 16 mph on tracks; 28 mph on wheels.
Suspension: Springs. hydraulic shock-absorbing mechanism, rubber tired rollers.
Wheels: Rubber tired. front and rear.
General arrangement: Similar to Mark II.
Dimensions: Length 21 ft.; width 9 ft.; height 8 ft. 10 in. on tracks; 9 ft. 6 in. on wheels.
Weight: 14 tons.
Engine: Armstrong-Siddeley, 8 cylinder, V-type, 90HP, air cooled.
Horsepower per ton: 6.4.
Obstacle ability: Trench 7 ft. 6 in.; stream 3 ft. 11 in.; vertical wall 23 in.
Fuel capacity: 118 gallons.
Special features: Wheel and track system. Change from wheels to tracks and vice versa was made by engine power, in about one minute, the crew not leaving the tank. The wheels were lowered for travel on roads and raised for travel on tracks. This vehicle was a modification of Vickers Mark IA.

Light Dragon Machine Gun Carrier.

Produced in 1926 by Vickers Armstrongs, Ltd. Total production, 1.

Crew: 3.
Armament: Two machine guns, one in forward turret, one in rear turret.
Armor: 0.25 in.
Maximum speed: 10 mph.
Suspension: Same as Vickers Mark I.
Tracks: Cast steel.
General arrangement: Driver in right front; engine in left front; gunners in front and rear.
Dimensions: Length 17 ft. 6 in.; width 9 ft.; height 6 ft.
Weight: 11.2 tons.
Engine: A E C Omnibus, 60 HP.
Horsepower per ton: 5.4.
Transmission: Planetary.
Obstacle ability: Trench 7 ft. 2 in.; stream 3 ft.; slope 30 degrees; vertical wall 2 ft. 7 in.
Fuel distance: 45 miles.
Special features: Driver and gunners in communication by laryngaphone.

Sixteen Ton Tank.

Produced in 1929 by Vickers Armstrongs, Ltd.

Crew: 4.
Armament: One 3 pounder (47 mm-1.85 in.) gun and one Vickers machine gun in one mount, and a pair of machine guns mounted in each of two smaller turrets.
Armor: 0.5 in. to 0.8 in.
Maximum speed: 31 mph.
Suspension: Similar to Vickers Medium tanks, but materially improved.
Tracks: Similar to Vickers Mark II.
General arrangement: Driver in front with machine gunner on each side; main turret in center; engine and final drive in rear.
Dimensions: Length 19 ft. 9 in.
Weight: 17.9 tons.
Engine: Armstrong-Siddeley, 180 HP, air cooled and probably horizontal.
Horsepower per ton: 10.
Transmission: Planetary.
Obstacle ability: Trench 9 ft.; slope 45 degrees; vertical wall 3 ft. 2 in.
Special features: Center of gravity farther forward than in previous tanks. Engine and transmission in a heat insulated compartment. Fuel tanks over tracks. Equipped with laryngaphone. Crew compartment protected against gas by filter and blower. Very steady gun platform and heavy fire power. Engine very accessible. This is considered to be, in most respects, a tank of very superior design. It appears to have more good features than any other tank up to this time.

Independent Tanks, Mark I and Mark II.

Produced in 1925 and 1928 by Vickers Armstrongs, Ltd. Total production, 2.

Crew: 10.
Armament: One 3-pounder (47 mm-1.85 in.) gun in central turret, one machine gun in each of the four small turrets (one machine gun for antiaircraft fire).

Armor: 0.78 in. to 1.0 in.
Maximum speed: 20 mph.
Suspension: Spring.
Tracks: Metal plates with depression in center.
General arrangement: Driver in front; gunners behind driver; engine and final drive in rear.

Plate 21.
1. Eighteen Ton Tank.
   (Wide World Photos.)
2. Vickers Convertible.
3. Independent Tank, Mark II.
(From *Royal Tank Corps Journal.*)

Dimensions: Length 30 ft. 6 in.; width 10 ft. 6 in.; height 9 ft.
Weight: 33.6 tons.
Engine: Armstrong-Siddeley, 12 cylinder, V-type, 380 HP, air cooled.
Horsepower per ton: 11.3.
Transmission: Planetary in Mark I; hydraulic in Mark II.
Obstacle ability: Trench 15 ft.; stream 4 ft.; slope 40 degrees; vertical wall 4 ft.; tree 30 in.
Fuel distance: 200 miles.
Special features: Built in accordance with the requirements of the British general staff for an independent tank. Hydraulic steering clutches and brakes. Laryngaphone and fire control indicators in each turret. An electric fan cooling system for crew compartment. A hollow bulkhead between engine and crew compartments. A small air cooled engine used to start the large engine. The Mark II was the same as the Mark I except that a hydraulic transmission was provided, cooling and ventilation were improved and there were some other minor changes.

Vickers Convertible.

Produced in 1929 by Vickers Armstrongs, Ltd. Total production, 1.

Crew: 3.
Armament: Two machine guns.
Maximum speed: 15 mph on tracks, 45 mph on wheels.
Suspension: Rollers and springs in movable track frames.

Tracks: Overlapping steel plates.
General arrangement: Engine in left front; driver in right front, gunners in center; final drive in rear.
Dimensions: Length 16 ft. 8 in.; width 7 ft. 6 in.; height 7 ft. 8 in.
Weight: 8.4 tons.
Engine: 135 HP, water cooled.
Horsepower per ton: 16.
Transmission: Sliding gear.
Fuel capacity: 48 gallons.
Special features: Rear wheels concealed within openings in the hull at rear when on tracks.

Vickers Armstrongs Six Ton, Alternative A.

Produced in 1930 by Vickers Armstrongs, Ltd. Total production, several.

Crew: 3.
Armament: Two cal. .303 machine guns or one cal. .303 and one cal. .50 machine gun.
Armor: 0.2 in. to 0.51 in.
Maximum speed: 21.75 mph.
Suspension: Leaf springs, pivoted bogies, and rubber tired rollers.
Tracks: Open plates without grousers. drop forged magnesium steel, width 9 in., pitch 3⅝ in. Track life 3,000 miles or more.
General arrangement: Final drive in front; crew in center; engine in rear.

Plate 22.
1. **Vickers Armstrongs Six Ton, Alternative A.**
   (Photo by Ordnance Dept., U.S.A.)
2. **Vickers Armstrongs Six Ton, Alternative B.**
   (Photo by Vickers Armstrongs. Ltd.)

Dimensions: Length 15 ft. ½ in.; width 7 ft. 11½ in.; heighth 6 ft. 9⅞ in.
Weight: 8.1 tons.
Engine: Special horizontal Armstrong-Siddeley, 4 cylinder, 87 HP, air cooled (using Roots blower); a large oil cooler is used.
Horsepower per ton: 10.7
Transmission: Sliding gear, 5 speeds forward, 2 reverse.
Obstacle ability: Trench 6 ft.; stream 3 ft.; slope 45 degrees; vertical wall, 2 ft. 5 in.
Fuel distance: 100 miles. **Fuel capacity:** 48 gallons.
Special features: Double sprockets. Equipped with Marconi short wave radio set and laryngaphone. The guns in each turret have a 240 degree field of fire. Engine, etc., very accessible. This design incorporates many excellent features. The short pitch of the track is very meritorious.

Vickers Armstrongs Six Ton, Alternative B.

Produced in 1930 by Vickers Armstrongs, Ltd. Total production, several.

Armament: One 47 mm (3 pdr-1.85 in.) gun and one cal. .303 machine gun in one mount.
Armor: 0.2 in. to 0.67 in.
Height: 7 ft.
Special features: Except as indicated, this vehicle is substantially the same as Alternative A.

British tank development has shown evidence of keen initiative and the application of sound principles of design.

CHAPTER XIII.

TANKS OF ALL COUNTRIES (Continued).

FRENCH.

Renault, Model B.S.

Produced in 1918 by Renault and Company. Total production, 200.

Crew: 2.
Armament: One short 75 mm (2.95 in.) gun.
Armor: 0.3 in. to 0.6 in.
Maximum speed: 6 mph.
Suspension: Spring suspension with bogies and rollers.
Tracks: Flat shoes with grousers on forward edge.
General arrangement: Driver in front; gunner in center; engine in rear.
Dimensions: Length 16 ft. 5 in.; width 5 ft. 8 in.; height 7 ft. 11 in.
Weight: 7.8 tons.
Engine: Renault, 4 cylinder, 39 HP, thermo syphon cooled.
Horsepower per ton: 5.
Transmission: Sliding gear, 4 speeds forward and 1 reverse.
Obstacle ability: Trench 5 ft.; stream 2 ft. 3 in.; slope 45 degrees; vertical wall 3 ft., tree 8 in.
Fuel distance: 35 miles. **Fuel capacity:** 20 gallons.
Special feature: Fixed turret.

Renault, M 1923.

Produced in 1923 by Renault and Company. Total production, 1.

Crew: 4.
Armament: One 75 mm. (2.95 in.) gun in front at right and one machine gun in turret.
Armor: 0.95 in.
Suspension: Spring with rubber tired rollers.
Tracks: Flat steel plates laterally flexible.
General arrangement: Driver left front; 75 mm gunner right front; machine gunner in center; engine in rear.
Dimensions: Length 19 ft. 10 in.; width 8 ft. 5 in.; height 6 ft. 8 in.
Weight: 14.3 tons.
Engine: Renault, 6 cylinder, 180 HP, water cooled.
Horsepower per ton: 12.6.
Special features: Tracks operated as in Delaunay Belleville tank.

Renault with Rubber Tracks.

Produced in 1924-1925 by Citroen Auto Company, supervised by M. Kegresse. Total production, about 100.

Crew: 2.
Armament: One 37 mm. (1.46 in.) gun or one machine gun.
Armor: 0.236 in. to 0.629 in.
Maximum speed: 10 mph.
Suspension: Kegresse with rollers.
Tracks: Kegresse rubber band, width, 12 in. Maximum track life about 900 miles.
General arrangement: Driver in front; gunner in center; engine and final drive in rear.
Dimensions: Length 18 ft.; width 6 ft. 5 in.; height 7 ft. 9 in.
Weight: 7.4 tons.
Engine: Renault, 4 cylinder, 39 HP, thermo syphon cooling.
Horsepower per ton: 5.3.
Transmission: Sliding gear, 4 speeds forward, 1 reverse.
Obstacle ability: Trench 7 ft. 4 in.; stream 2 ft. 3 in.; slope 38 degrees; vertical wall 2 ft. 9 in.
Fuel capacity: 20 gallons.
Special features: Rubber tracks increased speed and were very quiet. Tracks wear out quickly on rocky soil, but wear well on sandy soil. Two heavy 30 in. rollers attached to front of tank through leaf springs. Two 16 in. rollers in rear in lieu of tailpiece. Tested by U. S. Ordnance Department in 1925.

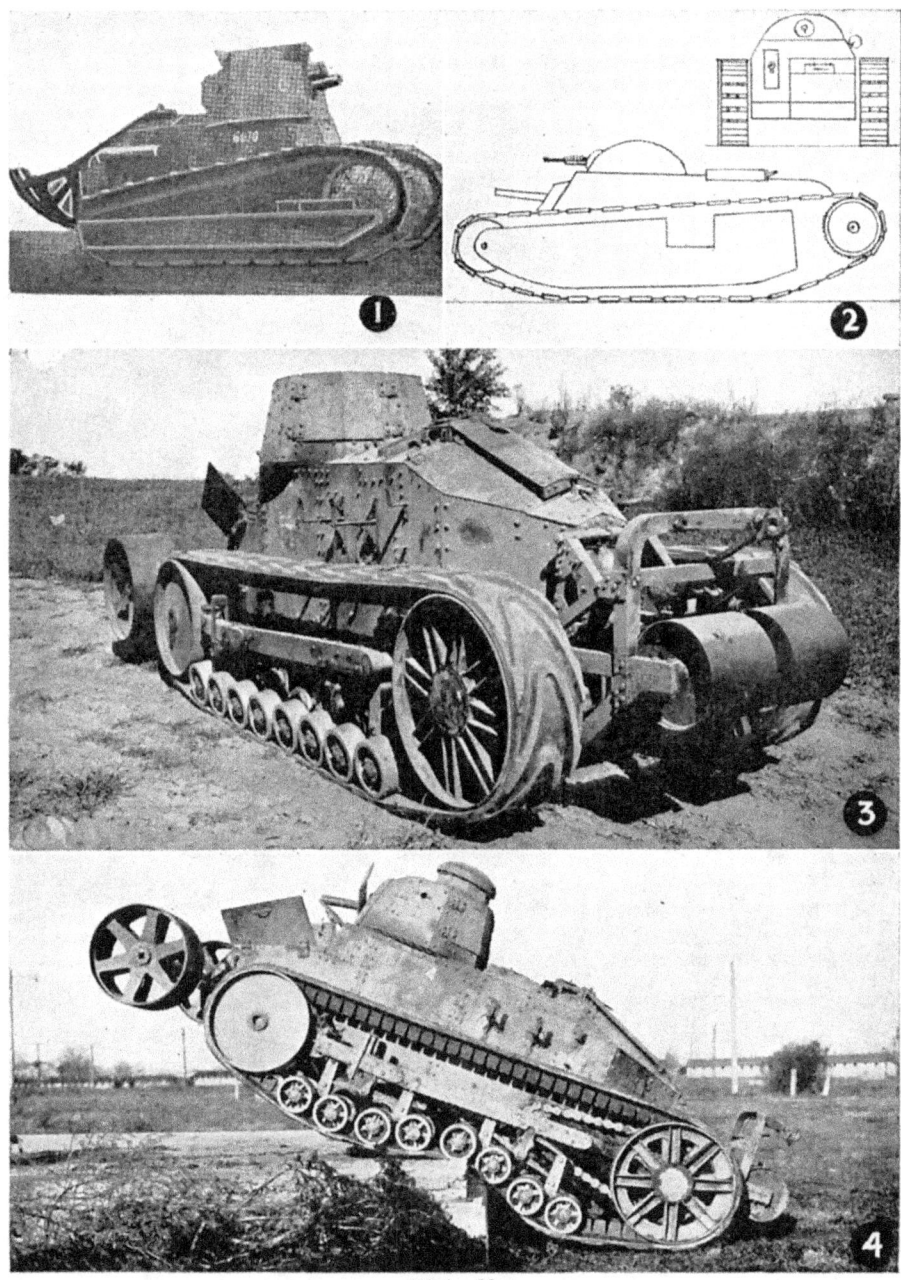

Plate 23.

1. **Renault Model BS.**
2. **Renault Model 1923.**
(From *Taschenbuch der Tanks* by Heigl.)
3. **Renault with Rubber Tracks.**
(Photo by Ordnance Dept., U.S.A.)
4. **Renault with Rubber Tracks.**
(Photo by Ordnance Dept., U.S.A.)

Renault, NC Model 1927.

Produced in 1927 by Renault and Company. Total production, 50.

Crew: 2.
Armament: One 37 mm (1.46 in.) gun or one machine gun.
Armor: 0.47 in. to 1.18 in.
Maximum speed: 11½ mph.
Suspension: Coil springs and bogies with rollers, one hydraulic shock absorber at the front at each side.
Tracks: Steel plates.
General arrangement: Driver in front; gunner in center; engine and final drive in rear.
Dimensions: Length 14 ft. 5 in.; width 5 ft. 7 in.; height 7 ft.
Weight: 8.69 tons.
Engine: Renault, 4 cylinder, 60 HP, water cooled.
Horsepower per ton: 6.9.
Transmission: Sliding gear, 6 speeds forward, 1 reverse.

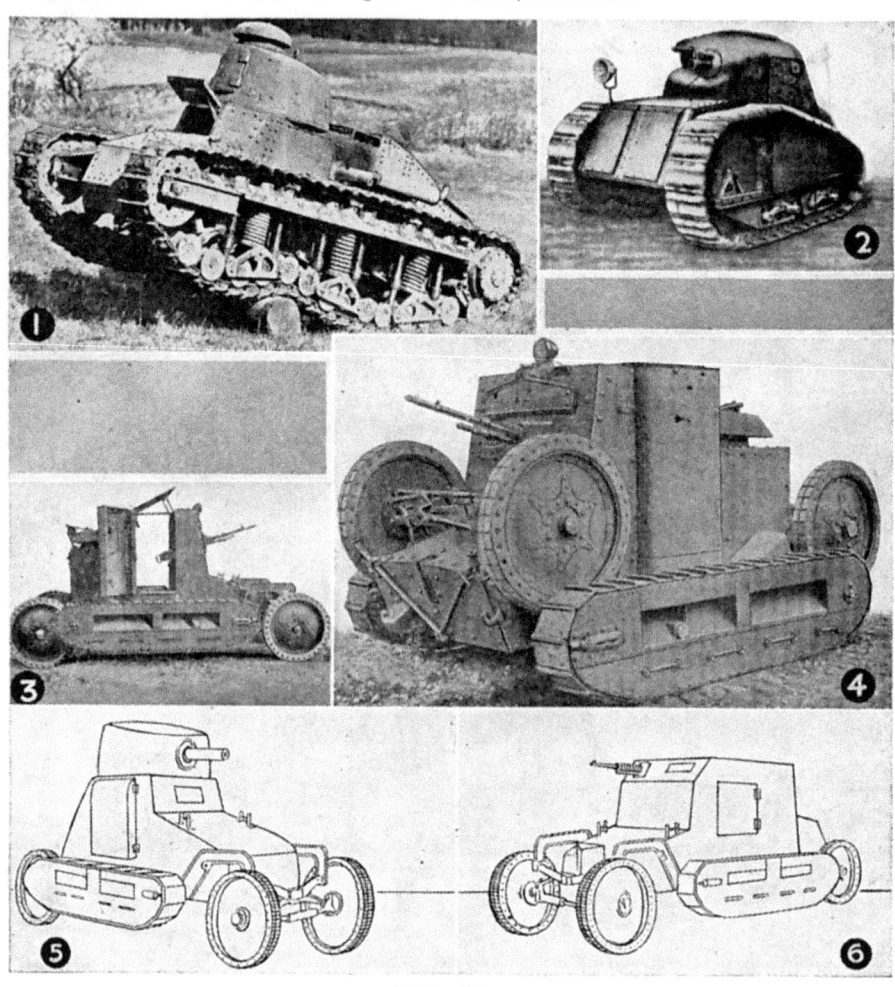

Plate 24.

1. Renault NC, Model 1927.
   (From *Taschenbuch der Tanks* by Heigl.)
2. Peugeot.
3. Chenilette St. Chamond, Model 1921. On wheels.
4. Chenilette St. Chamond, Model 1921. On tracks.
5. Chenilette St. Chamond, Model 1924.
6. Chenilette St. Chamond, Model 1926.

Obstacle ability: Trench about 7 ft.; stream 2 ft.; slope 45 degrees; vertical wall 2 ft.
Fuel distance: 70 miles. **Fuel capacity:** 53 gallons.
Special features: Center of gravity farther forward than in early Renault. Tailpiece removed. Sprockets and idlers small. Track is excellent and very quiet. Suspension exposed. Due chiefly to the removal of the heavy parts of suspension of early Renault, armor protection could be materially increased with a small increase in weight. By decreasing the height of the hull at the rear, the dead space from the center of the turret was reduced.

Chenilette St. Chamond, M 1921.

Produced in 1921 by St. Chamond Company. Total production, 2.

Crew: 1 man in one tank; 2 in the other.
Armament: One machine gun.
Mamimum speed: 3 mph on tracks, 15 mph on wheels.
Suspension: Rigid with rollers on tracks; springs on wheels.
Tracks: Flat steel plates with slight grousers.
General arrangement: Driver and gunner in front; engine in rear.
Dimensions: Length 11 ft. 10 in.; width 6 ft. 10.; height 6 ft. 4 in. (all on tracks).
Weight: 3.5 tons.
Engine: 2 cylinder, water cooled; mounted laterally.
Transmission: Sliding gear.
Special features: Changed from wheels to tracks in ten minutes by engine power, but change from tracks to wheels required running upon wooden blocks carried with tank. Crew dismounted for latter change. Driving of tank required the use of 5 pedals; difficult for one man. Slow and hard riding. Front wheels, when lifted interfered with vision. Steering wheel used on wheels; steering pedals on tracks.

Chenilette St. Chamond, M 1924.

Produced in 1924 by St. Chamond Company. Total production, 1.

Crew: 2.
Armament: One 45 mm (1.77 in.) gun.
Maximum speed: 3 mph on tracks, 15 mph on wheels.
Suspension: Similar to 1921 model.
Tracks: Flat steel plates with slight grousers.
General arrangement: Driver in front, gunner in center, engine in rear.
Dimensions: Length 11 ft. 10 in.; width 6 ft. 10 in.; height 7 ft. (all on tracks).
Weight: 3.6 tons.
Engine: 2 cylinder, water cooled; mounted laterally.
Transmission: Sliding gear.
Special features: Similar to 1921 models, but with 360 degree rotating turret.

Chenilette St. Chamond, M 1926.

Produced in 1926 by St. Chamond Company. Total production, 1.

Crew: 2.
Armament: One machine gun.
Dimensions: Length 11 ft. 10 in.; width 6 ft. 10 in.; height 6 ft. 6 in.
Transmission: Williams-Janney hydraulic.
Special features: Dismounting of crew not necessary in changing from wheels to tracks or reverse. Machine gun had limited field of fire. Transmission heavy for so small a tank.

Peugeot.

Produced in 1919 by Peugeot Motor Car Company. Total production, 1.

Crew: 2.
Armament: One 37 mm (1.46 in.) gun and one machine gun.
Suspension: Bogies supported by coil and leaf springs.
Tracks: Flat steel plates with double grousers.
General arrangement: Engine in front; crew in rear.
Engine: Peugeot, 4 cylinder.
Special features: Turret had rounded surfaces. Equipped with electric lights and starter.

Delaunay-Belleville.

Produced in 1920. Total production, 1.

Crew: 3.
Armament: One 37 mm (1.46 in.) gun or one machine gun in turret, two machine guns in front of tank.
Armor: 0.236 in. to 0.639 in.
Maximum speed: 12.5 mph.
Suspension: Spring with rollers.
Tracks: Flat steel plates laterally flexible.

General arrangement: Driver in front; gunners in center; engine in rear.
Dimensions: Length 16 ft. 5 in.
Weight: 14.3 tons.
Engine: Renault, 100 HP, water cooled.
Horsepower per ton: 7.

Plate 25.
1. Delaunay Belleville.
(From *Taschenbuch der Tanks* by Heigl.)
2. Char 1A.
3. Char IC.
4. Char 2 C.
(From *Taschenbuch der Tanks* by Heigl.)
5. Char 3C.

Transmission: Williams-Janney hydraulic.
Special features: Tracks operated flexibly in turning as in British Medium D.

Char Moyen.

Produced in 1925.

Armament: One 75 mm (2.95 in.) or one 155 mm (6.1 in) gun, and four machine guns.
Armor: 0.4 in. to 1.8 in.
Maximum speed: 10 mph.
Suspension: Spring.
Tracks: Kegresse half metal.
General arrangements: Driver in front; gunners in center; engine in rear.
Weight: 22 tons.
Special features: Resembles British Mark V.

Char 1 A.

Produced in December, 1917, by Société des Forges et Chantiers de la Mediterranée (La Seyne-Toulon). Total production, 1.

Crew: 6.
Armament: One 105 mm (4.1 in.) Schneider howitzer and one machine gun in turret, one machine gun at front of tank.
Armor: 0.63 in. to 1.38 in.
Maximum speed: 3.75 mph.
Suspension: Springs with rollers and bogies.
Tracks: Steel plates.
General arrangement: Driver in front; engine and final drive in rear.
Dimensions: Length 30 ft.; height 11 ft.
Weight: 50 tons.
Engine: Renault, airplane, V type, 12 cylinder, 240 HP.
Horsepower per ton: 4.8.
Transmission: Mechanical.
Obstacle ability: Slope 42 degrees.

Char 1 C.

Produced in 1919 by French Experimental Establishment. Total production, 1.

Crew: 10.
Armament: One 75 mm (2.95 in.) gun in forward turret, one machine gun in rear turret, and three machine guns in hull.
Armor: 0.51 in. to 1.77 in.
Maximum speed: 4 mph.
Suspension: Spring with rollers and bogies.
Tracks: Similar to Mark VIII.
General arrangement: Driver in front; crew in center and rear; engine in rear center.
Dimensions: Length 33 ft.
Weight: 74.4 tons.
Engine: Two, Renault, 6 cylinder, 250 HP each, with 2 electric generators and 2 motors.
Horsepower per ton: 6.7.
Transmission: Crochat-Colardeau, electric.
Obstacle ability: Trench 14 ft.
Special features: Armor intended to be impervious to 77 mm guns. The first operating test of this tank was made before the engines and dynamos were installed. Electric current was transmitted to the driving motors from an outside source through a cable. The cable and drum are shown in the illustration.

Char 2 C.

Produced in 1923. Total production, about 10.

Crew: 13.
Armament: One 75 mm (2.95 in.) or one 155 mm (6.1 in. gun, four machine guns and four reserve machine guns.
Armor: 0.4 in. to 1.8 in. (Said to be capable of withstanding 3 direct hits from a 75 mm gun.)
Maximum speed: 10 mph.
Suspension: Spring, with rollers.
Tracks: Armor plates with double grousers, 33 in. wide.
General arrangement: Driver forward; crew center and rear; engine rear center.
Dimensions: Length 33 ft. 7 in.; width 9 ft. 7 in.; height 13 ft. 3 in.
Weight: 75 tons.
Engine: Two, Daimler, sleeve valve, 300 HP each, with two electric generators and two motors.
Horsepower per ton: 8.
Transmission: Electric.
Obstacle ability: Trench 14 ft.; stream 6 ft.; slope 45 degrees; vertical wall 5 ft 7 in.; tree 31 in.
Fuel distance: 60 miles. **Fuel capacity:** 330 gallons.
Special features: Radio equipped. Both turrets equipped with stroboscopes. Very steady gun platform. Carried by rail on two special six wheel railway trucks. Superstructure prevents the firing of the main gun to the rear.

Char 3 C.

Produced in 1926. Total production, several.

Crew: 13.
Armament: One 155 mm (6.1 in.) gun, one 75 mm (2.95 in.) gun, and four machine guns.
Armor: 0.9 in. to 2.12 in.

Maximum speed: 7.5 mph.
Suspension: Spring, with rollers.
Tracks: Same as Char 2 C.
General arrangement: Same as Char 2 C.
Dimensions: Length 39 ft. 4 in.; width 9 ft. 7 in.; height 13 ft. 3 in.
Weight: 81½ tons.
Transmission: Electric.
Obstacle ability: Trench 17 ft. 3 in.; stream 6 ft.; slope 45 degrees; vertical wall 5 ft. 7 in.; tree 31 in.
Special features: Rear turret high to permit fire in diagonally forward directions.

Schneider.

Produced by Schneider and Company. Total production, 1.

Crew: 28.
Armament: Four 75 mm (2.95 in.) guns and nine machine guns.
Maximum speed: 5 mph.
Suspension: Spring.
Dimensions: Length 40 ft.; width 10 ft. 9 in.; height 9 ft. 8 in.
Weight: 141 tons.
Special features: Three turrets.

CHAPTER XIV.

TANKS OF ALL COUNTRIES (Continued).

GERMAN.

L. K. I.

Designed by Vollmer. Produced in 1918 by various manufacturers. Total production, 2.

Crew: 2 or 3.
Armament: Two machine guns.
Armor: 0.25 in. to 0.45 in.
Maximum speed: 8.5 mph.
Suspension: Leaf springs between hull and track frames.
Tracks: Steel plates with grousers, width 9.8 in., pitch 5.5 in.
General arrangement: Engine in front; driver in center; gunner in final drive in rear.
Dimensions: Length 16 ft. 7 in.; width 6 ft. 5 in.; height 8 ft. 3 in.
Weight: 9.4 tons.
Engine: 4 cylinder, 60 HP., forced water cooling.
Horsepower per ton: 6.4.
Transmission: Sliding gear, 8 speeds forward, 1 reverse.
Obstacle ability: Trench 6 ft. 6 in.; slope 41 degrees.
Fuel distance: 37 miles. **Fuel capacity:** 31 gallons.
Special features: Built around motor truck chassis to which tracks and hull were fitted. High interior temperature reduced by an exhaust fan. A German report on this vehicle states: "The L K vehicle, conceived on the spur of the moment, was a makeshift, due to the acute circumstances which restricted its design to the means at hand. The limited time available and the lack of material would not permit the development of a style of construction new in every particular or entirely suitable for the various specified requirements."

L. K. II.

Designed by Vollmer. Produced in 1918 by Daimler Motor Co. Total production, 2.

Crew: 2 or 3.
Armament: One 37 mm (1.46 in.) gun.
Armor: 0.25 in. to 0.45 in.
Maximum speed: 8.5 mph.
Tracks: Similar to L K I.
General arrangement: Engine in front; driver in center; gunner and final drive in rear.
Dimensions: Length 16 ft. 8 in.; width 6 ft. 5 in.; height 8 ft. 2 in.
Weight: 10.2 tons.
Engine: 4 cylinder, 60 HP, forced water cooling
Horsepower per ton: 5.8.
Transmission: Sliding gear, 8 speeds forward, 1 reverse.
Fuel capacity: 31 gallons.
Special features: This was an improved model of the L K I. An L K III was designed but never built. In this the engine was to be at the rear, and the armament was to be a 57 mm gun or a T U F machine gun in a rotating turret. A German report states: "A new design, the L K III, which followed later, contemplated placing the engine in the rear part to unite more closely the power unit and running gear, thereby leaving the front for the personnel."

A. 7. V. U. (Allgemeine Kriegsdepartement 7 Abteilung Verkehrswesen Wagen mit Umlaufende Ketten).

Produced in 1918 by Daimler Motor Company. Total production, 1.

Crew: 7.
Armament: Two 57 mm (2.24 in.) guns in sponsons and four machine guns.
Armor: 0.6 in. to 1.8 in.
Maximum speed: 7.5 mph.
Suspension: Coil springs, bogies and rollers.
Tracks: Solid plates with double grousers.

General arrangement: Driver in front; gunners in center; engine and final drive in rear.
Dimensions: Length 27 ft. 6 in.; width 15 ft. 5 in.; height 10 ft. 6 in.
Weight: 44 tons.
Engine: Two, Mercedes-Daimler, 4 cylinder, sleeve valve, each 150 HP, forced water cooling.
Horsepower per ton: 6.8.
Transmission: One sliding gear transmission for each engine, 3 speeds forward, 3 reverse.
Obstacle ability: Trench 13 ft.
Special features: Similar to British heavy tank design but with spring suspension and improved ventilation.

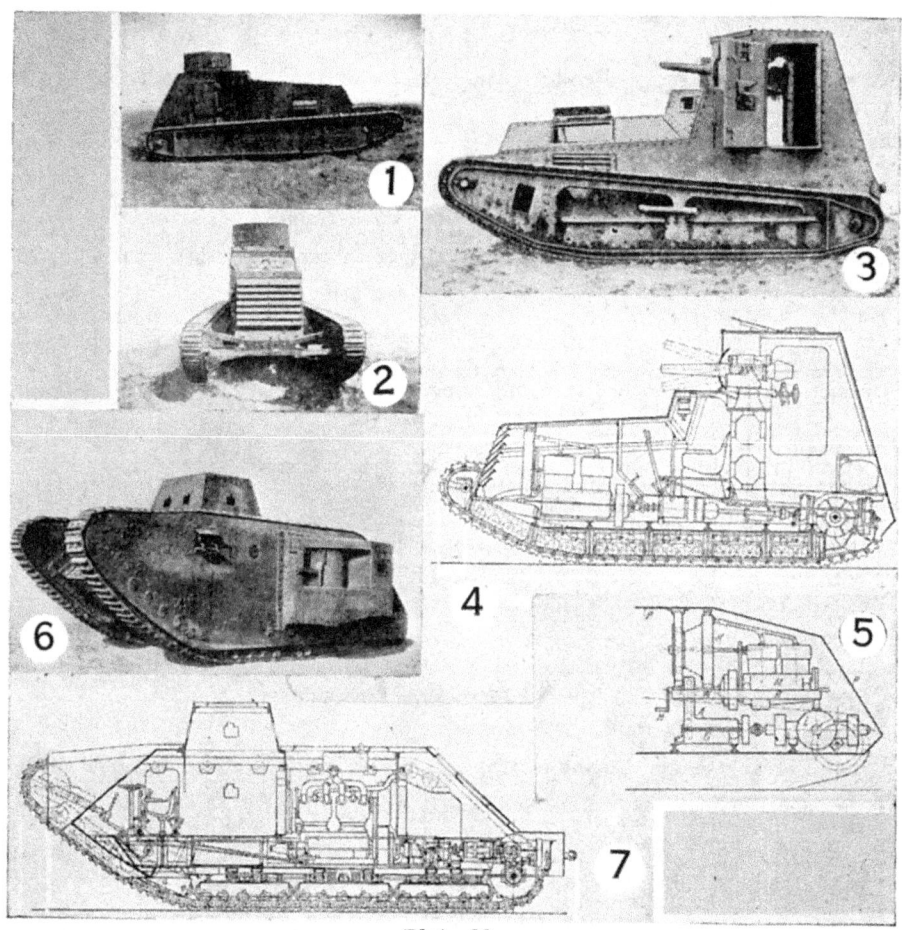

Plate 26.

1. **L K I, Side View.**
2. **L K I, Front View.**
3. **L K II.**
4. **L K II.** (From *Tanks* by Krüger.)
5. **Power Train of L K III.** (From *Tanks* by Krüger.)
6. **A 7 V U.**
7. **A 7 V U.** (From *Tanks* by Krüger.)

K Vehicle.

Produced in 1918 by Daimler Motor Co. Designed by Vollmer. Total production, 2.

Crew: 22.
Armament: Four 77 mm (3.03 in.) guns and six machine guns.

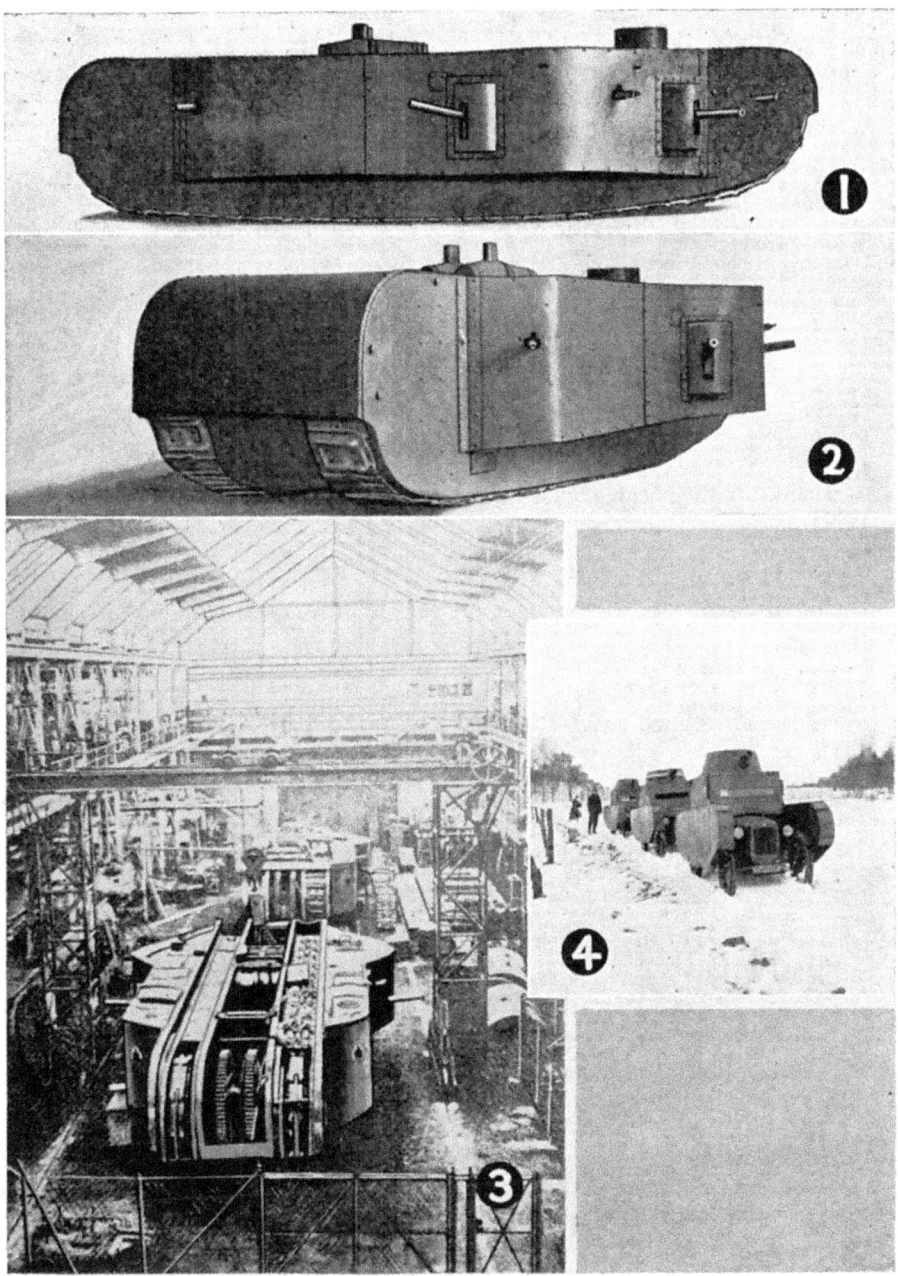

Plate 27.
1. K Vehicle.
(Photo by O. H. Hacker.)
2. K Vehicle.
(Photo by O. H. Hacker.)
3. Two K Vehicles under Construction in a German Factory.
4. German Post-War Tank Substitutes for Training in Tactics.

Armor: 0.39 in. to 1.18 in.
Maximum speed (estimated): 5 mph.
Suspension: 40 locomotive type springs.
Tracks: Orion. Rollers, carried as part of the tracks, rolled around upon stationary guides.
General arrangement: Driver in front; engine and final drive in rear.
Dimensions: Length 42 ft. 7 in.; width (hull) 10 ft. 2 in, (overall) 20 ft.; height 9 ft. 5 in.
Weight: 165.3 tons.
Engine: Two, Daimler, aircraft, sleeve valve, 600 HP each, forced glycerine cooling.
Horsepower per ton: 7.3.
Transmission: Sliding gear, 3 speeds forward and 1 in reverse.
Special features: Hull sprung to track frames. Divided into loads of 18 to 25 tons for transportation. Fitted with electric lights, elaborate fan cooling systems and submarine communication and control devices. Engines provided with electro-magnetic clutches.

These vehicles, which were nearing completion, were scrapped at the end of the war.

ITALIAN.

Fiat, Type 3000.

Produced in 1919 by Fiat Motor Company. Total production, 100.

Crew: 2.
Armament: Two machine guns in twin mount or one 37 mm (1.46 in.) gun.
Armor: 0.236 in. to 0.629 in.
Maximum speed: About 10 mph.
Suspension: Similar to French Renault.
Tracks: Similar to French Renault, 11 in. wide. Recently a new track of smaller plates has been tested.
General arrangement: Driver in front; gunner in center; engine and final drive in rear.
Dimensions: Length 13 ft. 9 in.; width 5 ft. 5 in.; height 7 ft. 2½ in.
Weight: 5.5 tons.
Engine: Fiat, 4 cylinder, 54 HP, water cooled; mounted laterally.
Horsepower per ton: 9.8.
Transmission: Sliding gear.
Obstacle ability: Trench 5 ft. 11 in.; stream 2 ft. 3 in.; slope 51 degrees; vertical wall 2 ft.
Fuel distance: 9 hours. **Fuel capacity:** 25 gallons.
Special features: Very low center of gravity. Some tanks fitted with device for cutting electrically charged wire. An improved Renault tank.

Fiat, Type 3000 B.

Produced in 1928 by Fiat Motor Company.

Crew: 2.
Armament: One 37 mm (1.46 in.) gun or one pair of machine guns.
Armor: 0.236 in. to 0.629 in.
Maximum speed: About 12½ mph.
Suspension: Similar to French Renault.
Tracks: Width 11 in.
General arrangement: Driver in front; gunner in center; engine and final drive in rear.
Dimensions: Length 13 ft. 10 in.; width 5 ft. 6 in.; height 7 ft. 3 in.
Weight: 6.2 tons.
Engine: Fiat, 4 cylinder, 63 HP.
Horsepower per ton: 10.2.
Transmission: Sliding gear; 3 speeds forward, 1 reverse.
Obstacle ability: Stream 35 in.; slope 40 degrees.
Special features: This tank has two mufflers, a water tight hull, a hand operated bilge pump, and improved eye slits.

Pavesi, M 1925.

Produced in 1925 by Pavesi Tractor Company. Total production, 1.

Crew: 2.
Armament: One machine gun or one 37 mm (1.46 in.) gun.
Armor: 0.236 in. to 0.629 in.
Maximum speed: 21 mph.
Suspension: All wheels sprung.
Wheels: 5 ft. 4 in. in diameter; 10 in. wide; permanently attached rubber grousers; brakes on all wheels.
General arrangement: Driver and engine in front; driver over engine; gunner in center; drive gear for rear wheels in rear.

Plate 28.

1. **Fiat, Type 3000.**
(From *Taschenbuch der Tanks* by Heigl.)
2. **Fiat, Type 3000.**
(From *Taschenbuch der Tanks* by Heigl.)
3. **Fiat, Type 3000 B.**
(Photo by O. H. Hacker.)
4. **Pavesi Tank, Model 1925.**
(From *Taschenbuch der Tanks* by Heigl.)
5. **Pavesi Tank, Model 1926.**
(From *Taschenbuch der Tanks* by Heigl.)

Dimensions: Length 13 ft. 11 in.; width 7 ft. 2 in.; height 6 ft. 9 in.
Weight: 5.5 tons.
Engine: 50 HP.
Horsepower per ton: 9.1.
Transmission: Sliding gear.
Obstacle ability: Trench 6 ft., if not more than 4 ft. 3 in. deep; stream 3 ft. 7 in.; slope 45 degrees; vertical wall 5 ft. 6 in.
Special features: Jointed hull facilitates sharp turns and allows wheels to follow ground conformation closely. Four wheel steering and drive. Fitted with a water pump to expel water while fording.

Pavesi, M 1926.

Produced in 1926 by Pavesi Tractor Company. Total production, 1.

Wheels: Rubber tired, detachable metal grousers.
Special features: Front part of hull increased in size. Otherwise substantially the same as first model.

Pavesi Tank Destroyer.

Produced in 1928 by Pavesi Tractor Company. Total production, 1.

Crew: 3.
Armament: One 75 mm (2.95 in.) gun.
Armor: 0.236 in. to 0.629 in.
Maximum speed: 15 mph.
Suspension: All wheels sprung.
Wheels: Rubber tired, detachable metal grousers.
Weight: 6.6 tons.
Special features: This vehicle is similar in many respects to the Pavesi tank.

Ansaldo.

Produced in 1930 by Ansaldo Motor Company. Total production, 1 or 2.

Crew: 3.
Armament: Either a short 47 mm (1.85 in.) gun or a long 37 mm (1.46 in.) gun in front of turret, and one machine gun in rear of turret.
Maximum speed: 27 mph.
Suspension: Spring.
Wheels: 4 ft. 10 in. in diameter, 1 ft. 4 in. wide; disc ribbed type, rubber and steel grousers.
General arrangement: Driver in front left side; engine in front right side; gunners in center and rear.
Dimensions: Length 15 ft.; width 8 ft. 6 in.; height 9 ft. 6 in.
Weight: 9.1 tons
Engine: Ansaldo, 110 HP, water cooled
Horsepower per ton: 12.1.
Transmission: Sliding gear; 4 speeds forward, 1 reverse.
Obstacle ability: Stream 5 ft.; slope 45 degrees; vertical wall 3 ft.
Special features: Cooling louver in front consists of series of concentric steel rings of increasing dimensions. Front wheels turn 40 degrees for steering and rear wheels 30 degrees from horizontal. Rear axle flexible instead of jointed hull. Water pumps for fording and gas filter standard equipment. Driver has cone stroboscope and gunner has periscope.

Fiat, Type 2000.

Produced in 1918 by Fiat Motor Company. Total production, 6.

Crew: 10.
Armament: One 65 mm (2.56 in.) gun in revolving turret, and seven machine guns located at the front, rear and sides.
Armor: 0.59 in. to 0.787 in.
Maximum speed: 4.5 mph.
Suspension: Pivoted bogies and rollers on leaf springs.
Tracks: Solid plates with longitudinal and lateral grousers, 1 ft. 5 in. wide.
General arrangement: Driver and final drive in front; engine under floor at rear end ; gunners in upper part, rear gunner over engine.
Dimensions: Length 22 ft. 6 in.; width 9 ft. 5 in.; height 11 ft. 5 in.
Weight: 44 tons
Engine: Fiat. 6 cylinder, 240 HP, water cooled.
Horsepower per ton: 5.5.
Transmission: Planetary.
Obstacle ability: Trench 9 ft. 10 in.; stream 3 ft.; slope 40 degrees; vertical wall 35 in.; tree 19 in.
Fuel distance: 40 miles. **Fuel capacity:** 132 gallons.
Special features: Driver normally uses periscope. Suspension has armor protection.

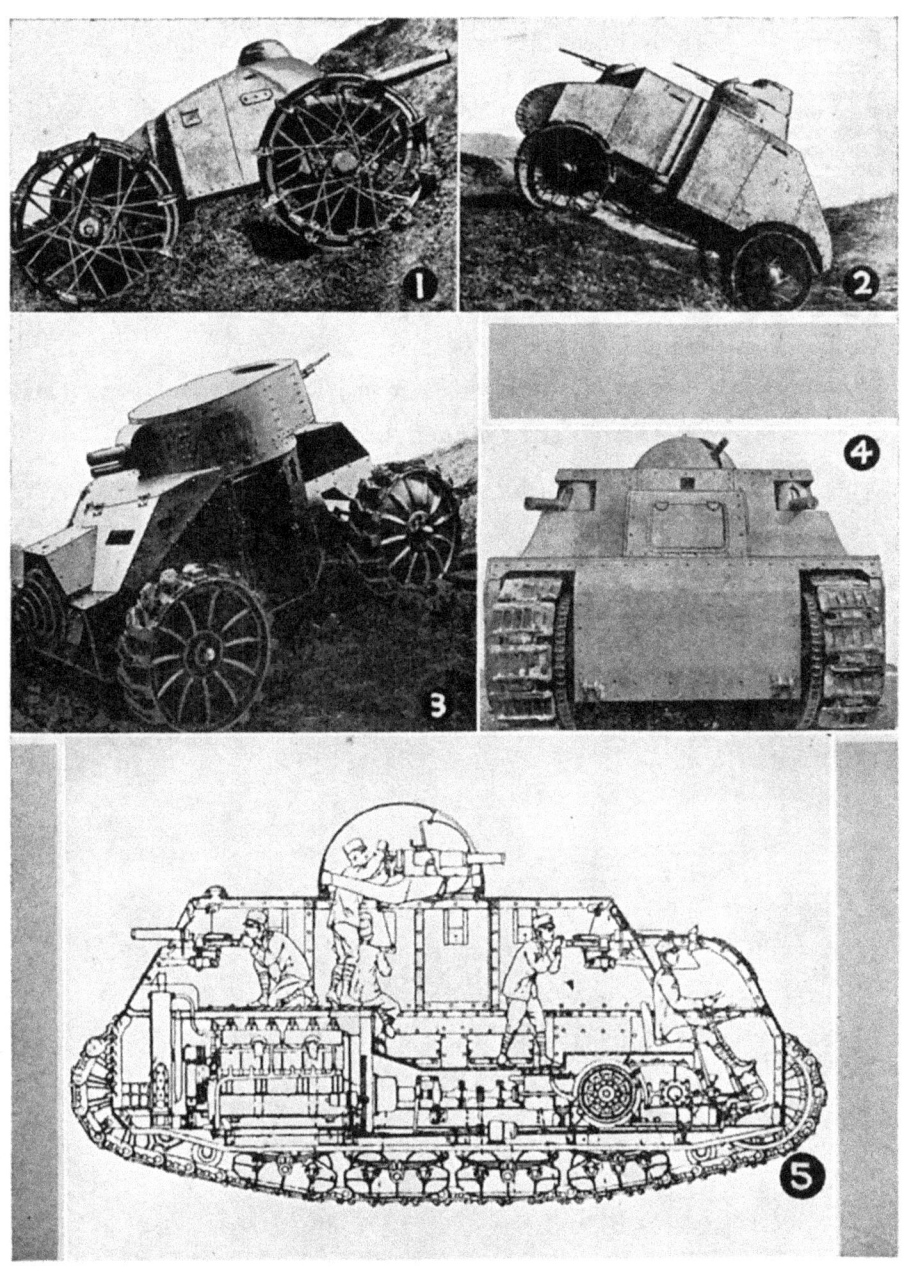

Plate 29.
1. **Pavesi Tank Destroyer.**
2. **Heavy Pavesi.**
3. **Ansaldo.**
(From *Taschenbuch der Tanks* by Heigl.)
4. **Fiat, Type 2000, Front View.**
(From *Taschenbuch der Tanks* by Heigl.)
5. **Fiat, Type 2000.**
(From *Taschenbuch der Tanks* by Heigl.)

Heavy Pavesi.

Produced in 1928 by Pavesi Tractor Company. Total production, 1.

Crew: 4.
Armament: Three machine guns.
Maximum speed: 15 mph.
Suspension: Spring.
Wheels: 5 ft. 4 in. in diameter, 14 in. wide.
General arrangement: Engine, driver, and 1 gunner in front; remainder of crew in rear.
Dimensions: Length 18 ft., height 11 ft. 2 in.
Special features: Equipped with water pumps for expelling water when fording.

G. L. 4.

Armament: One 75 mm (2.95 in.) gun and several machine guns.
Dimensions: Length 27 ft.; width 10 ft.; height 11 ft.
Weight: 38.5 tons.
Engine: Fiat, 8 cylinder, 200 HP.
Horsepower per ton: 5.2.
Obstacle ability: Trench 13 ft.; stream 3 ft. 6 in.; slope 40 degrees; vertical wall 3 ft.

Italian Light (Carden Loyd Type).

Produced recently.
This tank resembles the Carden Loyd Mark VI.

CHAPTER XV.

TANKS OF ALL COUNTRIES (Continued).

RUSSIAN.

Russian Renault.

Produced by Renault in France, modified in Russia. Total production, 116.

Crew: 2.
Armament: One 37 mm (1.46 in.) gun and one machine gun.
Armor: 0.256 in. to 0.6 in.
Maximum speed: 6.5 mph.
Suspension: Same as French Renault.
Tracks: Same as French Renault.
General arrangement: Same as French Renault.
Dimensions: Length 12 ft. 2 in. (without tail); width 5 ft. 4 in.; height 6 ft. 8 in.
Weight: 7.7 tons.
Engine: 4 cylinder, Fiat, 45 HP, water cooled.
Horsepower per ton: 5.8.
Obstacle ability: Trench 5½ ft.
Fuel distance: 30 miles.
Special features: Other features same as French Renault. Hotchkiss machine gun mounted on right of turret in ball mount; must be removed when firing 37 mm gun. Some tanks equipped with radio set and one machine gun.

Light Tank.

Crew: 2.
Armament: One 37 mm (1.46 in.) gun and one machine gun in front; 2 additional machine guns said to be mounted in rear.
Maximum speed: 10 mph.
Suspension: Spring.
Tracks: Similar to late Vickers Medium tanks.
General arrangement: Driver in front; gunners in center; engine and final drive in rear.
Weight: 8 tons.
Special features: Fitted with emitter for smoke or gas. Fuel tanks over tracks. Front idlers on exposed cross shaft in Mark I.

Eighty Ton Tank.

Produced in 1926 by Tank and Armored Car Works. Total production, 1.

Crew: 10.
Armament: Two 76 mm (2.99 in.) guns and 4 machine guns.
Armor: 0.5 in. to 1.6 in
Maximum speed: 6 mph
Weight: 80 tons.
Special features: Equipped with radio. Is gas proof. Has overhead tracks similar to the British Mark V but with a large turret and no sponsons.

Miscellaneous Notes Regarding Russian Tanks. In some of their British types of tanks, the Russians have provided for hermetic closing and the use of oxygen, thus facilitating travel through gassed areas.

The Russians have experimented with radio to a considerable degree. They have radio sets installed in Russian Renault, British Mark V, and British Medium A tanks.

JAPANESE.

Vickers Medium C.

Produced in 1928-1929 by Vickers Armstrongs, Ltd.

Crew: 6.
Armament: One 6 pounder (57 mm-2.24 in.) and four machine guns, one of the machine guns for antiaircraft fire.
Maximum speed: 18.6 mph.
Suspension: Same as Vickers Mark II.
Tracks: Same as Vickers Mark II, 16 in. wide.

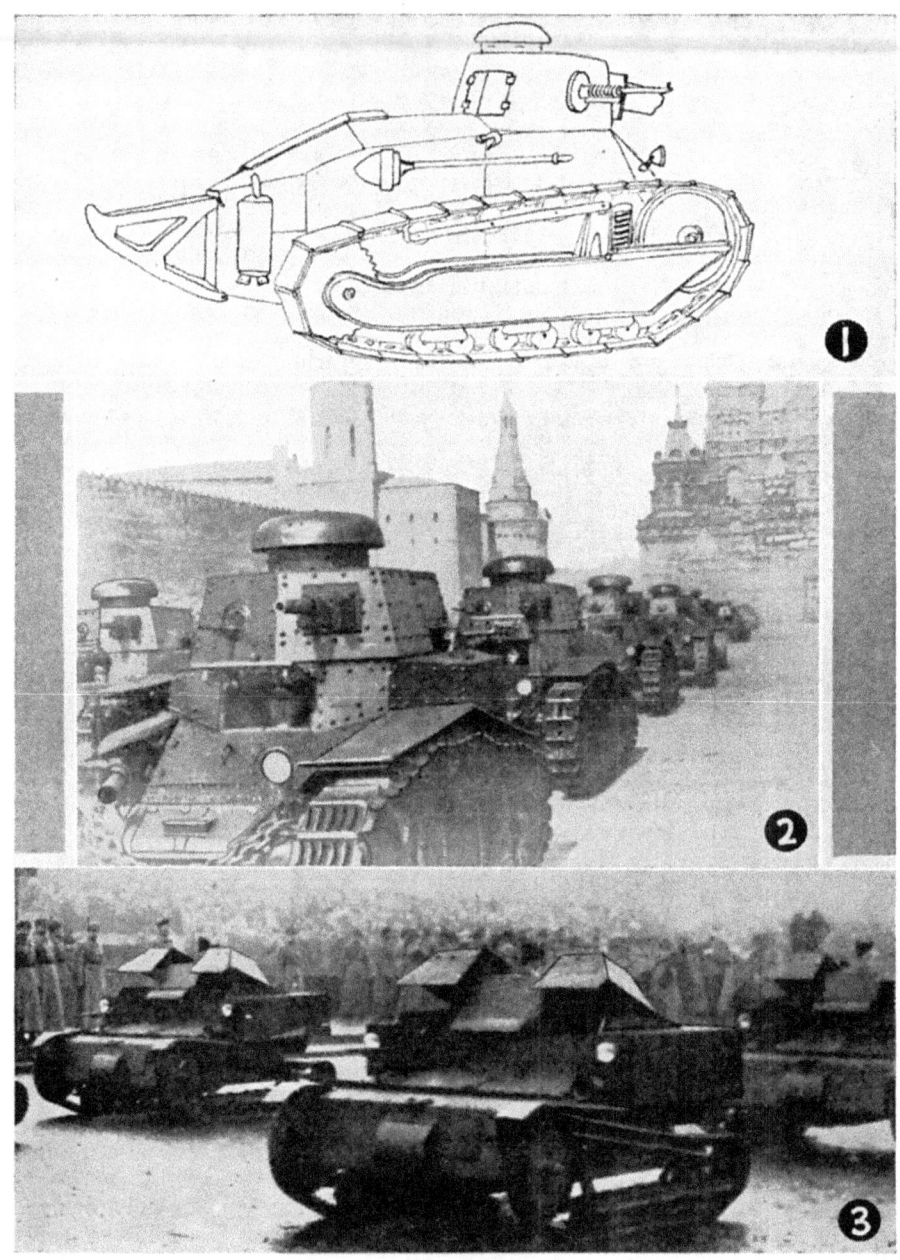

Plate 30.
1. **Russian Renault.**
2. **Russian Light Tank.**
3. **Carden Loyd Tanks in Russia.**
(Wide World Photos.)

General arrangement: Driver in front; gunners in center; engine in rear.
Dimensions: Length 17 ft. 6 in.; width 8 ft. 4 in.; height 7 ft. 10 in
Weight: 11.9 tons.
Engine: 110 HP, water cooled; later models Armstrong-Siddeley air cooled.
Horsepower per ton: 9.2.
Obstacle ability: Trench 6 ft. 6 in.; stream 2 ft. 5 in.; slope 45 degrees. vertical wall 2 ft.
Fuel distance: 144 miles. **Fuel capacity:** 70 gallons.

JAPANESE TANKS 149

Plate 31.
1. Vickers Medium C.
2. Vickers Medium C.
3. Radio Controlled Tank. Maj. Nagayama at the controls.
(Wide World Photos.)

Special features: Fire hazard reduced by bulkheads and outside fuel tanks. Good ventilation, vision, and roominess. Angles of impact well considered. Machine gun sponsons in hull may be swung inward to gain greater arc of fire. The original design of this tank placed the engine in front but Japan required that it be placed in rear, that the water cooled engine be replaced by an air cooled engine, and that the 3 pounder gun be replaced by a 6 pounder.

In a period of four years, the Japanese developed two radio-controlled tanks at a cost of about $50,000.00. One of these tanks is shown in Fig. 3, Plate 31. This development reached such a stage that the tank could, by remote control, be made to perform any one of a number different operations, these including such matters as acceleration, deceleration, changing direction, elevating and depressing the weapon, firing the weapon, discharging grenades, sounding a siren, etc.

Due largely perhaps to the hilly ground of Japan, the Japanese have found that their foreign-made tanks are in general, much underpowered.

At Shanghai, in the latter part of February, 1932, in connection with the Japanese drive to the westward against Kiangwan, the Japanese employed about 24 Medium and Renault NC tanks. The ground, however, in that area was unsuitable, being badly cut up by streams, ditches and rice paddies. As a result numerous parties were employed to prepare crossings and these parties suffered severely under Chinese fire. The tanks under such conditions could not be used vigorously and effectively. As the fight went on little effort was made to use them at all.

SWEDISH.
M 21.

Produced in 1921. Total production, about 12.

Crew: Combat tanks 4; command tanks 5.
Armament: *Male:* One 37 mm (1.46 in.) gun and one machine gun. *Female:* Two machine guns.
Armor: 0.39 in. to 0.59 in.
Maximum speed: 13 mph.
Suspension: Spring.
Tracks: Flat steel plates with slight grousers.
General arrangement: Engine in front; driver and crew in rear.
Dimensions: Length 17 ft.; width 6 ft.; height 8 ft.
Weight: 10.5 tons.
Engine: Daimler, sleeve valve, 4 cylinder, 55 HP, water cooled.
Horsepower per ton: 5.2.
Transmission: Sliding gear.
Obstacle ability: Trench 7 ft.; stream 24 in.; slope 40 degrees; vertical wall 3 ft.
Fuel distance: 60 miles.
Special features: Command tanks radio equipped, combat tanks with receivers only. Machine guns can be moved to several mounts. Designed by Vollmer, who built German war time tanks. Suspension protected by armor skirts.

Landskrona.

Produced in 1928 by Landskrona Company. Total production, 1.

Crew: 3.
Armament: One machine gun or one semi-automatic cannon.
Armor: 0.26 in. to 0.5 in.
Maximum speed: 16 mph on tracks; 35 mph on wheels.
Suspension: Spring.
Weight: 8 tons.
Engine: 8 cylinder, 115 HP.
Horsepower per ton: 14.4.
Transmission: Electric.
Obstacle ability: Trench 5 ft. 10 in.; stream 2 ft. 6 in.; slope 45 degrees
Fuel capacity: 33 gallons.

CZECHOSLOVAKIAN.
Light Tank.

Produced by Adamov Company. Total production, 1.

Crew: 2
Armament: One pair of light machine guns or one 20 mm (.79 in.) machine gun.
Armor: 0.31 in. to 0.5 in.
Speed: 10 mph on tracks; 30 mph on wheels

Plate 32.
1. **M 21, Swedish.**
2. **K H 50, Czechoslovakian.**
(From *Taschenbuch der Tanks* by Heigl.)
3. **Trubia, Spanish.**

Suspension: Spring.
General arrangement: Engine in front; driver in rear.
Dimensions: Length 8 ft. 6 in.; width 7 ft. 1 in.; height 7 ft. 4 in. on wheels.
Weight: 4.3 tons.
Engine: 50 HP.
Horsepower per ton: 11.6
Obstacle ability: Stream 2ft. 8in.
Special features: Tank has rotating turret.

K. H. 50-60-70 (Kolo Housenka).

Produced in 1924 by Skoda. Total production, 50.

Crew: 2 (3 when equipped with radio).
Armament: One 35 mm (1.38 in.) gun or one machine gun.
Armor: 0.25 in. to 0.5 in.
Maximum speed: 7.5 mph on tracks; 16-22 mph on wheels.
Suspension: Leaf springs and rollers.
Tracks: Steel plates 12 in. wide.
Wheels: Four, rubber tired.
General arrangement: Driver in front; gunner in center; engine in rear.
Dimensions: Length 14 ft. 8 in.; width 7 ft. 6 in.; height 7 ft. 5 in. on tracks. 8 ft. 4 in. on wheels.
Weight: 8.3 tons.
Engine: 4 cylinder, 50, 60, or 70 HP, water cooled.
Horepower per ton: 6, 7.2, or 8.4
Transmission: Sliding gear with two gear boxes.
Obstacle ability: Trench 5 ft. 11 in.; stream 2 ft. 7 in.; slope 45 degrees; vertical wall 1 ft. 8 in.; tree 12 in.
Fuel distance: 126 miles on wheels. **Fuel capacity:** 35 gallons.
Special features: Armored electric headlights in front. Crew must dismount in changing from wheels to tracks, or tracks to wheels. Ten minutes required for either change. The tank is run upon half-round wooden blocks. carried for the purpose. The wheels are then fixed in the proper position for either method of operation. Rear wheels lifted by engine power, front wheels by hand. Rear wheels also operate when on tracks. The design of this tank is not favorable for passage through wire entanglements. Triplex glass over all eye slits. Radio on command tanks only, with 30 mile radius. Wheels vulnerable. Designed by Vollmer who built German war time tanks.

SPANISH.
Trubia.

Produced in 1925. Total production, several.

Crew: 3.
Armament: One 37 mm (1.46 in.) gun or two machine guns.
Armor: 0.5 in. to 0.65 in.
Maximum speed: 12 mph.
Suspension: Spring.
Tracks: Flat steel plates with rollers attached.
General arrangement: Driver in front; gunners in center; engine and final drive in rear.
Dimensions: Length 17 ft. 8 in.; width 6 ft. 11 in.; height 7 ft. 10 in.
Weight: 8.9 tons .
Engine: 75 HP, water cooled.
Horsepower per ton: 8.4.
Transmission: Sliding gear.
Obstacle ability: Trench 7 ft.; stream 2 ft.; slope 40 degrees; vertical wall 15 in.
Fuel distance: 60 miles.
Special features: Small partly armored wheel in front for added length in trench crossing. Machine guns can be fired from several mounts. Tracks and rollers revolve around sprung track frames as in German K tank. Sprocket wheel, rollers, and part of tracks are protected by armor. This tank is in many respects similar to the French Renault but is intended to be an improvement over that model.

POLISH.
Cardosowitz.

Produced in 1925 by Renault and modified in Poland.

Crew: 2.
Armament: One 37 mm (1.46 in.) gun or one machine gun.
Armor: 0.236 in. to 0.629 in.
Maximum speed: 9 mph.
Suspension: Same as French Renault.
Tracks: 32 plates as received. In one modification 96 plates were substituted, in another 12 cables with grousers attached were substituted.
General arrangement: Driver in front; gunner in center; engine in rear.
Dimensions: Same as French Renault.
Special features: Other features same as French Renault.

CHAPTER XVI.

TANKS OF ALL COUNTRIES (Continued).

UNITED STATES.

Gas Electric.

Produced in 1918 by Holt Tractor Company and General Electric Company. Total production, 1.

Crew: 6.
Armament: One 75 mm (2.95 in.) mountain howitzer and two cal. .30 machine guns.
Armor: 0.25 in. to 0.63 in.
Maximum speed: 6 mph.
Suspension: Coil springs with rollers and bogies; two bogies on each side, forward bogie having 4 rollers, rear bogie 6.
Tracks: Flat steel plates with double grousers, width 15½ in., pitch 7½ in.
General arrangement: Driver and howitzer in front; driver above; machine guns in center at sides; engine in rear.
Dimensions: Length 16 ft. 6 in.; width 9 ft. 1 in.; height 7 ft. 9½ in.
Weight: 25 tons.
Engine: One Holt, 4-cylinder, high speed, 90 HP with electric generator, and an electric motor for each track, forced water cooling.
Horsepower per ton: 3.6.
Transmission: Electric.
Special features: Steered by varying current to driving motors and by braking. Ammunition racks in center. This was the first tank built in the United States.

Steam Tank, Track Laying.

Produced in 1918 by Corps of Engineers, U. S. Army. Total production, 1.

Crew: 8.
Armament: One flame thrower and 4 cal. .30 machine guns.
Armor: .5 in.
Maximum speed: 4 mph.
Suspension: Rigid, with rollers.
Tracks: Flat steel plates with grousers; width 24 inches; pitch 12½ inches. Upper part of each track supported by three rollers.
General arrangement: Driver and flame thrower in front; water and kerosene tanks in center; two machine guns in one sponson on each side; boilers and engines in rear.
Dimensions: Length 34 ft. 9 in.; width 12 ft. 6 in.; height 10 ft. 4½ in.
Weight: 50 tons.
Engine: Two 2-cylinder steam engines, one for each track, and one kerosene-burning boiler for each engine; total HP 500.
Horsepower per ton: 10.
Transmission: Sliding gear, 2 speeds forward, 1 reverse; large exposed gears and shafts in rear part of hull.
Special features: Steam was selected as motive power for this tank because the original plan was to utilize a steam jet from a 700 pound pressure as a carrier for the flame. The flame thrower in its final form used a 35-HP gasoline engine by means of which oil was put under a pressure of 1600 pounds per square inch. Through a small orifice in a nozzle, this was directed through a hole in the tank. The jet of flame, leaving the nozzle about the size of a lead pencil, became, at a range of 90 feet, a ball of flame 20 feet or more in diameter. The purpose of the flame thrower was to neutralize concrete pill-boxes. This was the second tank built in the United States.

Steam Tank, Three Wheeled.

Produced in 1918 by Holt Tractor Company. Total production, 1.

Crew: 6:
Armament: One 75 mm (2.95 in.) mountain howitzer and two cal. .30 machine guns.
Armor: 0.25 in. to 0.63 in.
Maximum speed: 5 mph.
Suspension: Rigid, on wheels.

Plate 33.
1. Gas Electric.
(Photo by Ordnance Dept., U.S.A.)
2. Ford Three Ton.
(Photo by Tank School, U.S.A.)
3. Steam Tank, Track Laying.
4. Steam Tank, Three Wheeled.
(Photo by Ordnance Dept., U.S.A.)
5. Skeleton Tank.
(Photo by Ordnance Dept., U.S.A.)

United States Tanks

Tracks: None.
General arrangement: Howitzer and its gunner low in front; driver above the gunner; engines in center; machine gun gunners and observers above engines; boilers and tanks in rear.
Dimensions: Length 22 ft. 3 in.; width 10 ft. 1 in.; height 9 ft. 10 in.
Weight: About 17 tons.
Engine: Two Doble 2-cylinder steam engines, combined HP 150.
Horsepower per ton: 8.8.
Special features: Hull carried at the front by 2 drive wheels 8 ft. in diameter and about 2 ft. wide, with grousers placed at opposing angles and with open spaces in the rim of the wheel between grousers. The rear end of the hull was carried by a swivelling drum-like trailer wheel to which was attached the tailpiece. One engine for each drive wheel. Doble type boilers condensing in Holt type radiator. This was the third tank built in the United States.

Skeleton Tank.

Produced in 1918 by Pioneer Tractor Company. Total production, 1.

Crew: 2.
Armament: One cal. .30 machine gun.
Armor: 0.5 in.
Maximum speed: 5 mph.
Suspension: Rigid. with rollers.
Tracks:. Width 12 in., pitch 11 in.
General arrangement: Driver in front, gunner in turret in rear; at each side from front to rear, radiator, engine and transmission.
Dimensions: Length 25 ft.; width 8 ft. 5 in.; height 9 ft. 6 in.
Weight: 8 tons.
Engine: Two Beaver, 4-cylinder, combined HP 100, forced water cooling.
Horsepower per ton: 12.5.
Transmission: Two speeds forward, one reverse.
Special features: Built with a view of securing a light vehicle capable of crossing wide trenches. Many structural members were pieces of iron pipe with standard plumbing connections.

Ford Three Ton.

Produced in 1918 by Ford Motor Company. Total production, 15.

Crew: 2.
Armament: One cal. .30 machine gun; arc of traverse 21 degrees; vertical arc 38 degrees.
Armor: 0.25 in. to 0.5 in.
Maximum speed: 8 mph.
Suspension: Leaf springs with bogies and rollers.
Tracks: Flat steel plates with grousers; width 7 in., pitch 7 in.
General arrangement: Driver and gunner in front; driver on right; engines and final drive in rear.
Dimensions: Length 13 ft. 8 in.; width 5 ft. 6 in.; height 5 ft. 3 in.
Weight: 3.1 tons.
Engines: Two, Ford, Model T, 4 cylinder, combined HP 45, forced water cooling.
Horsepower per ton: 14.5.
Transmission:. Planetary, 2 speeds forward, 1 reverse.
Obstacle ability: Trench 5 ft.; stream 21 in.; slope 25 degrees; vertical wall 20 in.
Fuel distance: 34 miles. **Fuel capacity:** 17 gallons.
Special features: Machine gun had limited traverse. Low and easily maneuvered. A fair accompanying tank except for limited fire power, defective cooling and ventilation, cramped quarters and other minor faults.

Mark I, Three Man.

Produced in 1918 by Ford Motor Company. Total production, 1.

Crew: 3.
Armament: One 37 mm (1.46 in.) gun and one cal. .30 machine gun.
Armor: 0.37 in. to 0.5 in.
Maximum speed: 9 mph.
Suspension: Leaf springs, pivoted bogies of three rollers each.
Tracks: Pressed plates with grousers; width 12 in., pitch 7 in.
General arrangement: Driver left front; machine gunner right front; 37 mm gun in center; engine and final drive at rear.
Dimensions: Length 16 ft. 6 in.; width 6 ft. 6 in.; height 7 ft. 9 in.
Weight: 7.5 tons.
Engine: Hudson, 6 cylinder, 60 HP, forced water cooling.
Horsepower per ton: 8.
Transmission: A complete transmission on each side, planetary, 2 speeds forward, 1 reverse.
Special features: Center of gravity too far to rear to negotiate obstacles satisfactorily. Track adjustment method unsatisfactory because idlers could not be moved independently.

Plate 34.
1. Mark I, Three Man.
2. Mark VIII.
(Photo by Tank School, U.S.A.)
3. Mark VIII.
(From a drawing by Ordnance Dept., U.S.A.)

United States Tanks

Mark VIII.

Produced in 1919 by Ordnance Department, U.S.A., the 6 pdr. guns, the armor plate, and various other parts being furnished by Great Britain. Total produced, 100.

Crew: 11.
Armament: Two 6 pounder (57 mm-2.24 in.) guns and five cal. .30 machine guns.
Armor: 0.236 in. to 0.63 in.
Maximum speed: 6.5 mph. Due to cooling difficulties, this tank cannot sustain a speed equal to that of the six-ton tank, M1917.
Suspension: Rigid, with rollers.
Tracks: Flat armor plates with slight grousers; width 26½ in., pitch 11.154 in.
General arrangement: Crew compartment in front; engine compartment in rear; driver at front of crew compartment.
Dimensions: .Length 34 ft. 2½ in.; width 12 ft. 5 in. (overall, sponsons not withdrawn); height 10 ft. 2½ in.
Weight:. 43.5 tons (including equipment).
Engine: Liberty, 12 cylinder, V type, 338 HP, forced water cooling.
Horsepower per ton: 7.8.
Transmission: Planetary, two speeds forward, two reverse.
Obstacle ability: Trench 16 ft.; stream 2 ft.; slope 40 degrees; vertical wall 54 in.; tree 18 in.
Fuel distance: 50 miles.
Special features: This tank has remarkable crushing and tractive ability and is intended for use against highly organized defensive works. Though seemingly suitable for the leading role, it is unsatisfactory due to slow speed, thin armor, and mechanical unreliability. An intratank telephone system enables the commander to communicate with his driver. 6 pdr. gunners, and mechanic. This tank uses the Autopulse electrical fuel pump.

Six Ton, M 1917.

(A copy of French Renault). Produced in 1918-1919 by various companies under supervision of Ordnance Department. Total production, 952.

Crew: 2.
Armament: One 37 mm (1.46 in.) gun or one cal. .30 machine gun.
Armor: 0.25 in. to 0.6 in.
Maximum speed: 5.5 mph.
Suspension: Coil and leaf springs, with bogies and rollers. Track frame is pivoted at rear end and sprung at front end.
Tracks: Flat steel plates with slight grousers; width 13⅜ in., pitch 9.84 in.
General arrangement: Driver in front; gunner in center; engine and final drive in rear.
Dimensions: Length 16 ft. 5 in.; width 5 ft. 10½ in.; height 7 ft. 7 in.
Weight: 7.25 tons, fully equipped; without equipment about 6.7 tons.
Engine: Buda, 4 cylinder, 42 HP, forced water cooling.
Horsepower per ton: 5.8.
Transmission: Sliding gear, 4 speeds forward, 1 reverse.
Obstacle ability: Trench 7 ft.; stream 2 ft.; slope 35 degrees; vertical wall 3 ft.; tree 8 in.
Fuel distance: 30 miles. **Fuel capacity:** 30 gallons.
Special features: Tail piece 2 ft. 7 in. long adds greatly to ability to negotiate obstacles. Noisy due to track bushings striking idlers and drive sprockets. Maintenance requires much labor. Suitable for the accompanying role, but slow.

Pilot Model, Six Ton, M 1917, A1.

Modified in 1929 by Engineering Department, Holabird Q. M. Depot. Total production, 1.

Maximum speed: 10.3 mph.
Dimensions: Length 17 ft. 3 ½ in.; width 7 ft. 10½ in.; height 7 ft. 7 in.
Weight: About the same as six ton M1917.
Engine: Franklin, 6 cylinder, 67 HP, air cooled, vertical draft.
Horsepower per ton: 9.2.
Transmission: Same as six ton M1917, except that case was reinforced.
Fuel distance: 50 miles. **Fuel capacity:** 29 gallons.
Special features: A six ton Model 1917 tank was modified by removing the Buda engine with clutch, radiator with fan, and minor parts; by enlarging the engine compartment slightly; and by installing a modified Franklin, air-cooled engine complete with clutch, and with fan mounted on crank shaft. A test was made on this tank of modified idlers which reduced the characteristic noise of the tank.

158   The Fighting Tanks Since 1916

Six Ton, M 1917, A1.

Modified in 1930-1931 by Ordnance Department. Total production, 7.

Maximum speed: 9 mph (governed).
Dimensions: Length 17 ft. 3½ in.; width 7 ft. 10½ in.; height 7 ft. 7 in.
Weight: 7.1 tons.
Engine: Franklin, 6 cylinder. 100 HP, air cooled, side draft.
Horsepower per ton: 14.2.
Transmission: Substantially the same as in Six Ton Tank.
Fuel distance: 50 miles. **Fuel capacity:** 29 gallons.
Special features: Substantially the same as pilot model except that engine and transmission were mounted upon a unit-supporting frame, and a more powerful engine was used. Due to the weakness of the old units still used, the engine

Plate 35.
1. **Six Ton, Model 1917.**
(Photo by Tank School, U.S.A.)
2. **Pilot Model, Six Ton M 1917 A-1. With rear parts removed and engine exposed.**
(Photo by Tank School, U.S.A.)
3. **Six Ton, Model 1917.**
(From a drawing by Ordnance Dept., U.S.A.)

speed was governed down from 3100 rpm to 2500 rpm. The pilot model, later modified, constituted one of these seven tanks. Twelve volt ignition and batteries were installed, the latter to be used with radio apparatus.

Medium A, M 1921.

Produced in 1921 by Ordnance Department. Total production, 1.

Crew: 4.
Armament: One 6 pounder (57 mm-2.24 in.) gun and one cal. .30 machine gun in one mount in main turret; one cal. .30 machine gun in upper turret.
Armor: 0.375 in. to 1.0 in.
Maximum speed: 10.1 mph.
Suspension: Bogies, rollers, and coil springs.
Tracks: Solid steel plates with grousers but grousers were hollow and constituted oil reservoirs for lubricating track pins.

Plate 36.

1. **Six Ton, M 1917 A 1.**
(Photo by Ordnance Dept., U.S.A.)
2. **Six Ton, M 1917 A 1, Engine and Transmission Sub-frame Assembly.**
(Photo by Ordnance Dept., U.S.A.)
3. **Medium, Model 1921 (Medium A).**
(Photo by Ordnance Dept., U.S.A.)

General arrangement: Driver in front; gunners in center; engine and final drive in rear.
Dimensions: Length 21 ft. 5 in., width 8 ft.; height 9 ft. 9 in.
Weight: 23 tons.
Engine: Murray and Tregurtha, marine, 6 cylinder, maximum HP 250, governed HP 170, forced water cooling
Horsepower per ton: 7.4 governed.
Transmission: Planetary and sliding, 4 speeds forward, 2 speeds reverse.
Obstacle ability: Trench 8 ft.; stream 3 ft.; slope 35 degrees, vertical wall 26 in.; tree 12 in.
Fuel distance: 50 miles.
Special features: The small upper turret carrying one machine gun revolved upon the larger turret as a base. A Liberty 338 HP engine was installed in this tank as an experiment and with it a maximum speed of about 25 mph was attained. A gear pump fuel system used. A rotary water-expelling pump was provided that was driven from the transmission.

Medium, M 1922.

Produced in 1922 by Ordnance Department. Total production, 1.

Crew: 4.
Armament: One 6-pounder (57 mm-2.24 in.) gun and one cal. .30 machine gun in one mount in main turret; one cal. .30 machine gun in upper turret.
Armor: 0.375 in. to 1.0 in.
Maximum speed: 15.7 mph.
Suspension: A chain type was first installed, the chain being replaced by a cable. The general nature of this suspension is shown in Fig. 3, Plate 47. This suspension was found to be unsatisfactory.
Tracks: Light wooden shoes within brackets pivoted at the center and also held by a coil spring at each side.
General arrangement: Driver in front; gunners in center; engine and final drive in rear.
Dimensions: Length 26 ft.; width 9 ft.; height 9 ft. 8½ in.
Weight: 25 tons.
Engine: Murray and Tregurtha, marine, 6 cylinder, maximum HP 250, governed HP 170, forced water cooling.
Horsepower per ton: 6.8 governed.
Transmission: Planetary and sliding gear, 4 speeds forward, 2 speeds reverse. Pneumatic control of transmission and brakes.
Special features: Tracks and suspension unsatisfactory. Tracks higher in rear than in front. Other features similar to Medium Tank, Model 1921.

Medium, T 1.

Recently renamed T 1-E 1. Has also been called Medium 23 Ton, T 1. Produced in 1925 by Ordnance Department. Total production, 1.

Crew: 4.
Armament: One 6 pounder (57 mm-2.24 in.) gun and one cal. .30 machine gun in one mount in main turret and one cal. .30 machine gun in upper turret.
Armor: 0.375 in. to 1.0 in.
Maximum speed: 11.3 mph
Suspension: Bogies, rollers, and coil springs.
Tracks: Forged steel skeleton type with grousers, width 18 in., pitch 8⅛ in.
General arrangement: Driver in front; gunners in center; engine and final drive in rear.
Dimensions: Length: 21 ft. 6 in.; width 8 ft.; height 9 ft. 7½ in.
Weight: 22 tons.
Engine: Special Packard, 8 cylinder, V type, 200 HP, forced water cooling.
Horsepower per ton: 9.1.
Transmission: Planetary and sliding gear, 4 speeds forward and 1 reverse.
Obstacle ability: Trench 8 ft.; stream 2 ft.; slope 35 degrees; vertical wall 31 in.; tree 12 in.
Fuel distance: 50 miles. **Fuel capacity:** 95 gallons.
Special features: This tank has an unusually high ground pressure, it has not as great horsepower per ton as is desirable, and it has a number of lesser faults. In spite of all this, it is a good tank and is fairly well suited to a definite tactical mission—that of the leading role. A Liberty 338 HP engine has recently been installed in place of the special Packard.

Medium, T 2.

Produced in 1930 by James Cunningham, Sons and Company. (Ordnance design). Total production, 1.

Crew: 4.
Armament: One 47 mm (1.85 in.) gun and one cal. .50 machine gun in one mount in turret; one 37 mm (1.46 in.) gun and one cal. .30 machine gun in one mount in the hull.

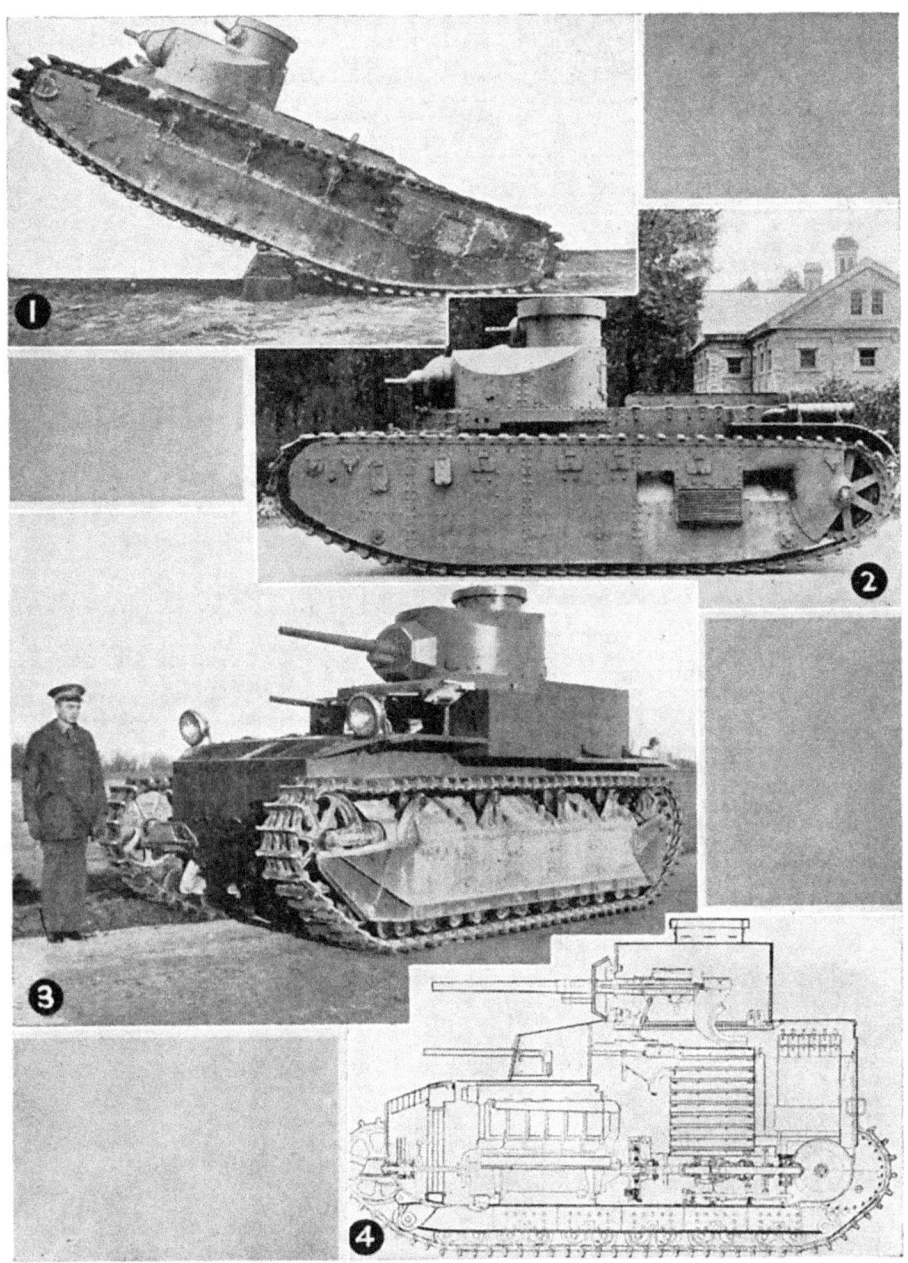

Plate 37.
1. Medium, Model 1922.
(Photo by Ordnance Dept., U.S.A.)
2. Medium, T 1.
(Photo by Ordnance Dept., U.S.A.)
3. Medium, T 2.
(Photo by Ordnance Dept., U.S.A.)
4. Medium, T 2.

Armor: 0.25 in. to 0.875 in.
Maximum speed: 25 mph; governed speed 20 mph.
Suspension: Bogies, rollers, and coil springs. Front and rear rollers, which are slightly elevated, are unsprung.
Tracks: Cast steel skeleton type with long integral grousers, width 15 in.; pitch 5¼ in.
General arrangement: Driver in left front; engine in right front; gunners in center; ammunition space and final drive in rear.
Dimensions: Length 16 ft.; width 8 ft.; height 9 ft. 1 in.
Weight: 15 tons.
Engine: Modified Liberty, 12 cylinder, V type, 318 HP, forced water cooling. HP decreased through reduced compression.
Horsepower per ton: 21.2.
Transmission: Cotta sliding gear, 4 speeds forward, 1 reverse.
Obstacle ability: Trench 6 ft.; stream 48 in.; slope 35 degrees.
Fuel distance: 90 miles. **Fuel capacity:** 94 gallons.
Special features: Gunners interfere with each other. Guns mounted in hull have limited traverse. One cal. .30 machine gun substituted in 1931 for one 37 mm gun and machine gun; one cal. .30 AA machine gun added. The steering brakes of this tank are operated through a vacuum booster. An experimental gunner's seat is attached to the gun cradle, and the weight of the receiver, the gunner and gunner's seat, is balanced by counterweights (over 600 pounds) mounted in front of the gun mount trunnion pins. A Sperry electric-driven gyroscopic direction indicator has been installed.

One Man Tank, Experimental.

Track Development Chassis, T 1. Produced in 1928 by James Cunningham, Sons and Company. (Ordnance design). Total production, 1.

Crew: 1.
Armament: One cal. .30 machine gun.
Armor: 0.125 in.
Maximum speed: 19.5 mph.
Suspension: Rear wheels slightly sprung (coil springs); unsprung front wheels drive; wheels of aluminum; solid rubber tires.
Tracks: Each track includes two flexible steel bands 4½ in. wide, lined with commercial belting; steel grousers outside, guiding lugs on inner surface; total track width 10 in.
General arrangement: Transmission in front; engine in center; man in rear; man straddles rear part of engine.
Dimensions: Length 8 ft. 7 in., width 4 ft. 9 in., height 5 ft. 1½ in.
Weight: 1.5 tons.
Engine: Ford, Model A, 4 cylinder, 42 HP, forced water cooling.
Horsepower per ton: 28.
Transmission: Sliding gear, 3 speeds forward, 1 reverse.
Special features: The experimental development of tracks was an important purpose of this construction.

Light Tank, T 1.

Produced in 1927 by James Cunningham, Sons and Company. (Ordnance design). Total production, 1.

Crew: 2.
Armament: One 37 mm (1.46 in.) gun and one cal. .30 machine gun in one mount.
Armor: 0.25 in. to 0.375 in.
Maximum speed: 20 mph.
Suspension: Rollers, bogies, and equalizing links; no springs.
Tracks: Cast steel skeleton type with grousers, width 12 in., pitch 6.5 in.
General arrangement: Engine in front; driver in center; gunner and final drive in rear.
Dimensions: Length, 12 ft. 6 in.; width 5 ft. 10½ in.; height 7 ft. 1½ in.
Weight: 7.5 tons.
Engine: Cunningham, 8 cylinder, V type, 105 HP, forced water cooling.
Horsepower per ton: 14.
Transmission: Cotta type sliding gear, 3 speeds forward, 1 reverse.
Obstacle ability: Trench 6 ft.; stream 20 in.; slope 30 degrees; vertical wall 20 in.; tree 10 in.
Fuel distance: 65 miles.
Special features: The series of light tanks, of which this was the first, embodied numerous improvements in design over earlier U. S. tanks. The tanks were not, however, deemed satisfactory due to the general arrangement and the inability of the suspension to cope adequately with vibration and shocks. This particular tank was later converted into a cargo carrier and, from that, into a motorized reel for artillery.

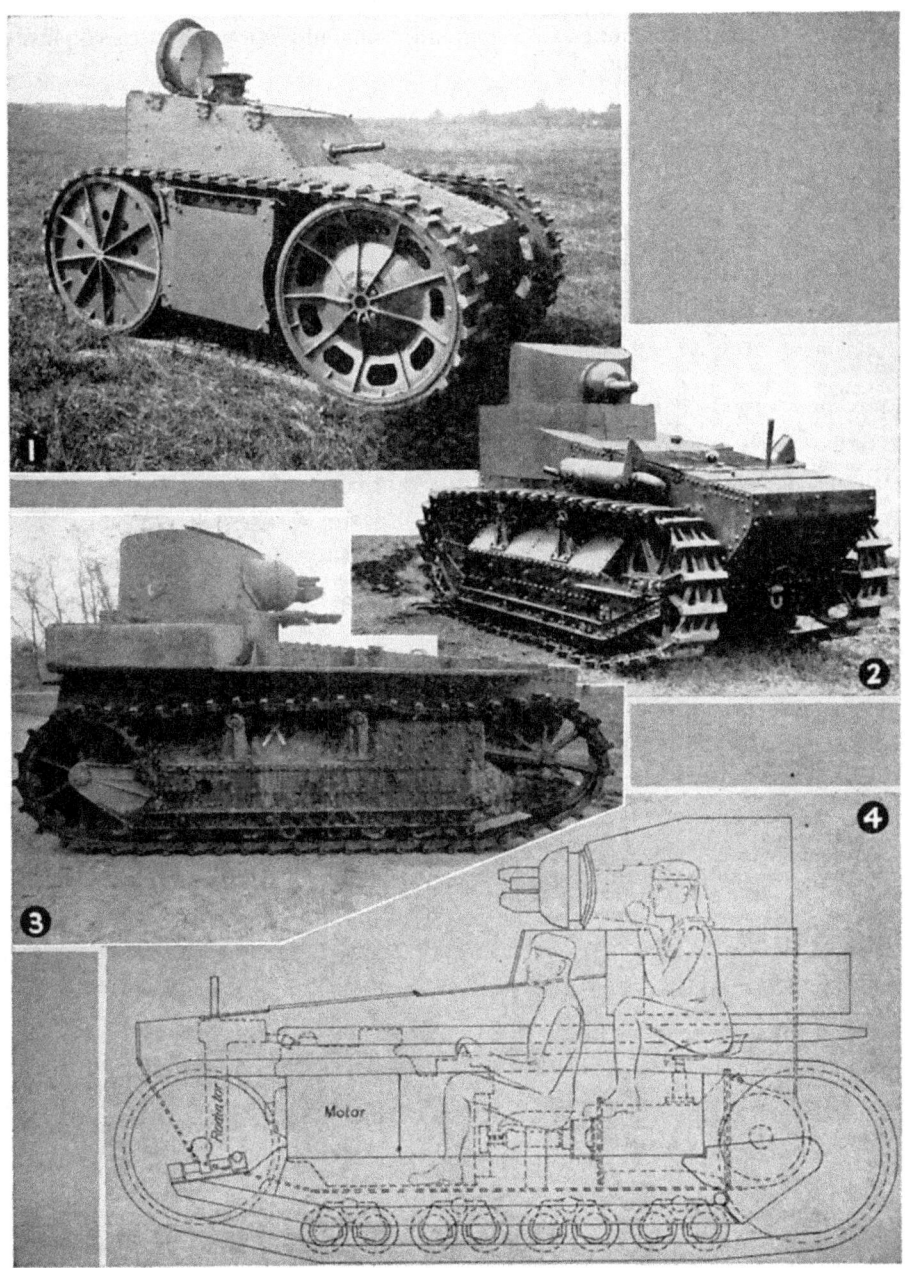

Plate 38.
1. **One Man Tank, Experimental. Track Development Chassis T 1.**
(Photo by Ordnance Dept., U.S.A.)
2. **Light Tank, T 1.**
(Photo by Ordnance Dept., U.S.A.)
3. **Light Tank, T 1-E 1.**
(Photo by Ordnance Dept., U.S.A.)
4. **Light Tank, T 1-E 1.**

Light Tank, T 1-E 1.

Produced in 1928 by James Cunningham, Sons and Company. (Ordnance design). Total production, 4.

Dimensions: Length 12 ft. 8½ in.; width 5 ft. 10½ in.; height 7 ft. 1⅝ in
Weight: 7.5 tons.
Special features: This type differed only slightly from the T 1. The projection of the body beyond the front extremities of the tracks was eliminated. The fuel tanks were placed above the tracks. The air circulation was altered somewhat. This type has a maximum speed of 18 mph. The skeleton type of track used on these tanks picked up barbed wire, thereby stalling the tank.

Light Tank, T 1-E 2.

Produced in 1929 by James Cunningham, Sons and Company. (Ordnance design). Total production, 1.

Crew: 2.
Armament: One 37 mm (1.46 in.) semiautomatic gun and one cal. .30 machine gun in one mount.
Armor: 0.25 in. to 0.625 in.
Maximum speed: 16 mph.
Suspension: Same type as T 1.
Tracks: Same type as T 1, width 13 in., pitch 6.75 in.
General arrangement: Same as T 1.
Dimensions: Length 12 ft. 8½ in.; width 6 ft. 2 in.; height 7 ft. 6½ in.
Weight: 8.9 tons.
Engine: Cunningham, 8 cylinder, V type, 132 HP, forced water cooling.
Horsepower per ton: 14.8.
Transmission: Same type as T 1 but high gear ratio reduced and intermediate increased.
Obstacle ability: Trench 6 ft.; stream 20 in.; slope 27 degrees; vertical wall 22 in.; tree 10 in.
Fuel distance: 75 miles.
Special features: Minor defects existing in T 1-E 1 were corrected, the armor was made heavier, and the engine horsepower was increased.

Light Tank, T 1-E 3.

This was a modification of one of the T 1-E 1 tanks effected by the Ordnance Department in 1930. Total production, 1.

Crew: 2.
Armament: Same as T 1-E 2.
Armor: 0.25 in. to 0.625 in.
Maximum speed: 21.9 mph.
Suspension: Rollers, bogies, vertical coil springs within hydraulic shock absorbers. The shock absorbing action works only in one direction.
Tracks: Same as T 1-E 1.
General arrangement: Same as T 1-E 1.
Dimensions: Same as T 1-E 1.
Weight: 8.5 tons.
Engine: Same as T 1-E 2.
Horsepower per ton: 15.6.
Transmission: Same as T 1-E 1.
Obstacle ability: Trench 6 ft.; stream 20 in.; slope 35 degrees.
Fuel distance: 75 miles. **Fuel capacity:** 50 gallons.
Special features: The improved suspension materially improved the riding qualities. This was the only important modification. A supplementary floor was installed in the gunner's compartment flush with the top of the transmission.

Light Tank, T 1-E 4.

Produced in 1932 by Ordnance Department. Total production, 1.

Crew: 4.
Armament: One 37 mm (1.46 in.) semiautomatic gun M 1924 and one caliber .30 machine gun in one mount.
Armor: 0.25 in. to 0.625 in.
Maximum speed: 20 mph.
Suspension: Pivoted semieliptic springs and bogies; rollers with dual solid rubber tires.
Tracks: Forged steel plates, mostly solid, giving surface for rollers, but open at edges for double sprocket teeth; no grousers; width 13¼ in., pitch 4 in.
General arrangement: Final drive, transmission and driver in front; gunner in center; engine in rear.
Dimensions: Length 15 ft. 5 in.; width 7 ft. 2¾ in.; height 6 ft. 6¾ in.
Weight: 8.6 tons.
Engine: Cunningham, 8 cylinder, V type. 140 HP, forced water cooling.
Horsepower per ton: 16.3.

UNITED STATES TANKS 165

Plate 39.
1. Light Tank, T 1-E 2.
(Photo by Ordnance Dept., U.S.A.)
2. Light Tank, T 1-E 3.
(Photo by Ordnance Dept., U.S.A.)
3. Light Tank, T 1-E 4.
(Photo by Ordnance Dept., U.S.A.)

Transmission: Cotta type sliding gear, 3 speeds forward, 1 reverse.
Obstacle ability: Trench 7 ft.; slope 30 degrees.
Fuel capacity: 50 gallons.
Special features: This tank is a modification of the T1-E1 design. The engine, power train, and final drive are reversed, thus placing the engine in rear and the final drive in front. There are adjustable air louvers in the fire screen.

Christie, M 1919.

Produced in 1919 by Front Drive Motor Company. Total production, 1.

Crew: 3.
Armament: One 6 pounder (57 mm-2.24 in.) gun in main turret and one cal. .30 machine gun in upper turret.
Armor: 0.25 in. to 1.0 in.
Suspension: Rubber tired wheels, center wheels only sprung.
Tracks: Removable; flat steel plates 15 in. wide, pitch 9¾ in.
General arrangement: Driver in front; gunners in center; engine and final drive in rear.
Dimensions: Length 18 ft. 2 in.; width 8 ft. 6 in.; height 9 ft.
Weight: 13.5 tons.

Plate 40.
Light Tank, T 1-E 4. (Photo by Ordnance Dept., U.S.A.)

Engine: Christie, 6 cylinder, 120 HP. forced water cooling.
Horsepower per ton: 8.9.
Transmission: Sliding gear. 4 speeds forward, 4 reverse.
Fuel distance: 35 miles on tracks, 75 miles on wheels. **Fuel capacity:** 59 gallons.
Special features: Tracks carried above the wheels when driving on wheels. Small pyramidal lugs on inside of plates for driving and guiding tracks. Center wheels raised when running on wheels. 15 minutes to change from wheels to tracks or vice versa.

Christie, M 1921.

Rebuilt from the 1919 tank in 1921 by the Front Drive Motor Company. Total production, 1.

Crew: 4.
Armament: One 6 pounder (57 mm-2.24 in.) gun in front and one cal. .30 machine gun on each side.
Armor: 0.25 in. to 0.75 in.
Maximum speed: 7 mph on tracks; 14 mph on wheels.
Suspension: Wheels with double rubber tires; front wheels sprung with coil springs, center wheels on pivoted bogies.
Tracks: Removable; flat steel plates with grousers and driving lugs; width 15 in.; pitch 9¾ in.
General arrangement: Gunners in front; commander and driver in center (driver at left); engine and final drive in rear.
Dimensions: Length 18 ft. 2 in.; width 8 ft. 6 in.; height 7 ft. 1 in.
Weight: 14 tons.
Engine: Christie, 6 cylinder, 120 HP, forced water cooling. Mounted laterally.
Horsepower per ton: 8.6.
Transmission: Sliding gear, 4 speeds forward, 4 reverse; a complete transmission on each side.

Plate 41.

1. **Christie, Model 1919.**
(Photo by Ordnance Dept., U.S.A.)
2. **Christie, Model 1921.**
(Photo by Ordnance Dept., U.S.A.)
3. **Christie Chassis, Model 1928.**
(International News Photos, Inc.)

Obstacle ability: Trench 7½ ft.; slope 40 degrees.
Fuel distance: 60 miles on tracks, 100 miles on wheels. **Fuel capacity:** 67 gallons.
Special features: Drive wheels unsprung; crew compartment small; maneuverability poor.

Christie Chassis, M 1928.

Produced in 1928 by U. S. Wheel Track Layer Corporation. Total production, 1.

Crew: Undetermined.
Armament: Undetermined.
Armor: 0.5 in.
Maximum speed: 42.5 mph on tracks, 70 mph on wheels.
Suspension: Essentially the same as in the later 1931 tank.
Tracks: Forged steel plates with U-shaped driving lugs attached to each alternate plate, width 10 in., pitch 10 in.
General arrangement: Crew compartment in front; driver in left rear part of crew compartment; engine and final drive in rear.
Dimensions: Length 17 ft.; width 7 ft.; height 6 ft.
Weight: 8.6 tons.
Engine: Liberty, 12 cylinder, V type. 338 HP, forced water cooling.
Horsepower per ton: 39.3.
Transmission: Sliding gear, 4 speeds forward, 1 reverse.
Obstacle ability: Trench 7 ft.; stream 5 ft.; slope 37 degrees; vertical wall 28 in.
Fuel distance: 75 miles on tracks, 115 miles on wheels. **Fuel capacity:** 35 gallons.
Special features: This was the forerunner of the 1931 tank and although lighter in weight was similar as a vehicle to the later tank in all essential features.

Christie, M 1931.

Convertible Medium Tank, T3. Produced in 1931 by U. S. Wheel Track Layer Corporation. Total production, 7.

Crew: 2.
Armament: One 37 mm (1.46 in.) gun and one cal. .30 machine gun in one mount.
Armor: 0.25 in. to 0.625 in., exclusive of 0.188 in. inner hull of nickel steel.
Maximum speeds: 40 mph on tracks, 70 mph on wheels.
Suspension: Four large weight-bearing wheels on each side, each with dual rubber tires and mounted upon a pivoted arm upon which bears a long adjustable coil spring. The liberal compression amplitude gives each of these wheels an independent maximum vertical movement of about 14 inches.
Tracks: Forged steel plates, each alternate plate having a driving lug integral therewith; width 10¼ in., pitch 10 in. When on wheels the tracks are carried on shelves at the sides of the tank.
General arrangement: Driver in front; gunner in center; engine and final drive in rear.
Dimensions: Length 17 ft. 10 in.; width 7 ft. 4 in.; height 7 ft. 3 in. (The height varies slightly with the adjustment of the suspension springs.)
Weight: 10½ tons.
Engine: Liberty, 12 cylinder, V-type, 338 HP, forced water cooling.
Horsepower per ton: 32.2.
Transmission: Sliding gear, 4 speeds forward, 1 reverse.
Obstacle ability: Trench 7 ft.; stream 3½ ft.; slope 35 degrees; tree 8 in.
Fuel distance: On tracks 170 miles, on wheels 250 miles. **Fuel capacity:** 89 gallons.
Special features: Six of these seven tanks have chain drive from sprocket to rear road wheel when on wheels; the other has gear drive instead. Thirty minutes is required to change from tracks to wheels or vice versa. Two additional chassis of this type were purchased by Russia. In February, 1931, one of these U. S. tanks made a cross-country run of 141 miles at an average speed of 21.1 mph with no mechanical difficulties. Due to the time required for the operation, the change from wheels to tracks should be made before there is any prospect of coming under hostile fire. Since this type of tank is durable and very fast on tracks, the early change from wheels to tracks is not objectionable.

Christie Light Tank, M 1932.

Produced in 1932 by U. S. Wheel Track Layer Corporation. Total production, 1.

Crew: 3.
Armament: Undetermined. (Can carry one cannon and one or more machine guns.)
Armor: 0.375 to 0.5 in. (Thicker armor may be installed.)
Maximum speed: 120 mph on wheels; 60 mph on tracks.
Suspension: Similar to 1931 model but with a maximum vertical movement of 24 inches; wheels of duraluminum with pneumatic tires.
Tracks: Steel plates; width 11 in.; pitch 7 in.; track pins ⅜ in. diameter.
General arrangement: Cannon in front; crew in front center; engine and final drive in rear.

Dimensions: Length about 22 ft.; width about 7 ft.; height about 5 ft. 8 in. (without turret.)
Weight: About 5 tons.
Engine: Hispano-Suiza, V-type. 12 cylinder, 750 HP, forced water cooling.
Horsepower per ton: 150.

Plate 42.
1. Christie, Model 1931. Mr. Christie in tank.
(International News Photos, Inc.)
2. Christie, Model 1931. Mr. Christie on top.
(Photo by *Elizabeth Daily Journal*, Elizabeth, N. J.)

Mr. J. Walter Christie, inventor of Christie tanks, can be seen in Figs. 1 and 2, Plate 42. Sixty-seven years old in 1932, Mr. Christie has spent his life in the automotive industry, his various positions including apprentice machinist, automobile racing driver, and designing engineer. He has made numerous successful automotive inventions, but, in recent years, has concentrated his efforts and resources upon the development of fighting tanks.

Plate 43.
1. Christie, Model 1931. Tracks partly removed.
2. Christie, Model 1931.
(Photo by Ordnance Dept., U.S.A.)
3. Christie, Model 1931. Interior.
(Photo by Ordnance Dept., U.S.A.)

Transmission: Sliding gear; 3 speeds forward, 1 reverse; has a power take-off for operating proposed flying propeller.
Obstacle ability: Can jump across a 12-foot trench.
Fuel capacity: 89 gallons.
Special features: Very light construction throughout. In this design it was contemplated that this vehicle could be carried up by a special airplane carrier and later released close to the ground. Cross-flow horizontal radiators above the engine.

Combat Car, T 2.

Formerly Convertible Armored Car T 5. Produced in 1931 by Ordnance Department. Total production, 1.

Crew: 3 or 4.
Armament: One cal. .50 machine gun and one cal. .30 machine gun in one mount in turret; one cal. .30 machine gun for use at front of crew compartment or for AA fire, one Thompson submachine gun; all guns air cooled.
Armor: 0.25 in. to 0.50 in.

Plate 44.

1. **Christie Light Tank, Model 1932.**
(Photo by Fieldman, Elizabeth, N. J.)
2. **Combat Car, T 2.**
(Photo by Ordnance Dept., U.S.A.)
3. **Christie Light Tank, Model 1932.**
(Photo by Fieldman, Elizabeth, N. J.)

Maximum speed: 30 mph on wheels; 20 mph on tracks.
Suspension: Springs and rubber tired wheels.
Tracks: When on wheels the tracks are carried on shelves at the sides of the car.
General arrangement: Driver and one gunner in front; gunner at center in turret; engine and final drive in rear.
Dimensions: Length 14 ft. 9 in.; width 6 ft. 3 in.; height 7 ft. 5 in.
Weight: 8½ tons
Engine: Continental, radial, 7 cylinder, 165 HP, air cooled
Horsepower per ton: 19.4.
Transmission: Sliding gear, 4 speeds forward, 1 reverse.
Obstacle ability: Slope 35 degrees.
Fuel distance: 125 miles on wheels, 100 miles on tracks. **Fuel capacity:** 50 gallons.
Special features: Operates on wheels on the road and on tracks across country. The steering brakes are operated through an hydraulic booster. An improved model of this car, T-3, is under consideration.

CHAPTER XVII.

TANKS POSSESSED BY THE VARIOUS COUNTRIES.

Although a considerable amount of experimental work in tank construction is conducted in the smaller countries of the world, these countries rely chiefly, for the tanks that they require, upon the standard types purchasable from the larger countries. The following list, except as to a considerable number of items not available for publication, shows the approximate numbers of the different types possessed by the various countries.

Country	Number	Type
Albania	2	Fiat, Type 3000
Australia	4	Vickers, Mark II
Belgium	75	Renault
Bolivia	....	
Brazil	....	
Chile	....	
China	15	Renault
	20	Vickers or Carden Loyd
Czechoslovakia	11	Renault
	1	Light tank, Wheel and Track, 4.3 tons
	50	K H 50
Denmark	1	Fiat, Type 3000
		Renault (modified)
Estonia	10	Renault.
	4	Mark V
Finland	32	Renault
France	25	Carden Loyd
	4	Chenilette, St. Chamond
	2000-3000	Renault
	about 100	Renault with Kegresse tracks
	....	Renault, N C M 27
	1	Renault, M 1923
		Mark V Star (obsolete)
		Char, Moyen
		Char, 2 C
		Char, 3 C
Great Britain	300	Carden Loyd
	10	Mark 1 A and new light tanks
	10	Vickers Armstrongs, 6 ton
	100	Vickers, Medium
		Sixteen Ton
	2	Independent
Greece	6	Fiat, Type 3000
Holland	1	Renault
Hungary		
Ireland	1	Vickers, Medium C
	1	Mark V
Italy	25	Carden Loyd
	100	Fiat, Type 3000
	2	K H 50
	4	Pavesi
	2	Ansaldo
	10	Fiat, Type 2000
Japan	1	Chenilette, St. Chamond
		Renault
		Renault, N C M27
	1	K H 50
	1	Experimental light tank
	1	Experimental heavy tank
	5	Medium, Mark A
	2	Vickers, Mark I
		Vickers, Medium C
		Mark V
Yugoslavia	50	Renault
	50	Renault, N C M27
Latvia	1	Fiat, Type 3000
	7	Renault
	2	Medium B
	10	Mark V
Lithuania	16	Renault
	2	Fiat, Type 3000
Persia	2	Renault
Poland	10	Carden Loyd
	100	Renault

	70	Renault, N C M27
	25	Char 2C or Char 3C
	60	Cardosowitz
	25	Mark V
	3	Mark V, Star
	5	A 7 V
Rumania	75	Renault
Russia	20	Carden Loyd, Mark VI
	1	Fiat, Type 3000
	20	Renault
	116	Russian Renault
	70	Russian light tanks
	40	Vickers, Mark II
	2	Christie, M 1931
	8	Medium, Mark A
	25	Mark V
	8	80-Ton
Spain	20	Renault
	5	Trubia
	5	Schneider
Sweden	1	Renault
	12	M-21
Switzerland	2	Renault
United States	3	Ford
	940	Six Ton
	7	Six Ton, with Franklin engine modification
	5	T 1 series, light
	3	T 1 series, medium
	1	Medium, T 2
	7	Christie, M1931
	100	Mark VIII

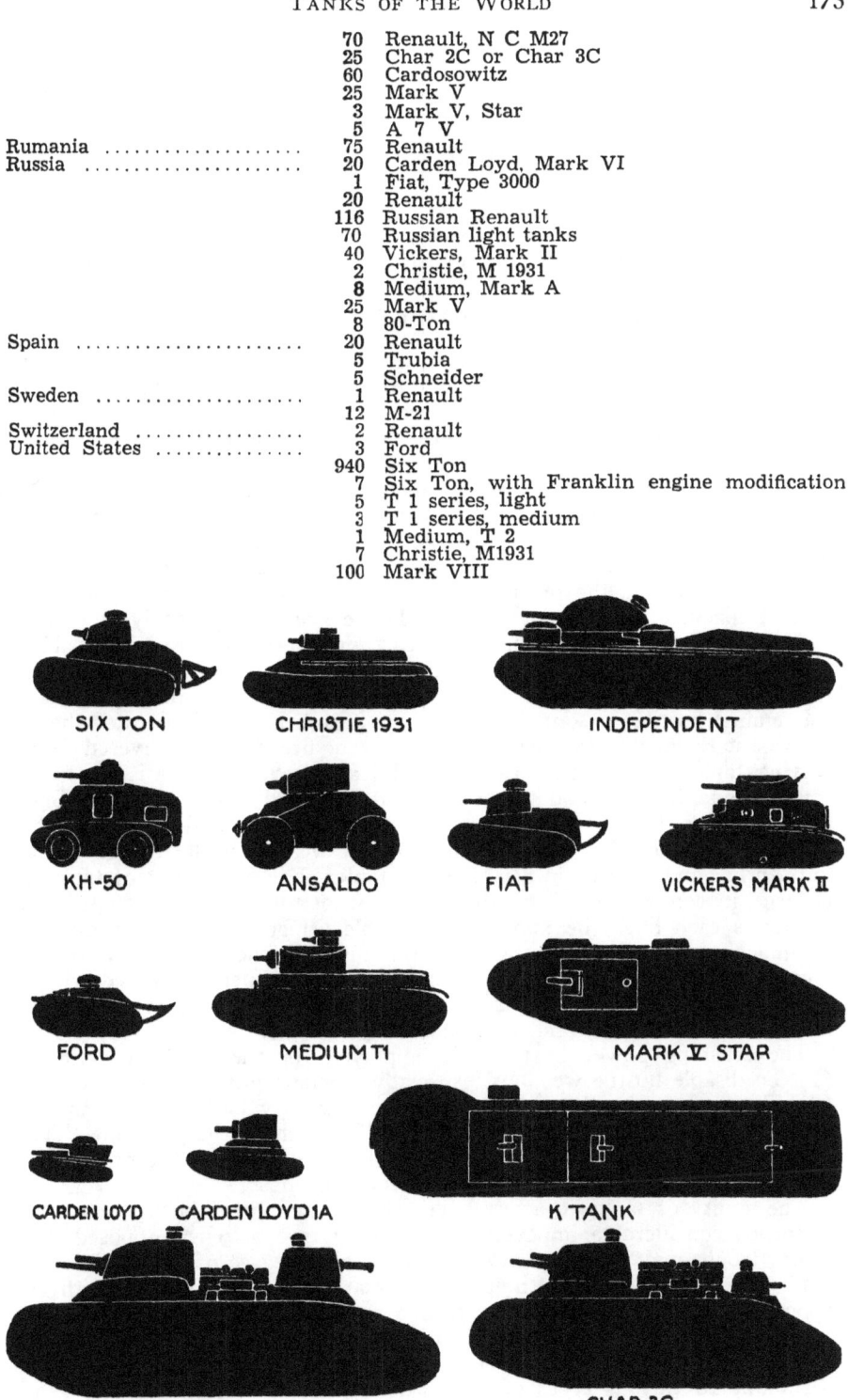

Plate 45.
Shapes and Relative Sizes of Various Tanks in Outline.

CHAPTER XVIII.

POWERS AND LIMITATIONS OF TANKS.
CONSIDERATIONS GOVERNING THEIR EMPLOYMENT.

POWERS.

The combat power of a tank includes its fire power, its crushing power, and its adverse effect upon the morale of opposing troops. The ability of a tank to employ its combat power effectively is dependent upon its speed, its mechanical reliability and endurance, its armor protection, its ability to negotiate difficult ground, the training of its crew, and several lesser factors.

The fire power of a tank is dependent upon the type and design of the tank. In many old and in nearly all modern types, the fire power includes one or more antitank weapons and one or more small caliber machine guns. The antitank guns are, for the most part, cannons and, in tanks of the World War period, the prime purpose of such cannons was the neutralization of machine gun emplacements. To be prepared properly for general combat use, a tank should have at least one antitank weapon capable of penetrating, at reasonable ranges, the armor of probable types of hostile tanks. All tanks should be equipped with one or more machine guns for use against hostile personnel. Fire from a tank is most accurate when the tank is stationary. However, fairly accurate fire can be delivered from a moving tank provided the crew is well trained, the ground is not unduly rough, the tank has a good suspension, and the speed is not excessive. Should a stop for the delivery of fire be justified, the stop should be of very brief duration, as otherwise some hostile antitank weapon within range is likely to open fire and succeed in disabling the tank. The fire power of tank units is large in comparison with other types of organizations. Assuming that the tanks each have the minimum desirable armament of one cannon and one machine gun, a company with the minimum number of tanks would have 9 cannons and 9 machine guns. A heavy tank regiment equipped with our old heavy tank would have 162 cannons and 405 machine guns, exclusive of its spare machine guns and its replacement sections.

The crushing power of tanks is employed for the following purposes: (1) To disable hostile weapons, especially machine guns and antitank guns. (2) To crush paths through obstacles, particularly barbed wire, and thus facilitate the advance of foot troops. (3) To crush hostile personnel. This is most practicable when lightly covered emplacements are located or when fast tanks come suddenly upon hostile foot troops in mass.

The ability of tanks to affect adversely the morale of hostile troops is a factor of considerable importance. Infantry troops, when opposed to infantry troops may be expected to display a high degree of courage, for they think that they have something approximating an even chance with their opponents. But the best of foot troops opposed to tanks are likely to have an utterly helpless feeling that will disintegrate morale and organization.

For information as to the speed, armor, ability to negotiate obstacles, etc., pertaining to various types of tanks, the reader is referred to the detailed descriptions in previous parts of this book.

In any comprehensive study of the powers of tanks from the tactical viewpoint, consideration must be given to numbers and method of employment. One hundred tanks have a tactical value materially greater than ten times the tactical value of ten tanks. If tank tactics were such that normally tanks were sent individually against hostile antitank guns, the combat value of tanks would be trivial. But, by their sudden use in mass by surprise in an intelligently selected locality, they can, through numbers and speed, overwhelm the local resistance and, through their own success, do much to insure the attainment of a strategic objective of great importance. It is this decisiveness of their action in numbers that so greatly augments their value. In this connection, it is perhaps worth while to consider the plight of the antitank gunner. Suppose he is installed in a very good location with a field of fire to the front for a distance of 500 yards. If a lone hostile tank appears out of the woods or over the crest, his morale may well be high for he will have an excellent chance to destroy his enemy's machine. But suppost, instead of one, a swarm of thirty or more appears headed in his direction at an average speed of 15 miles per hour or more. What will be the thoughts of the gunner then? In a minute or less he will find himself surrounded by those hostile monsters. If he opens fire at once and has good luck he may stop one, two, or three of the tanks. But such fire will most likely draw attention to his position, in which case he will himself be destroyed within two or three minutes. Confronted with such a situation, will not a considerable number of gunners hold their fire in the hope that they may thus avoid detection and so prolong their existence for a time? If the gunners do not sacrifice themselves, the task of the tanks is made that much easier; if they do, the tank casualties that they are able to cause will be inadequate, the majority of the tanks will get through, and the purpose of the tank attack will be accomplished. Moreover, we must not assume that all antitank gunners will always be ready and on the alert. Often they will lose many precious seconds. Nor can we assume that such a field of fire will always be usable. Other agencies will also frequently cooperate with assaulting tanks by putting down smoke upon known or suspected locations of hostile antitank weapons thus diminishing the advantages of such defensive agencies.

In dwelling upon the situation that occurs when tanks confront antitank guns, let us not overlook the great value of tanks in destroying small caliber machine guns and thus allowing friendly foot troops to advance without suffering terrific casualties. While a few such guns may mow down advancing foot troops in prohibitive numbers, they themselves are in turn easy prey to tanks. Well constructed tanks are practically immune to their fire. It is a simple and easy matter for a tank to rush toward such a gun, destroy or disperse its crew by fire, and destroy the weapon by crushing.

Although it is generally admitted that tanks possess offensive power to a formidable degree, the statement is often encountered that they have little or no holding power. This is in turn sometimes misconstrued so as to create the impression that tanks have little or no defensive power. It is perfectly true that the use of tanks as stationary blockhouses is a grossly unsuitable procedure. If they were habitually so used, most of them would normally soon be destroyed by hostile shell fire. But mobile defensive power is something else and it is possessed by tanks to a high degree. The purpose of defense is to prevent the advance of the enemy beyond a certain point or line. What does it matter by what method that purpose is accomplished? What does it matter whether the advance of the enemy is blocked by the occupation of a defensive position or whether he is vigorously hit by offensive action each time he advances beyond predetermined limits? Tanks are not used in the former method, but, when the purpose is of sufficient importance, they may be used in accordance with the latter, and to excellent effect. In some cases an enemy may go around or avoid a defensive position, but he will find it much more difficult to go around or avoid a highly mobile offensive force of suitable strength. In a large general defense, more or less localized offensive action is necessary if collapse and defeat are to be avoided. The availability of an efficient means for insuring the success of such an offensive effort (counterattack) may be of most vital importance.

LIMITATIONS.

The tank, being a complicated machine that must undergo long-continued and rough usage, is subject to wear and breakdown. It requires servicing at fairly frequent intervals. From time to time it requires overhaul. It is, therefore, to be expected that, at times, the number of tanks in action or ready for action will fall short of plans and desires. After a strenuous day of tank activity, much time may be required for servicing and repairs before a satisfactory number of tanks is again ready for combat. If we ignore maintenance requirements and attempt to operate tanks for an unduly long period without respite, we merely kill the goose that lays the golden egg. The relation of maintenance time to operation time varies greatly with tanks of different periods. This is to be expected since the same is true in the case of the automobile. Let us compare the automobile of 1901 with that made thirty years later. In the old days it was exceptional if one could travel any considerable distance without mechanical breakdown. Today one may expect to go thousands of miles without giving the automobile any attention other than to supply fuel, water and lubricants. The tanks of the World War period were crudely built in many respects. The present-day tank is very greatly improved, but in its development it still lags much behind the present-day automobile in mechanical endurance. If tank organizations are equipped with tanks of the World War type the ratio of maintenance personnel must be materially greater than if the units are equipped with modern tanks. To summarize, the necessary maintenance requirements interfere materially with operations and this is true to a much greater degree with the older types of tanks.

Tanks are, in large measure, deaf, dumb and blind. The noise of tank operation interferes greatly with the hearing of the crew, and this condition is, of course, much aggravated in battle by the fire of weapons. Up to the present time it has been found very difficult to provide the tank with means to enable it to communicate satisfactorily with other tanks and other agencies. The tank is therefore in large measure dumb. The vision of a crew within a tank is curtailed to a material degree due to the necessity of the observation being effected through slits in the armor. This interference is aggravated when the tank is travelling over rough ground. In most cases overhead observation by the crew is completely blocked. This deaf, dumb and blind condition, which still greatly handicaps direction and control after the launching of the attack, will probably be much alleviated by future developments.

The operations of tanks are limited and delayed by obstacles, the most serious of which are water courses. For details regarding obstacles and the ability of tanks to overcome them, see Chapters II to XVI and Chapter XXVI.

A tank in motion presents a conspicuous target on the battlefield. However, when concealment is not available, if it moves speedily, it is a difficult target for artillery to hit, even by direct aiming. This is particularly true of the ordinary field pieces of the World War variety. The tank has the particular advantages that it is not likely to be seriously injured except by a direct hit, that a single tank is a small artillery target, that individual tanks themselves occupy only a very small part of the ground covered by a group as a whole, and that the speed of tanks makes them difficult for the artillery to keep up with by any kind of fire other than direct. It appears, therefore, that fast tanks are not ordinarily in great danger from artillery fire, especially when that fire is in the form of a barrage or observed concentration. Tanks generally are supposed to be and should be invulnerable to small arms fire (such as cal. .30 A.P.). Some heavily armored tanks are invulnerable to weapons much more formidable. Tanks, like foot soldiers, may often take advantage of defilade and thus save themselves from being hit. Like warships, they obtain some advantage from camouflage painting. Camouflage is more successful, however, when the tank is not in motion.

Another limitation on the employment of tanks, in addition to the inherent weaknesses of the vehicles themselves, is the fatigue and strain upon the crew which may cause an otherwise efficient vehicle to operate to much poorer advantage.

Tanks in assault often require supporting fire of one kind or another if they are to avoid an undue number of casualties and render satisfactory service.

PRINCIPAL CONSIDERATIONS GOVERNING THE EMPLOYMENT OF TANKS.

Role of the Tank, in General.

The obvious and primary purpose of the armored, motorized, fighting vehicle that we call the tank is to facilitate offensive action. This offensive action may be part of a general attack or part of a general *defense*. If wars could be won by engaging only in passive defense, there might be no justification for building tanks. A gun, protected by earth and concealed by vegetation, is a much more difficult target for the enemy to discover and attack than a similar gun carried about by a conspicuous hulk of steel. In passive defense, the stationary, concealed weapon has advantages. But wars cannot thus be won. Though we may assume the passive defensive temporarily to excellent purpose, there is sure to come a time when, if we are to win, we must move upon the enemy and strike him. When that time comes and we must advance, is it better to expose to hostile fire our bare skins, or a half inch or more of armor plate? At that time, is it better to close with the enemy at the rate of the foot soldier on the battlefield, or at a rate of 10, 20 or more miles per hour? The answers are obvious. The armor of a battleship will not stop all hostile projectiles, but it will stop many of them. So it is with the tank.

The development of the tank toward perfection in the desirable qualities of protection, high mobility, and mechanical reliability, is transforming it into an offensive necessity. The intense interest shown by almost every country in the world indicates that the truth of this statement is generally recognized.

Tanks, properly employed, will accomplish two important purposes. They will augment the rapidity and success of the attack, and they will greatly lessen the number of casualties in the ranks of the side that uses them. The tank, like a good medicine, is valuable for its proper function but is not a panacea. It will not win the war alone. It will not supersede the artillery, infantry or cavalry. However, it is rather obvious that, if tanks and armored cars are available in very liberal numbers, the requirement for horsemen will be somewhat lessened.

Specific Uses for Tanks.

The following possible uses for tanks should be noted:

(1) To assist, by close and continuous cooperation, infantry bat-

talions in the attack. This is called the **accompanying role** (U. S. terminology).

(2) To exploit a success attained in the earlier phases of an attack or to counterattack in defense. This may be considered the **reserve role**. It may include participation in pursuit.

(3) To break the way in mass for a main effort against a hostile flank or for the main effort of a penetration. This is called the **leading role** (U. S. terminology).

(4) To participate in the work of security forces and detachments (advance, flank and rear guards).

(5) To assist cavalry units in the accomplishment of their missions.

(6) Within a relatively independent organization, known as an armored force or mechanized force, to: (a) Attack the hostile flank or rear (large reserves, etc) in coordination with an attack launched by the main force. (b) Participate in the exploitation of a success attained by the main force. (c) Delay and harass hostile columns (such as approaching reenforcements). (d) Meet and defeat a hostile armored force. (e) Seize an advance or flank point of strategic importance in order to deny temporarily its occupation by the enemy or to defend it pending the arrival of other troops. (f) Execute reconnaissance in force and counterreconnaissance. (g) Make raids for such purposes as interrupting communications, destroying bridges, munitions, etc.

(7) To participate in landing operations.

Principles of Tank Employment.

If excellent tanks were available in huge numbers, it would be appropriate and desirable to apportion them to many of the uses just enumerated. But, for many reasons, we cannot expect to have tanks in the desired numbers. To establish appropriate priorities, therefore, the various needs must be studied comparatively. The basic principle that we should concentrate our strength at the time and place where a decision is likely to be obtained, applies to the use of tanks at least as fully as it does to other arms.

The supreme purpose in war is to win. The most decisive means is a successful attack made in pursuance of a sensibly selected strategic purpose. Insofar as generalship is concerned, the success of an attack depends, more than anything else, upon the adequacy of measures taken to insure a far-reaching success in the area of the main effort. The modern tank is the assistant de luxe for insuring such success. It is rather clear from these considerations that, if tanks are frittered away on projects of secondary importance, the strength for the main effort of the important attack may be weakened sufficiently to change what might have been a decisive success into a wasted effort. Such a dissipation of these tactically valuable vehicles is a most serious blunder and the one most likely to be made in connection with the employment of tanks.

To secure the most effective results, tanks must be employed, in general, in accordance with the following:

(1) Simultaneously, in large numbers.

(2) Massed adequately in a decisive locality (but not precluding their use at other places).

(3) On a sufficiently wide front.

(4) In depth sufficient to have the requisite driving power. (This implies suitable reserves.)

(5) On terrain selected for its suitability for tank operations.

(6) With such secrecy and mobility as will effect surprise to the maximum practicable degree.

(7) With due regard for the strategic and tactical requirements of the particular situation.

Tanks should always be guarded against unnecessary loss in every practicable way. The principal safety precautions to be observed in their employment, are as follows:

(1) The time and place selected should be such that the enemy's antitank defense will not, in that area and at that time, be at highest strength and efficiency.

(2) The attack should be launched so suddenly and swiftly that the enemy will have no adequate time in which to readjust his defense.

(3) Supporting smoke should be put down by suitable agencies so as to protect the tanks at those particular points where it is expected that they will be in the greatest danger from hostile antitank guns.

(4) In driving through especially dangerous zones tanks should be massed more densely than normally and move at a relatively high speed. By their numbers, power, and speed, they utterly overwhelm the local defense. By the hopelessness of the defensive task and of their situation, the morale of the defenders will be shattered to a considerable degree and, in many instances, they will not exert even the defensive power of which they are capable.

(5) If the enemy is in a compact defensive position but with only one flank protected by an impassable obstacle and he is liberally equipped for defense against tanks, he may, by threat of complete envelopment, be forced to extend his line before our tanks are used. Having thus caused him to attenuate his defense, the main element of the tanks may be plunged through at a selected place with but little losses.

Through the application of the foregoing principles, and assisted by the supporting fires of other branches, tanks are enabled to lead and escort other troops in a deep and expeditious penetration of the hostile position with a minimum number of casualties among the escorted troops.

Under appropriate conditions, the use of tanks, or an armored force consisting largely of tanks, independent of other types of troops, may be fully justified.

CHAPTER XIX.
TANK TYPES AND USES.

In considering the uses planned for various types of tanks by the nations of the world, interest centers in Great Britain, France, and the United States. Russia ranks second in progress in equipping her army with modern tanks, and should be included in the list but for the fact that little information is available regarding her developments and plans. Great Britain has been the leader since the World War in this development, both mechanically and tactically. France has modernized, in a considerable measure, her wartime tanks, and has built some that are completely new. She attaches very high importance to the value of tanks, and maintains more of them ready for service than any other nation. In tactical employment, it is believed that she has made only moderate progress, due in a measure to the economic impossibility of scrapping the many tanks already in existence and building new ones to take their places. The United States, in proportion to her wealth and prominence as a world power, has done little along the line of tank development. The nations not mentioned that have tank organizations have not, so far as known, carried on any developments of unusual interest.

War Types.

The tanks used in the World War have been described and illustrated in earlier chapters. Except for the British Whippet (Medium A), and the German A-7-V (the latter of which was not well designed for cross-country travel), they were all decidedly slow. During that time, the accompanying role was the only mission given any extensive consideration. On the relatively few occasions when tanks became separated from infantry troops and advanced deeply and alone into the hostile position, such procedure was considered accidental and undesirable. In the early instances of tank employment, they were used in a piecemeal and scattered fashion. Late in the war, as a result of early experience, they were used in greater numbers, simultaneously, and better concentrated, but still for the accompanying role only. The British, in building long and heavy tanks, sought ability to cross wide trenches, a high degree of crushing power, and a liberal amount of armament per tank. the French in adopting the light Renault type, indicated their desire for an inconspicuous target and a moderate cost per tank .It was their belief that the Renault with its tailpiece, had sufficient trench-crossing ability.

Very Small Tanks.

The one-man tank idea has been experimented with in England, France and the United States. It now seems to have no friends or supporters. The principal difficulty is that one man cannot perform, simultaneously, in a satisfactory way, the duties of driver, navigator, gunner, and observer. Then too, a well-constructed vehicle for one man would not be much cheaper than a similar and much more efficient vehicle for two men.

The British seem more opposed than anyone else to having the advance of tanks delayed for the purpose of keeping them near to advancing foot troops: hence the Carden-Loyd machine gun carrier, which is in reality a tank, although no longer so considered in England. Although provision is made in this tank for firing the weapon from the vehicle, the primary purpose is to take machine guns and ammunition forward promptly and rapidly after the tank advance, to positions where the weapons can be fired normally from the ground.

Light Tanks.

Within the field of the normal light tank (a vehicle that has a crew of 2 or 3 men and uses its armament without dismounting it), France has developed a number of machines, the most promising of which seems to be the NC M27 and modifications thereof. The NC M27 is a modification of the old Renault, but it has a greatly improved suspension and thicker armor. So far as known, France still plans to use her light tanks for the accompanying role only.

Great Britain has developed new fast light tanks, including the Vickers Armstrongs 6 ton and the Carden Loyd Mark 1A. She is planning to use her fast light tanks primarily, if not exclusively, in combination with fast medium tanks as part of a large armored force. There is little, if anything, to indicate that Great Britain attaches much importance to the accompanying role. She gives much consideration to the employment of her armored force (or forces) for the following purposes:

(1) Attacks against the hostile flank or rear in coordination with the attack by the main forces of all arms. (2) Independent assignments, including delaying actions, reconnaissance in force, raids, and combat with hostile armored forces. (3) A frontal penetration when necessary, the armored force plunging on deeply without reference to the advance of the foot troops following. (4) Participation in exploitation.

Great Britain attaches importance to the use of small fast tanks in a scouting role to serve her medium tanks. This is perhaps partly due to the fact that her medium tanks are somewhat sluggish in comparison with the scouts. However there are advocates of the view that the requirements for a scout tank and a light combat tank are substantially the same, to wit: high speed, armor protection against small weapons, armament to include one antitank weapon and at least one machine gun, size and weight the minimum consistent with other requirements. Occasions will undoubtedly arise where the crew of a tank serving in a scouting capacity and not equipped with an antitank weapon, will wish that their tank were so equipped. One of the chief merits of the small scout tank is its low cost, both initial and for upkeep. Commanders of large infantry units may find a small tank very useful for reconnaissance purposes.

The United States regards its M1917 light tank as usable for the accompanying role under favorable circumstances and in the absence of a more efficient type in adequate numbers. The usefulness of this tank can be materially increased by certain alterations. The quantity production is contemplated of a somewhat heavier and much faster tank (Christie type) that will carry one antitank weapon and one or more machine guns. The United States view is that such a vehicle will serve, to a considerable extent, as a general-purpose tank. It will be appropriate for the accompanying role, in which case it will utilize its speed in advancing rapidly by bounds; it will be suitable for the leading role in open warfare and in lightly stabilized situations; and it will be of first importance in a mechanized force, which will operate in conjunction with and under the control of the cavalry. One of the reasons why such a vehicle is favored as a general purpose tank, is the economy that may be effected and the quantity production that may be facilitated if efforts are largely concentrated upon one type. The United States' interest in scout tanks and heavy medium tanks has been less; although it is recognized that, for use in the leading role against a stabilized position, a moderately fast, heavily armored, heavy-medium tank is very appropriate and desirable. According to the American view, it is appropriate that tanks in the leading role should utilize their potential mobility and not retard their advance for the purpose of remaining near the advancing infantry troops.

Medium Tanks.

The typical British medium tank now in the hands of troops, is the Vickers Mark I to Mark II A. inclusive. The new 16 ton tank is probably intended to supersede the present medium tanks in due time. It is believed to be the British view that the medium tank is habitually employed as a part of an armored force. The medium tanks are regarded as expensive and precious. Their high tactical value is habitually reserved for major efforts.

The French medium tanks are practically nonexistent so far as known, with the possible exception of the Char Moyen, an experimental medium tank of the same fire power and armor as the Char 2 C, but weighing about one third as much.

The French recognize the leading tank role to a certain extent, but only at the pace of and close to the advancing foot troops. For this, they favor the use of very large and heavy tanks ("breaking-through tanks," they call them), that can withstand the fire of small antitank guns, that have liberal fire power, and that they expect to be necessarily slow.

The *close support tanks* of the British Army include: the very light Carden Loyd mounting a Vickers 47 mm gun or a Stokes mortar, the self-propelled mount for 18 pounder gun, and a medium tank armed with a mortar. The British seem to look with increasing favor upon the use of smoke for this purpose. This sort of supporting vehicle does not seem to have been developed in any other country to any considerable extent.

Heavy Tanks.

The British are not wholly convinced of the need for any tank larger than the 16 ton but, recognizing the possible need for a heavier vehicle, have designed and made two, the so-called Independent types. The French have built the Char 2 C and Char 3 C for "break-through" purposes. The Italians have built the G L 4. In general, however, the heavy type of tank seems to lack ardent supporters and relatively few have been constructed.

If there is ever justification for a heavy tank, it must be based upon a combination of unusual fire power and thick armor. The weapons must be able to put out of action any hostile tank, and the armor must be able to resist all light weight antitank guns. The large and heavy tank would of course have some advantages in an assault against extremely well organized defenses. But being very expensive, relatively slow, and difficult to transport by rail or across rivers, it seems to have a good many disadvantages. Moreover, it is believed that its tracks would always be vulnerable to the fire from much lighter tanks. Its fire power and its armor would not protect it long once it had been immobilized by a broken track.

A few very large and heavy tanks have been constructed by certain countries. Very little has been published regarding them. It is believed that such tanks would be prohibitive in cost either in peace or war, that they would introduce difficulties as to transportation and river crossings, and it is doubtful if they would be of greater value than the lighter types of heavy tank.

Amphibious Tanks.

The British have been interested in the past in amphibious tanks. They have built several and are prepared to build others if it should be found desirable to do so. One such vehicle has been built in the United States, the Christie. Tanks of this nature offer one possible means of effecting landings on hostile shores. The advantage possessed by such tanks in effecting rapid crossings of streams is obvious.

CHAPTER XX.

TANK DESIGN

The design of a satisfactory tank for combat purposes is a complicated problem. Some requirements, if given undue weight will conflict to a serious extent with other very important requirements. If a start is made with a minor consideration as the fundamental requirement, or if careless deductions are made on the way, the plan for a tank is very likely to prove wholly unsatisfactory. In this chapter the purpose is to discuss many of the factors that must be given consideration in tank design, and to show how these matters may be taken up in such a sequence and in such a way that they will lead from the first fundamental to a final plan for a highly satisfactory combat vehicle.

Relatively Unimportant General Considerations.

Weight. In tank design, the rule regarding weight is simple. It must always be the minimum practicable. Not being the most important factor, the final weight will be dependent upon other factors. If it is planned to carry a tank on a certain type of motortruck or flat car, the tank must not be too heavy to be so carried. If it is deemed necessary that a proposed tank be able to pass over a certain type of bridge, then the tank must not be too heavy for such bridge. It should be borne in mind that any unnecessary weight is always objectionable, if for no other reason, because it unnecessarily and adversely affects mobility.

Size. Size is another general consideration of secondary importance, and the rule again is that it should be the minimum practicable. A tank must be large enough to accommodate the necessary armament, crew, parts, equipment, and supplies. It must have sufficient length to be able to cross obstacles. Unnecessary size means a larger target and greater weight, both of which are undesirable.

Shape. Shape is the relation between length, width and height. This is partly dependent upon other factors, but shape, itself, is entitled to some consideration. If the tank is too long in proportion to its width, steering will be unduly difficult. If it is too wide in proportion to its length it may, due to its shortness, have insufficient ability to negotiate obstacles and may make an unstable gun platform, or its excessive width may adversely affect traffic when the tanks are traveling upon roads. If it is too high in relation to its other dimensions, it presents an unnecessarily conspicuous target and its center of gravity will probably be too high in relation to its length and width, making the tank top heavy and liable to overturn on rough ground, with disastrous results. It is, of course, especially likely to overturn if its center of gravity is high in relation to its width. An excessive height is likely to interfere with passage under bridges when tanks are being transported by truck or by rail.

The First Fundamental of Design.

Tactical Purpose. The tactical purpose is the first fundamental—the basic factor for initial consideration. Its determination initiates the determination of the type of tank. When the tactical purpose has been determined, from it are deduced the **armament**, the **speed**, the **armor**, and characteristics of lesser importance. From these, further details are deduced, and so the type and ultimate design are determined. The tactical purposes are considered in Chapters XVIII, XIX, and XXVIII.

Armament and Associated Matters.

Limitations on Fire Power. It is desirable that a tank have the maximum practicable degree of fire power. However, weapons, and the men and ammunition that must be with them, occupy space and have weight. The size and weight of the tank must not be unduly expanded in order to add *unnecessarily* to the fire power. When more than a minimum complete armament is planned, it must be after a careful consideration of it in relation to the resulting size and weight of the tank. The proposed armament must also be considered in relation to the proposed type of superstructure. If a rotating turret is employed, a flexibility of fire is afforded that permits the use of a minimum number of weapons. When there is *not* a rotating turret, the rigidity of the superstructure and the space available within it permits the mounting of more weapons, and the absence of the flexibility of the turret tends to require it in order that the field of fire may be as complete as practicable.

Scout Tank. If the tank is to be merely a light, inexpensive, two-man scout tank, a single cal. .30 machine gun may suffice, or, at most, a cal. .50 machine gun.

Light Combat Tank. If the tank is to be a completely armed and equipped vehicle for assault purposes, of minimum size and weight, the armament should be one antitank weapon and one or two ordinary machine guns.

Leading Tank. A tank intended for the leading role, especially if it is to be used against a stabilized situation, must be fairly large (for crushing and the negotiation of obstacles) and it must be heavily armored. It is therefore appropriate to give it greater fire power than the light combat tank. For such a tank, having a rotating turret, a suitable armament might be one antitank gun and three or four machine guns within the turret and perhaps one or two machine guns outside the turret. Without a rotating turret, such a tank might have one or two antitank guns and four or five machine guns. A tank built for this role should be able to fire machine guns to the rear and flanks while an antitank gun is firing forward. Small turrets, in addition to a main turret, are provided in some cases.

Heavy Tank. In this chapter no heavier type of tank is considered than that mentioned in the preceding paragraph.

Nature of Antitank Weapons in Tanks. No effort is made to prophesy what types of antitank weapons will be found most desirable for use in tanks. There are, however, some general requirements that may be stated. The weapons must be able to penetrate, at reasonable ranges, the armor of the hostile tanks that will most likely be encountered. In this there should be some margin of safety, for there may be little or no opportunity to change armament after the enemy has thickened his armor. The large caliber machine gun has some advantage as an antitank weapon since it can usually register upon a target more quickly than a cannon. On the other hand the projectiles of the machine gun may penetrate without doing any damage worthy of consideration, whereas, if a shell can be made to explode within the armor, the neutralization of the tank is practically certain. Since it is easily practicable to carry larger weapons in large tanks than in small ones, we may properly expect the larger tanks to have weapons of greater penetrating ability.

Mounting of Weapons. Weapons in tanks must be so mounted as to permit the needful traverse and elevation and yet so shielded that an absolute minimum of hostile bullet splash will be permitted to enter, regardless of the direction in which the guns are pointing or the direction from which the fire is coming. Such a weapon, of light or medium weight, should be mounted upon horizontal and vertical trunnions so that the gunner may easily point the gun in any direction within the necessary limits. Guns not mounted in rotating turrets must often be placed in sponsons in such a way as to permit a large horizontal arc of fire. A considerable angle of depression is desirable so that the gunners may fire at hostile soldiers close to the tank.

In the case of antitank guns of light or medium weight, for use in tanks, it is believed best to mount an ordinary machine gun, parallel to such a gun, in the same mount. This enables the gunner, when observing a prospective target, to decide upon the kind of fire he deems most appropriate and initiate it almost instantly.

Antiaircraft Weapons. For a time there seemed to be a prevalent notion that a tank without an antiaircraft weapon ready for instant use was seriously deficient. This view appears to be unsound and has few defenders today. The fire from a tank is and should be primarily, if not wholly, for delivery upon terrestrial objects. The disadvantages of general purpose tools often outweigh their advantages. Such is believed to be the case when an attempt is made to combine a combat tank with an antiaircraft weapon. The tank must have complete overhead cover. The antiaircraft gunner needs complete overhead and lateral vision. Let us assume that such a gunner is within a tank and he has a door in the top of fairly large size and conveniently located. Although it is probably not true, let us assume that he does not occupy an excessive amount of valuable space within the tank. If the tank is traveling, it is believed that the motion of his mount combined with his limited vision would render such a gunner's efforts completely futile. On the other hand, if the tank were stationary, it would be much better for the gunner to be outside where he might have an appropriate field of view. Hence the deduction that, if a tank must carry an antiaircraft machine gun, it should be for the purpose of mounting it somewhere outside, either on a bracket attached to the tank or on an entirely separate mount. While it would be unwise to discourage defense against hostile aircraft, it is believed that the best defense for tank units can be furnished by attached antiaircraft units. Next best after that is the carrying of machine guns for use against aircraft from external mountings only.

Crew. The number of men composing the crew of a tank depends more upon the armament than anything else. As previously indicated, if there is to be a specially light tank for scouting purposes, a two man crew is deemed suitable. In the case of the light combat tank a crew of three is preferred. That provides one driver, one gunner, and a tank commander. The gunner operates one antitank weapon and one machine gun. The tank commander commands, observes, navigates, and, if necessary, may assist the gunner or operate the second machine gun if the tank has one. Leading tanks of the types indicated on the preceding page would require crews of from four to six. Tanks of great size and weight would naturally have more armament and more space within and would, in consequence, utilize still larger crews.

Speed and Associated Matters.

Speed. Speed depends upon a number of factors, the most important of which is power. The developments in motive power have made possible and practicable much greater speeds than formerly. High speed in a tank traveling on a smooth straight road is not a primary objective, but it is an indication that the tank has sufficient power per ton to enable it to perform creditably as a vehicle on the battlefield under the various situations that are likely to arise. If a tank traveling up a ten per cent slope while under fire from an antitank weapon can maintain a speed of 15 or more miles per hour, it has a better prospect of continuing its usefulness than one that can maintain only 3 miles per hour under such conditions. This is only one illustration of the value of superior mobility on the battlefield. A dozen others might be cited. Superior mobility requires a very material reserve of power over that needed for average conditions. This can be had only by providing ample horsepower per ton. Assuming a properly designed transmission, if there is a reserve of power appropriate for the demands of the battlefield, there is available the ability to run at high speed on the road. Moreover, an ample reserve of power contributes toward reliability, and protects the engine against excessive depreciation, thereby reducing the frequency of overhaul. And, after all, high road speed facilitates strategic mobility. Fast-moving vehicles require the use of a road for a much shorter time than slow-moving vehicles. If a road is cleared for combat vehicles when the circumstances are right, mobility may be enhanced for a particular purpose that will contribute greatly toward success. So high road speed as an objective should not be disregarded after all.

It is believed that a fast general purpose tank requires a minimum of 25 horsepower per ton. A heavily armored tank for the leading role might do very well with 20 or perhaps even somewhat less.

Power Train and Transmission. The power train includes all such parts, as clutches, transmission, and reduction gears, by means of which the power of the engine is transmitted to the tracks or the road wheels. All these parts should be strong enough to withstand the strains that the operation of the engine is likely to put upon them, with a liberal factor of safety.

The transmission of a tank requires a wider variation of gear ratios than that of an automobile. The speed range of the tank may be considered as from one-half mile per hour to sixty miles per hour. While any given type may not have so much as that, its variation is necessarily large, and it is believed that any tank should have at least four speeds forward. Tanks of low engine power require even more, if they are to pull well in low gear and make fairly good speed when the going is favorable.

As for automobiles, the sliding gear transmission has been found best for tanks, although the planetary or epicyclic type has been used in a number of the heavier tanks. As in the commercial automotive world, a transmission for a tank would, in theory, be better if it had a multitude or an infinite number of gear ratios. Although a number of such types have been built, they have, for the most part, been found unsatisfactory in practice.

In some tanks electric and hydraulic types of transmissions have been tried and are said to be satisfactory, but no such type has come into general use.

Suspension.

The suspension of a tank is the means of support between its tracks or its road wheels and its body. It is the keystone of vehicular efficiency. The lack of an efficient suspension makes high speed impossible, and adversely affects durability, comfort, and accuracy of fire. Since high power in the engine cannot be used unless the suspension is efficient, it may be truly said that power and suspension are the indispensables upon which speed depends.

To be efficient, a suspension must absorb the vibrations, bumps and shocks caused by the roughness of the ground; it must handle the body of the tank gently and cause it to move as nearly as practicable in a straight line in spite

Design 187

of the rough ground that it may be passing over; it should reduce bouncing and rocking as much as practicable.

In designing a suspension, to accomplish the desired purpose, some or all of the following agencies are employed:
(1) A fairly large number of points of support (wheels or rollers).
(2) Equalization (bogies, levers or cables).
(3) Elasticity (coil springs, leaf springs, rubber tires or rubber buffers).

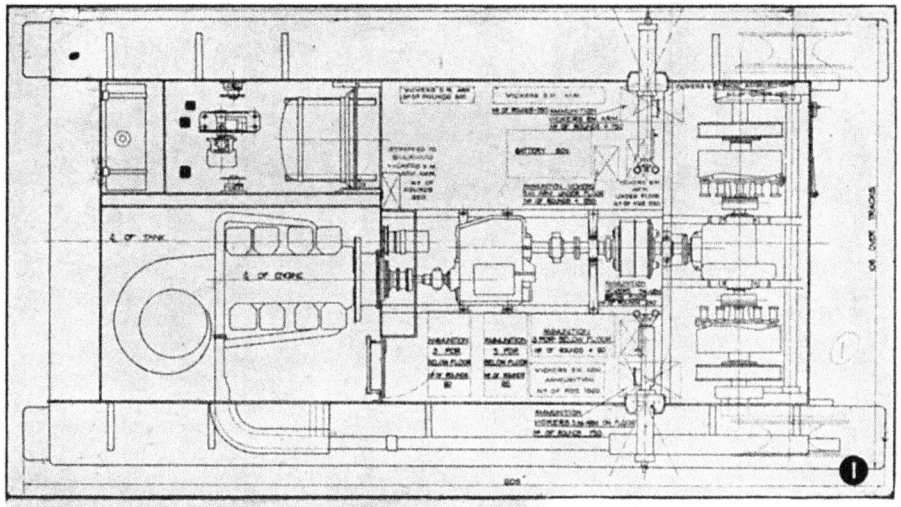

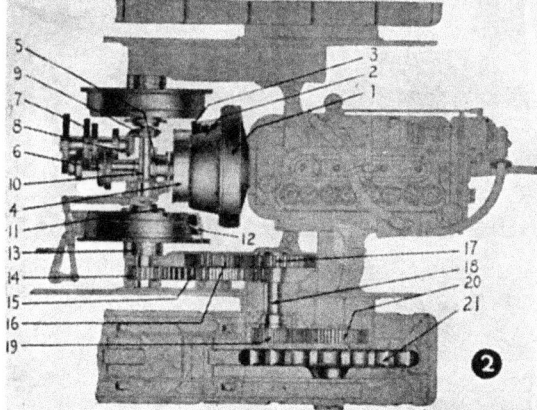

1. Flywheel
2. Main clutch
3. Driving member, main clutch
4. Fan belt drive
5. Second alignment joint
6. Driving gears on main shaft
7. Driving gears on countershaft
8. Bevel pinion
9. Bevel gear on transverse shaft
10. Transverse shaft
11. Driven member, steering clutch
12. Driving member, steering clutch
13. Oldham coupling
14. Driving gear, 1st pair reduction gears
15. Driven gear, 1st pair reduction gears
16. Driving gear, 2nd pair reduction gears
17. Driven gear, 2nd pair reduction gears
18. Shaft (integral part of 19)
19. Driving gear, final reduction gears
20. Driven gear, final reduction gears
21. Drive sprocket

Plate 46.
1. Vickers Medium, Power Train with Engine in Front and Final Drive in Rear.
2. U.S. Six Ton, Power Train with Engine and Final Drive in Rear.
(From a drawing by Ordnance Dept., U.S.A.)

(4) Dampening (shock absorbing devices to reduce bouncing and rocking).

An automobile has only 4 points of support, as have also the Italian four-wheel tanks, which have no tracks. This latter type of vehicle is apparently satisfactory for its purpose. It is intended for use in mountainous country and is not operated at high speed.

A tank should have a more efficient suspension than is needed in an automobile because a tank must be able to travel over very rough surfaces at a rapid rate. Four points of support are considered inadequate in a track-

188   THE FIGHTING TANKS SINCE 1916

laying type of tank. From 8 to 12 points are deemed necessary if the suspension is to have a reasonably high degree of efficiency.

Some suspensions employ a combination of equalization and elasticity. Others use one without the other. The principles involved are, of course,

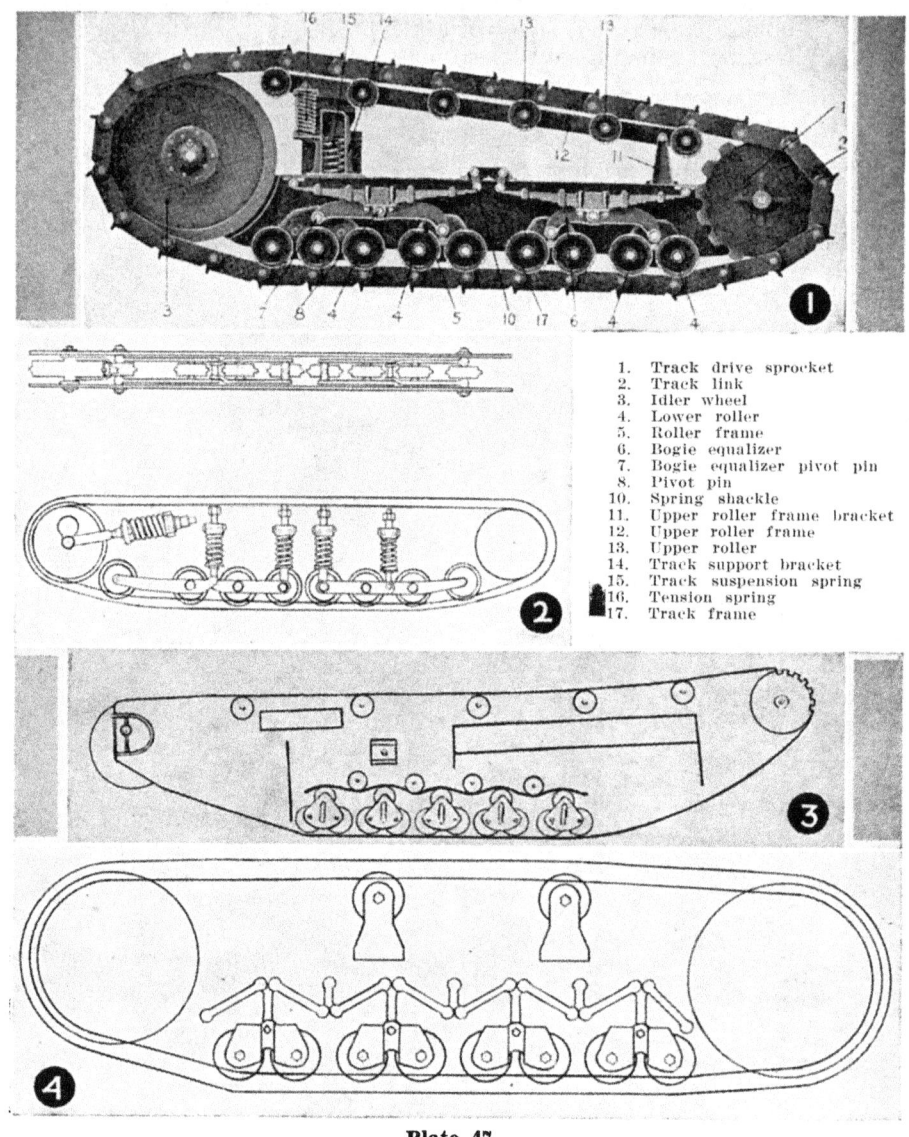

Plate 47.
1. Suspension of U.S. Six Ton, M1917.
(From a drawing by Ordnance Dept., U.S.A.)
2. Suspension of U.S. Medium, T 1.
3. Cable Type of Suspension, U.S. Medium, M1922.
4. Link Type of Suspension (Equalizing), U.S. Light Tank T1-E1.

entirely different. In the equalization system, a sudden rise that the tank must pass over is divided into a series of several sudden rises of lesser amplitude, each of which may be a fairly sharp jolt. In the elastic system, such a rise is temporarily completely absorbed, provided the rise is not so great as to exceed the compression amplitude. As the rise is absorbed, the springs immediately set to work to raise the tank at a smooth, steady and relatively slow rate. If the rise in the ground is passed over sufficiently soon, the compression may be released to normal before the tank has risen more than a fraction of the height of the rise in the ground.

In either an equalization or an elastic system there is a maximum amplitude in the motion of wheels or rollers that rise and fall beneath, and in relation to, the body of the tank. This amplitude is of vital importance in suspension efficiency.

Suppose a tank, with an equalizing suspension, and going fairly fast, strikes a log. If the height of the log exceeds the amplitude of motion in the suspension, the body of the tank will immediately receive a very severe jolt. If the height of the log is *less* than the amplitude of motion, the suspension will transform what might be one severe jolt into several jolts of less severity.

It should be clear from the foregoing that an elastic suspension cushions the jolts, whereas an equalizing suspension merely multiplies them and decreases their severity. A search has failed to disclose any equalizing suspension, uncombined with elasticity, in any tank of any army, that has proven even fairly satisfactory.

Rubber tires are very desirable in the suspension of a tank because they dampen out small vibrations and because they contribute greatly toward silence in operation.

Most tank suspensions that use springs are of only mediocre efficiency and are unsuitable for a rapid pace over moderately rough ground. This is primarily due to the fact that the compression amplitude is only a very few inches and hence markedly inadequate.

In the suspension of the Christie tank, M 1931, a wheel may rise eleven inches above normal, and may exert pressure three inches below normal. The 1932 Christie has a compression amplitude of 24 inches. This is by far the greatest known in tank suspensions up to the present time.

Of the combination suspensions, one of the most efficient known, is that of a British light tank (the Vickers Armstrongs 6 Ton). This suspension, although it has a small compression amplitude uses rubber tires and derives a maximum benefit from the equalization principle. Taken altogether, this is a very good suspension. And yet it seems appropriate at this time to make a comparison that reflects unfavorably upon it.

Suppose that both this 6 ton tank and the Christie tank are driven across an eight-inch log. Observing the side of the British tank, it may be noted that there are two main axles, one forward and one rear. Due to the small compression amplitude, when the front axle of the 6 ton tank passes directly over the log, the front part of the tank must have been raised about four inches. But, when the Christie tank passes over the same log, each wheel, successively and individually, rises eight inches and promptly lowers again to normal, while, at no time during the passage, is the front or the rear part of the tank raised more than about one inch. The crux of the comparison is thus very apparent. Although the jolt is cushioned in either case, it is very much more of a jolt in the case of the 6 ton tank than in the case of the Christie.

All of this leads to the maxim that the compression amplitude is the keystone of suspension efficiency.

Shock absorbers, so universally used in automobiles, have appeared but seldom in tank design. Tanks need the relief that they afford no less than automobiles. It seems a certainty that the tanks of the future will not be without them.

The essentials of tank suspension construction, by means of which comfort and a relatively steady gun platform are secured, have now been ade-

Plate 48.
1. Spring and Hydraulic Suspension, U.S. Light Tank T 1-E 3.
 (Photo by Ordnance Dept., U.S.A.)
2. A Six Ton Tank Nosing Over a Rise and Exposing Its Belly.
 (Photo by Tank School, U.S.A.)
3. Operation of Christie Type Suspension.

quately considered. There remains only the matter of unusual constructions intended to facilitate strategic mobility. It has been recognized for a number of years that the endless track is superfluous and undesirable for travelling long distances on good roads; but, since the early types of tanks moved upon steel rollers that were plainly unsuitable for road contact, a problem was created, and efforts were made in different directions to solve it.

In England, France, and Czechoslovakia, experimental tanks were built that had tracks and rollers for combat use, and entirely separate wheels for road use. This involved *two complete separate suspensions* for each tank. In some of these vehicles provision was made for conversion by raising and lowering the wheels very quickly through the use of engine power. Space interference was the main factor that made the success of this type of vehicle an impossibility. Having two complete independent sets of running gear, neither was satisfactory. The superfluous load, of one kind or another, carried at all times, was of course, objectionable. There was an exceptionally large number of parts to get out of order or become damaged. So far as known, this type of vehicle no longer has supporters. It has been definitely abandoned by the British.

In the Christie effort toward the same objective, large rubber-tired wheels are used both for running on the road and running on the tracks. There is, of course, but one suspension. It becomes merely a case of using or not using the tracks. The tracks are carried upon shelves on the outside when the vehicle makes long trips by road. When the tracks are off, the rear road wheels are driven by chains or gears from the sprocket wheel and the front wheels are steered as in an automobile. When the tracks are on, they are driven by the sprocket wheels and the front wheels are locked straight to the front. Although it requires something like a half an hour to put on or take off the tracks, the Christie system is by far the best wheel and track combination that has yet been developed.

While it is expected that the wheel and track construction will be a decided success for vehicles weighing in the vicinity of ten tons, it is an interesting question as to whether success might also be anticipated if the size were enlarged so as to double or treble the weight. The first trouble that appears is the axle load. When the Christie machine travels without tracks, most of the weight is placed on the front and rear axles in order to facilitate steering. When we think of transporting at a high rate of speed, on ordinary roads, the larger part of something like 25 tons on two axles, we see immediately that the idea of taking off tracks for road travel has its limitations.

It should not be thought from the foregoing that the Christie type of suspension is unsuitable for a 25-ton tank. On the contrary, it would probably be a very satisfactory type, provided that the vehicle were always operated upon its tracks.

Armor, and Other Protection Features.

Armor. We now come to the third of the controlling factors that are deduced from the tactical mission—armor. Any combat tank should have armor to protect it at least against any sort of projectiles from any weapon that a single soldier can carry about in its complete form. Accordingly all tanks must be protected against armor piercing cal. .30 bullets. Otherwise, if tanks were vulnerable to the fire from individual foot soldiers, it is obvious that their cost in conjunction with their vulnerability would bar them from the realm of the practical. It appears at present that a half inch of high grade armor plate will furnish complete protection against cal. .30 armor piercing ammunition at any range.

The armor of a tank is not of the same thickness throughout. A study is made, in connection with the probable directions of impact, of the planes in which the various plates of armor are installed. Where the normal impact is probable and where personnel or vulnerable parts are located behind the armor, the maximum thickness is used. Materially thinner plate is used on top and bottom surfaces. The forward part of the bottom surface is usually thicker than the center and rear portions due to the danger of exposing the forward part of the bottom surface to fire when nosing up over an obstacle while advancing against resistance. (See Fig. 2, Plate 48.)

We have considered, in a general way, the minimum thickness of armor for tanks. The maximum is the thickest and most effective armor that the other factors of design will allow. A square foot of armor plate one-half inch thick weighs approximately 20 pounds. Thick armor therefore materially increases the weight. This in turn requires a stronger and heavier suspension. It also increases the ground pressure per square inch which in turn may cause trouble when soft ground must be passed over. With a given engine power, increased weight reduces the horsepower per ton. If we have less horsepower per ton and increased weight to pound the suspension when passing over rough ground, we lose in mobility. We cannot armor against all hostile fire. No matter how thick armor we use, the enemy can employ a gun that will pierce it. It is here that judgment must be exercised in adjusting the conflicting requirements of speed and armor. The British incline to high speed and thin armor; the French to thick armor and slow speed. The minimum thickness at vital points in French tank specifications is now 1.18 inches (30 mm). In studying this problem, the probable types of hostile guns should be considered and the highly important factor of tactical purpose should always be kept in mind.

The armor of the tank must have openings, particularly for the engine cooling draft, vision, and ventilation. Some sort of an additional closable opening that would permit the driver to see the ground when crossing obstacles might be worth while. All openings should be cleverly designed and so located as to minimize the chance of entrance of undesirable matter such as bullets, grenades, earth, water, and gas.

Bullet Splash. Ordinary bullets will splash particles of metal through remarkably thin cracks. While such splash will not ordinarily produce any severe wounds, unless the eye is struck, it will cause annoying cuts and burns and will interfere materially with the efficiency of the crew. To overcome this, great care is exercised in the design of door and pistol port closures, turret races, gun mounts, and all other necessary openings. By skilful design and careful workmanship such places as door closures and turret races can be so made that, by requiring the splash to make numerous right angle turns on its way coming in, complete protection is afforded. There are various ways by which the splash coming in through the observation slits can be reduced, but it is believed that one of the best contributions to this end is the use on the inside of pieces of bullet-proof laminated glass which are quickly and easily replaceable when damaged. Exterior flanges or ridges that divert splash from vulnerable openings was a war-time expedient that possesses a considerable degree of merit.

Speed. Armoring against hostile fire is the most important phase of protection, but speed is also a factor that cannot be ignored. Even when a tank is attempting to travel in a straight course, as it moves over the usual irregular terrain, it is, from the viewpoint of the hostile gunner, an irregular course. But the tank, instead of attempting to maintain a straight course, will often intentionally pursue an irregular course. If it is then travelling at a rate of 15 miles per hour or more, it will be a difficult target to hit. If several tanks are advancing toward a hostile gunner at a rapid pace, he, having time for but few shots, cannot expect to disable many of them. The speed of modern tanks cuts the time factor so ruthlessly from the viewpoint of the hostile artillery that the danger from indirect artillery fire is normally very slight except when the tanks come under such fire accidentally. We thus see that high speed affords considerable protection.

Gas Protection. There has been a great deal of discussion in various parts of the world about protecting tanks against gas, but in some countries equipped with tanks very little has been done to solve the problem. In a few foreign countries the type of equipment most generally discussed has been installed experimentally. This scheme comprises the following:

(1) The crew compartment and the engine compartment are suitably separated so that the cooling draft, if contaminated, will not affect the crew.

(2) All openings in the shell of the crew compartment (gun mounts, etc.) are made as tight as practicable so that a slightly increased air pressure can be built up within.

(3) By means of a power-operated blower, air from without is put through a gas protection filter and delivered to the crew compartment. Simultaneously, air escapes from the crew compartment through the various cracks and openings that it is not practicable to close completely.

Without such protection as outlined above, the crews must wear gas masks. But these will not protect against the vesicant effects of mustard. Morever, the wearing of gas masks interferes with the efficiency of the crew. These disadvantages pertaining to the use of gas masks can be avoided if a satisfactory device such as outlined above is developed.

Fire Protection. The crew and the tank must be protected against unnecessary fire hazards.

The fuel tanks should be completely inclosed with the maximum thickness of armor. This protects the crew and engine compartments against fuel leakage and consequent fire hazard since bullets penetrating one thickness of armor will probably be stopped by the second thickness.

When the engine is stopped, the flow of fuel should be stopped automatically and positively so that no leakage will occur in the vicinity of the engine.

There should be a closable opening in the partition between compartments so that the crew may extinguish a fire in the engine compartment.

The tank should be equipped with the type of fire extinguisher most effective in putting out gasoline fires.

Heat and Fumes. The crew should be protected from unnecessary heat and fumes from the engine, exhaust pipes, etc. This requires that there be a suitable partition between the crew compartment and the engine compartment. This protection is also involved in the general arrangement of the tank, as will be brought out later.

Space Required.

After having determined the **armament, speed,** and **armor,** and before determining the necessary horsepower, the **space** requirements must be considered thus to make possible an intelligent estimate as to the probable weight.

Crew Compartment. The size of the crew compartment is determined first, and, in so doing, the armament and crew receive first consideration. There must be space enough so that the guns will not interfere with each other and so that the gunners operating them will not interfere with one another. It should be borne in mind that a gunner in the crouching posture necessary for firing occupies much more space than he would if standing erect. There must be space to mount the guns so that it will be practicable for the gunners to assume suitable firing positions. The floor space below the gunners must be level and free from obstructions, for otherwise firing efficiency will be seriously reduced. Since the members of the crew must live in the tank for long periods of time during travel and combat, their comfort should receive thoughtful attention. Fairly comfortable seats should be provided for their use while travelling. Dangerous projections on the interior of the compartment must be minimized and, at those points where particularly needed, protective cushions should be placed. Before the dimensions of the crew compartment are finally determined, consideration should be given to the space requirements for the following: suitable amounts of ammunition conveniently arranged for use; tank equipment (spare parts, accessories, tools, etc.); radio instruments and other communication means; personal equipment, food, and water; and gas protection mechanism, if contemplated.

Engine Compartment. The size of the engine compartment should be determined next. A type, power, and size of engine must be assumed and adequate space therefor allowed. (Theoretically, at least, this is subject to correction after a later weight computation shall have been made.) Consideration must also be given to the space requirements for the following: the cooling system; the transmission, steering clutches and other parts of the power train; batteries and generators or whatever is necessary to furnish the power needed for radio and other electrical purposes. The requirements of accessibility should not be overlooked as they affect the space requirement.

Determination of Horsepower.

Having decided upon the **armament,** the approximate **speed** (and the corresponding horsepower per ton), the **armor,** the **size** of the hull, the general design of power train units, the track, and the suspension system, an approximate **weight computation** is in order. When the approximate weight has been determined, the required **horsepower** of the engine is found by comparing the tentative weight of the tank with the horsepower per ton desired. This completes the outstanding essentials of the design.

Engines and Cooling.

Gasoline engines similar to those used in automobiles have predominated so far in tanks. The Diesel type, since it produces no electrical interference with radio, since it is so economical in the use of fuel, and since it reduces the fire hazard, is worthy of consideration for tank use. It has already been installed in some foreign tanks. Most British tanks use air cooling. Air cooling possesses important advantages over water cooling for tank use. These advantages are of much less importance in the case of the automobile. In any tank, whether air or water cooled, there must be maintained a large stream of air for engine cooling purposes. This air must enter through openings at one place and emerge through openings at another place. In action, it is probable that bullets or fragments thereof will ricochet into such openings. If a water-cooling radiator is used it is very likely to be punctured. A tank, that thus loses its water in action, loses its mobility almost immediately and stands an excellent chance of being put out of action completely by hostile fire. Even if the radiator of such a tank in action is not punctured by fire, it is very common for water to leak out otherwise, or it may be boiled away. If a tank in action needs urgently to have the water replenished, there is no assurance that there will be water available near at hand, or that water that is available will not seriously injure the parts of the cooling system. There is no assurance that the crew will know where to find water. If they succeed in driving to water, there is no assurance that they will be able to live outside the tank while effecting replenishment. In winter the danger of freezing is a serious disadvantage of water cooling, more so in the case of tanks than automobiles. The importance of the advantages of air cooling is obvious. The United States has found air cooling practicable in its Franklin-engined tanks, armored cars, and trucks. It has been used in the majority of airplanes all over the world. The British will use nothing else in their latest tanks. Apparently therefore, it is desirable and practicable, and we may expect to find it extensively used in our own future tanks.

Tracks.

In order to have the necessary ability to pass in and out of ravines and trenches, a tank must have support materially in front of the front of the body and likewise in rear of the rear of the body. Since the tank normally moves forward, the forward part of the tank will be the part that will need to be lifted out when crossing a trench. The forward extension should therefore be the tracks; at the rear, a tailpiece may be substituted if deemed desirable. An example of a satisfactory tailpiece is found in our so-called six ton tank.

The tracks of a tank must be of sufficient width to give an adequate supporting surface in relation to the aggregate weight of the vehicle. When a tank is moving on hard ground, the ground pressure per square inch is not important. But when it must pass over soft ground, the ground pressure may be the determining factor in whether it will get mired or pass through successfully. Due to the varying shapes of tanks, the amount of submergence affects the length of track in contact with the ground much more in some cases than in others. Hence a slight submergence (one or two inches) does not permit the ground pressure to furnish its best comparative index regarding the ability of tanks to pass through mud. The using arm in the United States has therefore selected a submergence of ten inches as its basis for comparison in this respect. Naturally, in designing tanks, efforts are made to get a satisfactorily low ground pressure. Many tanks, at a submergence of ten inches, have a ground pressure of about five pounds per square inch. This is very satisfactory. A pressure of six pounds per square inch is deemed fairly satisfactory.

In the building of large tanks, there is a general consideration that adversely affects ground pressure. It is best emphasized through a comparison with ships. Suppose, after building a small vessel having one inch of armor, we wish to build a larger vessel, of the same shape, with thicker armor. In the case of the ship, the factor of flotation (the displacement) and the weight of the vessel both vary with the cube of the length, and we may therefore build everything larger in proportion, thicken our armor, and not disturb our relative factor of flotation. But like procedure in the case of the tank brings different results. Suppose we wish to build a second tank increasing the size and the thickness of armor in an exactly similar way. The weight, as before, increases in proportion to the cube of the length. But the area of the tracks supporting the weight increases only in proportion to the square of the length. It is therefore true that, if the second tank is larger than the first tank in every way in the same ratio, the number of pounds of ground pressure per square inch in the case of the second tank will be greater than in the case of the first tank and in exact proportion to their relative lengths. This explains why it is that, in the case of large tanks, the width of the tracks is greater in relation to the width of the tank than is true in the case of small tanks. It also indicates that, while a low unit ground pressure is easy to attain for small tanks, we may find much difficulty in the case of a large tank unless sacrifices of other features are made in the interest of light weight. This is one of a number of factors tending to indicate that **very large tanks are less desirable than those of moderate size.**

In most tanks, the tracks are made of steel plates (or links), each of which is attached to its adjacent plate by means of a steel pin. The length and width of tracks have been discussed—there remains for consideration the pitch. The pitch is the length of the track plate as measured from pin center to pin center. The pitch should be as short as practicable; otherwise there is much unnecessary banging of part against part with resulting greater wear and greater noise. A good example of an excellent short-pitch track is that of the Vickers Armstrongs 6 ton tank. If this short-pitch track operated against rubber to the same extent as the Christie track, it would be still more quiet. And, if the 1931 Christie track had as short a pitch as the British track mentioned, it also would be more quiet in operation. In the case of the Christie track, however, the lugs integral with the alternate track plates must be strong and rather long to fulfill their dual purpose of guiding the track and driving it. The pitch of that track cannot, therefore, be very greatly reduced without altering the general design.

The old types of tanks used simple types of toothed sprocket wheels for driving, the teeth entering between adjacent track pins. There was usually a great amount of friction and wear at these points of engagement. The Christie sprocket wheel relies for its driving contact upon four rollers which are located between the two tires but somewhat closer to the wheel center. These rollers turn comparatively easily. In operation, the lugs of the track successively plunge in between adjacent rollers. The rotation of the rollers is effective to a great degree in minimizing the friction. In the case of the British track mentioned, each track plate has a centrally located lug, not so large or tall as the Christie lug and simpler in construction, that serves merely to steer the track on its proper course as the lug passes between the pairs of tires on the rollers and wheels. The sprocket wheel of this tank has a double row of sprocket teeth for driving. Each plate of the track engages two of these teeth, one on each side. Although this system does not involve any roller bearings, by a skillful design of the sprocket wheel and the track, the friction is very greatly reduced.

The tracks of tanks are subjected to very great strains and hard wear (not to mention gunfire). They should be made of the best material—of steel that is durable, tough and of proper hardness. Some tracks are hardened castings—some are forgings. The track pins are usually especially hard in order to minimize wear. Steel bushings are used in the track plates in some cases for contact with the pins. When worn, they can be replaced with relatively slight expense and trouble.

Grousers. Grousers are lateral ridges that move with the tracks for the purpose of preventing or reducing the slippage of the track on the ground when traveling through mud, up a steep slope, etc. Grousers are integral or detachable. An integral grouser is part of the track plate. The majority of such grousers are low (not more than an inch in height). Detachable grousers are high grousers that are clamped temporarily to the tracks by means of screws. They are put on only when swamps, slippery hills or the like must be negotiated. Slow tanks need grousers badly when climbing slippery hills. Fast tanks rush the hills and usually have less need of grousers.

Ability to Negotiate Obstacles.

Tanks cross trenches and ditches by one or the other of two processes, which are: (1) *Spanning process*. To span a trench is to cross it without sinking into it to any material degree. (2) *In and out process*. In this case the tank noses down into the trench and then climbs out the other side.

The size and shape of the obstacle determines which process occurs. Most obstacles of this nature are crossed by the latter method, and therefore *in and out* efficiency in the design of a tank is more important than spanning efficiency. If a trench is very deep and has walls that are almost vertical, it cannot, in most cases, be crossed by a tank at all unless the tank can span it. Under the most favorable circumstances, a tank can span a trench the width of which is slightly less than one-half the track length of the tank.

To increase the trench-crossing ability (particularly by the *in and out* process), the center of gravity of a tank should be as low as practicable. This is so because, whenever a tank is tipped forward or backward, its center of gravity is displaced in an unfavorable direction (downhill) in relation to

Plate 49.
Method of Determining Location of Center of Gravity of a Tank.
(Photo by Ordnance Dept., U.S.A.)

the ground covered by its tracks. With the center of gravity at the longitudinal center of the tank, the situation is most favorable for *spanning*. With the center of gravity several inches in advance of the longitudinal center, the situation is most favorable for *in and out* crossing. The latter is true because a center of gravity so located favors the tipping of the rear end up out of the trench when the tank is emerging. That is the critical time of the passage. There is relatively no difficulty in nosing into a trench or in lifting the nose after it gets in. Some tanks, having centers of gravity in rear of the centers of the tanks, can back through trenches that they cannot cross by forward motion. It follows from the foregoing, that the center of gravity of a tank should be low and as close as practicable to a point several inches in front of the longitudinal center.

The ability of a tank to climb over a vertical wall is closely related to its ability to climb out of a trench. In both cases, the critical and difficult part is the lifting of the tail. Assuming the idlers to be in front and the sprockets to be in the rear, a high idler and a low sprocket favor the successful passage of walls. Although a rear sprocket should be as low as practicable for this reason, its height is determined by the vertical amplitude of the suspension action. If the wheels or rollers will rise, for example, eight inches in passing over a sharp elevation, then the height of the lower part of the periphery of the sprocket should be not less than eight inches above the lower part of the periphery of such wheels or rollers, in their normal position. If the sprocket is lower than this it will unnecessarily transmit shocks from rough ground to the body of the tank.

It seems to be a habit among tank users to report that this tank and that can climb a 45 degree slope. It is believed that most of such reports are based upon an inadequate investigation. If a tank can maintain a steady speed up a 40 degree slope when the soil is dry, firm and altogether exceptionally favorable, it is a better hill climber than most of them. Of course many tanks can rush up to a very short 45 degree slope and ascend it successfully, largely through their own momentum. Such a performance is no indication of the actual climbing ability of the tank. If steep hills are slippery, long grousers are very helpful. Fast tanks seldom have to make steady climbs of steep slopes, for they can usually rush them at such high speeds that the momentum will assist them all the way up. It is of course desirable that tanks have the ability to climb 45 degree slopes, but, where that figure appears in the data of this book, there is no assurance that it can be accepted at full face value.

The hulls of most tanks admit water rather freely, in which case the fording ability of the tank is dependent upon the height of the lowest of such parts as elements of electrical system, fan, carburetor, etc. If the hull is water-tight, deeper water can naturally be forded. If the hull is not water-tight, the fan is low, and the stream is not wide, an increased depth can be negotiated by protecting the ignition system with canvas. The water spray kicked up by the fan will thus not stall the tank.

Tanks have a limited ramming ability. Many of them can knock down trees, brick walls, etc., provided such objects are not too strong. Such activity is of course likely to damage the tank, depending in a measure upon the strength and arrangement of the forward parts. Such use of a tank is to be discouraged except where necessary. When it is done contact should be made with the obstacle very slowly if circumstances will permit. For accomplishing the purpose, reliance should be placed upon weight and power rather than upon shock effect.

The ability of a tank to get through barbed wire is chiefly dependent upon the presence or absence of such angles (especially in the track) as are likely to hook and hold the wire. When the wire has become hooked, it becomes a question of the amount of wire and its strength against the power of the tank. Powerful tanks can often keep going unless the wire involved is very great in quantity. In designing tracks great care should be exercised to see that there are no places unnecessarily shaped so that they can catch wire. Very small and light tanks attempting to pass through well constructed wire entanglements, may become tipped at such an angle as to stall or overturn.

In determining the clearance under the hull in tank design, an important consideration is the ability to straddle objects of moderate height without becoming bellied. This calls for a high floor. On the other hand the desirability of a low center of gravity requires that the floor be low. A reasonable compromise between these requirements must therefore be effected.

General Arrangement.

The general arrangement of the tank and its contents is a matter that has not as yet crystallized to a degree comparable with that of the automobile. In the automobile the engine and driving axle are normally at opposite ends of the vehicle due to the fact that the body is rigid with respect to the engine, but liberally sprung with respect to the axle. This condition is quite different in the tank, where the driving axle and sprockets are normally sufficiently elevated so that they may be made rigid with respect to the body.

The various general arrangements of tanks include the following:

(1) Driver in the front, gunners in the center, engine and final drive in the rear.

(2) Engine and driver in the front, gunners in the center, ammunition and final drive in the rear.

(3) Engine in the front, driver in the center, gunners and final drive in the rear.

(4) Final drive in the front, driver in the front center, gunners in the rear center, and engine in the rear.

The general arrangement is governed by mechanical or utilitarian considerations or by a combination of both. A mechanical consideration of sufficient cogency might well govern over all other considerations. However, it is believed that there is no such cogent mechanical consideration. The facts that the engine, drive sprocket and body are all rigid with one another, the ease with which satisfactory provision may be made so that the driver may operate his transmission and engine controls at a distance, and the satisfactory mechanical operation that has been demonstrated in the various types, all so indicate. Reaching the conclusion that the mechanical considerations are not highly important in this matter, we should examine the utilitarian requirements with considerable care.

The principal utilitarian considerations are:

(a) The driver must have a good view to the front and flanks including a close view of the immediate foreground.

(b) The gunners must have an all-around view and the best practicable gun platform.

(c) To secure prompt and effective fire, the gunners require flat and unobstructed standing space.

(d) For the sake of comfort and efficiency the crew should be as free as practicable from engine heat.

(e) The efficiency of the crew should be interfered with as little as practicable by dust and oil mist.

(f) To facilitate the negotiation of obstacles, the center of gravity must not be excessively far from, and should be as close as practicable to, the preferred location, which is low and slightly forward of the center of the tank.

Let us now see how the various general arrangements affect the stated utilitarian requirements. We shall consider the arrangements successively and in the same order as previously enumerated.

(1) (Driver in the front, gunners in the center, engine and final drive in the rear.)

The position of the driver in front is most favorable as to (a) (driver's view), and is not unfavorable as to (b) (gunner's view and platform), (c) (gunner's standing space), (d) (heat), (e) (dust and mist) or (f) (center of gravity).

The position of the gunners (or gunner) in the center is entirely favorable as to (b) (gunner's view and platform) especially so as to the gun platform since the pitching motion is much less there than at either end. This position is not unfavorable as to (a) (driver's view), (c) (gunner's standing space), (d) (heat), (e) (dust and mist) or (f) (center of gravity).

The position of the engine and final drive at the rear is particularly favorable as to (c) (gunner's standing space), (d) (heat) and (e) (dust and mist); it is not unfavorable as to (a) (driver's view) or (b) (gunner's view and platform), but it is ordinarily somewhat unfavorable as to (f) (center of gravity). However it is entirely practicable to extend the tracks somewhat farther to the rear of the engine than is usual and thus move the center of gravity relatively forward to a satisfactory location. The utilization of somewhat heavier armor in the front of the tank will also contribute to the favorable placing of the center of gravity.

(2) (Engine and driver in the front, gunners in the center, ammunition and final drive in the rear).

The position of the driver in front is favorable and in no way unfavorable. (See previous discussion.)

The position of the engine in front is favorable as to (f) (center of gravity), but is unfavorable as to (a) (driver's view), (b) (gunner's view and platform), (c) (gunner's standing space), (d) (heat) and (e) (dust and mist). When the engine is in front, its heat reaches the crew much more than when it is in rear. The cooling draft of the engine kicks up much dust and some oil mist. The dust interferes with vision and also, as the tank moves forward into the cloud, the dust and the oil mist settle on the lens of the telescopic sight and on the glass protecting the observation slits, thus still further interfering with vision.

The position of the gunners (or gunner) in the center is favorable and in no way unfavorable. (See previous discussion.)

The location of the engine and the final drive at opposite ends is favorable as to (f) (center of gravity), not unfavorable as to (a) (driver's view), (b) (gunner's view and platform), (d) (heat) or (e) (dust and mist), but is markedly unfavorable as to (c) (gunner's standing space). It necessitates the transmission of power through the center part of the tank and above the floor. The shaft with its housing or other protection interferes with the gunner's use of his feet.

(3) (Engine in the front, driver in the center, gunners and final drive in the rear.)

The position of the engine in front is on the whole unfavorable. (See previous discussion.)

The position of the driver in the center is unfavorable as to (a) (driver's view); otherwise it is neither favorable nor unfavorable except as it interferes with the desirable placing of other elements. The driver, from this position cannot get a sufficiently close view of his immediate foreground.

The position of the gunners (or gunner) in the rear is unfavorable as to (b) (gunner's view and platform) as regards gun platform. In this position they are subjected to an undue amount of pitching motion. Since this position of the gunners requires the engine to be further forward, it is also unfavorable as to (d) (heat) and (e) (dust and mist). Placing the gunners at the rear appears to offer no specific advantages whatever.

The location of the engine and final drive at opposite ends is favorable as to (f) (center of gravity) but unfavorable as to (c) (gunner's standing space). (See previous discussion.)

(4) Final drive in front, driver in the front center, gunner in the rear center, and engine in the rear.)

The final drive in front is exceptional. It causes the track to have more earth adhering to it as it comes in contact with the idler and less as it comes in contact with the sprocket. This may be a slight advantage.

The position of the driver slightly to the rear of the forward position is unfavorable as to (a) (driver's view) although only to a slight degree.

The position of the gunners in the rear center is substantially the same as the center, and hence favorable. (See previous discussion.)

The position of the engine at the rear is favorable. (See previous discussion.)

The location of the engine and final drive at opposite ends is favorable as to (f) (center of gravity) but markedly unfavorable as to (c) (gunner's standing space). (See previous discussion.)

It appears from the foregoing study that the least objectionable general arrangement, from the utilitarian viewpoint, is: driver in front, gunners in the center, engine and final drive in the rear. The deduction that this arrangement is preferable seems to accord with experience, since there are more tanks with this arrangement than with any other.

General Arrangement. Cargo Carrier.

We must not conclude that, because a certain general arrangement is best for a combat tank, the same arrangement is best for a cross-country cargo carrier. In fact, it is not. As a matter of economy, a chassis of the same type as built for a tank may be made to serve for a cargo carrier, but, if a special chassis is to be built to meet the needs of the cargo carrier, the following requirements should be considered:

(1) The driver should not be at the center or the rear. The cargo may be stacked high. The driver must see where he is going. He should be located at the front.

(2) A handy and suitable opening should be provided somewhere for the loading and unloading of the cargo. The tracks are a fairly high barrier and hence preclude the use of a side of the vehicle for this purpose. There are therefore left for consideration only the front end and rear end.

(3) It is already apparent that the driver should be in front. His control levers, etc. will therefore interfere with the use of the front end for loading and unloading. If the engine is placed at the rear, both ends will be blocked. If the engine is placed at the front, the rear end will be clear and wholly available for loading and unloading. The reasons that make the front end objectionable for the engine in the case of the tank do not apply in the case of the cargo carrier. Therefore the engine and the driver should both be at the front end.

(4) The only question remaining is the desirable location of the final drive. In a track-laying vehicle it is mechanically of negligible consequence whether the drive sprocket wheel is at the front or at the rear. If it is placed at the rear of a cargo vehicle with the engine in front the result will be that the drive shaft must run through the cargo space above the floor. This is obviously undesirable and it is wholly unnecessary since the drive sprocket wheel can readily be placed at the front rather than the rear.

Conclusion: In building a chassis for use as a cross-country cargo carrier, the power plant, final drive, and driver should all be placed in the front, thus leaving as much as possible of the center and rear clear for cargo and the rear end completely clear for loading and unloading.

Superstructure.

In the superstructure of the tank, there should be not less than two doors through which the crew may enter or leave. These doors should be so located with reference to each other that, if the tank is overturned, the simultaneous blocking of both doorways will be highly improbable.

A rotating turret in the superstructure has advantages and disadvantages. The main advantage is that if a tank that has one is traveling in a given direction, a cannon mounted in the turret can be fired at a target in any direction. On the other hand, with one or two cannons mounted in a tank without such a turret, it might be necessary to make a wide and undesirable change of direction of travel in order to permit the fire of a cannon to be brought to bear upon a given target. The chief disadvantages of the rotating turret are: (1) Most of the turrets now built cannot be rotated through a wide arc within the very few seconds that might at times be desirable. (2) Rotating turrets may become damaged in such a way as to prevent their rotation. (3) If it has the necessary strength and armor, the rotating turret is likely to add somewhat to the weight. (4) A rotating turret should turn freely under all circumstances, yet without such looseness in the race as to permit vibration. This is an important problem confronting the designer, and one that has not always been successfully solved.

There is the question as to whether these disadvantages are serious and insuperable. It seems improbable that there is any formidable barrier to the employment of power for turret rotation, supplemented of course by facilities for emergency hand operation. With high speed rotation, the value of the turret would be materially enhanced.

To facilitate ease of operation, a turret should be balanced with respect to its axis of rotation.

If it becomes impossible to rotate a turret, fire can still be brought to bear upon a given target by turning the tank—a procedure that might be necessary if there were no rotating turret in the first place.

As far as weight is concerned, it would seem that the economy resulting from fewer guns and men by the employment of a rotating turret would probably offset the added weight of the rugged and complicated construction required for the rotating turret. And, furthermore, a little weight more or less is not of much importance if it is foreseen in the design computation.

Strength, Reliability, Lubrication.

A tank, possessing, as it must, mobility in the face of rough ground and obstacles, is necessarily subjected to rather violent bumps and stresses. Its parts must have the strength to stand such treatment without much danger of mechanical trouble being caused thereby. To be able to ram walls and trees with safety to itself, the forward parts of a tank must be exceedingly strong. The design of a tank should be such that it will be durable and trouble-proof in spite of long wear and rough treatment. To be durable and reliable, a tank must have an excellent lubrication system. Not only must there be proper and thorough lubrication, but the system should be so designed that the work of servicing may be accomplished with the greatest celerity and ease.

Accessibility. Accessibility is a factor of importance that should never be overlooked in design. It should be easy to get at all lubrication points, engine and transmission support bolts, drainage cocks, spark plugs, wiring, magnetos, carburetors, batteries, adjustments of fan belts, power train units, etc. Assemblies should be removable with the greatest practicable ease and with a minimum disturbance of other parts.

Fuel Distance. The fuel distance should be as great as practicable. To be specific, in modern tanks, for travel on tracks over very favorable ground, it should not be less than about 100 miles. The oil supply should outlast the fuel with a margin of safety under all normal circumstances.

Location of Fuel Tanks. It is desirable that the fuel tanks be so located as to insure the maximum protection from hostile fire and, if punctured, the minimum fire hazard to the crew. The fuel tanks of the Mark VIII tank are especially well located in both respects. However, other considerations will sometimes make the preferred location not available.

Maneuverability and Ease of Control. A tank should be able to stop, start and turn suddenly and quickly without harm resulting therefrom. The controls should be handily located, easy to understand, operable with slight effort, effective and reliable in their application.

Silence in Operation. A tank should be as silent in its operation as practicable. Especially should it be largely free from characteristic noises that would distinguish it from other types of motor vehicles. The important noises to be reduced come from the tracks, sprockets, idlers and exhaust. To hide the flash and subdue the noise, efficient mufflers are imperative.

Lights. Suitable lights for moving at night should be considered a necessary part of the tank. Additional lights dimmed to such a degree as to be visible for only a very short distance are also desirable. The latter are useful when brighter lights would create a danger of detection by the enemy.

CHAPTER XXI

TANK EQUIPMENT AND ACCESSORIES

Tank equipment and accessories are classified and discussed under the following headings:

(1) **Communication:**
    (a) *Interior:* Voice, touch, speaking tubes, visual, telephone.
    (b) *Exterior:* Semaphore, flags, pyrotechnics, lights, sound, pigeons, radio telephone, radio telegraph.

(2) **Armament:** Weapons, sights, shellbags, shell ejection, machine gun belt loaders, oil, cleaning rods, tools, spare parts, ammunition racks, target designating devices.

(3) **Protection:**
    (a) *Against bullet splash:* Doors, gun mounts, turret races. Vision devices: Eye slits, laminated glass, periscopes, stroboscopes, geoscopes, goggles and masks.
    (b) *Head protection:* Helmets, eye slit pads.
    (c) *Against observation:* Camouflage nets and brackets.
    (d) *Fire extinguishers.*
    (e) *Against gas:* Gas masks for crew, devices for crew compartment filtration and ventilation, oxygen containers, engine protection.
    (f) *Smoke projection.*
    (g) *Against mines.*
    (h) *First aid supplies.*

(4) **Vehicular operation:** Speedometers and odometers, map boards, direction indicators, lights, bilge pumps, grousers, unditching gear, fascines.

(5) **Engine accessories:** Priming and starting, tachometer and governor, air coolers and oil coolers. Oil pressure, fuel quantity, and temperature guages; fuel feed mechanism.

(6) **Vehicular maintenance:** Tools, spare parts, trouble lights.

(7) **Miscellaneous:** Personal equipment of the crew and racks therefor, rations, water, log book, instruction book, canvas water buckets, funnels for water, oil and gas, canvas engine covers, replacement glass for vision devices, pigeon baskets, towing cables, jacks, extra track plates, axes, picks, shovels and grapnels.

Interior Communication.

In small tanks where the members of the crew are close to each other, special communication equipment has, as a rule, not been furnished. The tank commander has relied upon the voice and a few touch signals for giving directions to the driver. These methods are not wholly satisfactory because the tank commander, who is usually also the gunner, is too much occupied in observing, in loading and operating his weapon, and in turning his turret, to reach down and give the driver a signal for each change in direction or speed. Speaking tubes have been tried on medium and heavy tanks but this method has not been very satisfactory on account of the noise in the tank.

Visual interior communication systems have been advocated, to be so arranged that orders could be transmitted to the driver and other members of the crew by a series of lights installed at each station, the lights and combinations of lights to convey certain messages. Another system for accomplishing the same result by means of indicators at each station, operated by means of connecting wires from the commander's station, has been tried on some British tanks.

Neither system is entirely satisfactory because they are hand operated and tend to divert the attention of the man for whom the message is intended from other important matters, such as observing and operating his weapon.

The German heavy K tank used communication equipment similar to that used by the German submarines.

An intratank telephone system, (laryngaphone) may be used in tanks equipped with batteries. It is probably the best solution as it

Plate 50.
1. Transmitter and Receiver of U.S. Intratank Telephone.
2. British Laryngaphone Set.

leaves the hands free to perform the assigned tasks and does not take the attention of the commander or the members of the crew from their work. As used on the U. S. Mark VIII tank, the transmitter is a small button shaped device held against the larynx by means of an elastic band which is fastened around the throat. The receivers are held against the ears by a band passing under the chin and over the head. When the operator speaks the vocal cord sets up a vibration in the transmitter directly instead of through the air, as with the ordinary telephone transmitter. A switch box is located at each station, the commander having a switch box to which all stations are connected. Each phone set is connected to the switch box for that station by a long cord and plug so that it will not be necessary to remove the set if the operator has to leave his station for any purpose. By closing the proper circuit, the commander may speak to the mechanic, to either of the six pounder gunners, or to the driver. As originally installed, he could also speak to the guide, who wore a head set connected by means of a long cord to the driver's switch box. This part of the system has been discontinued in favor of subdued light signals given by the guide by means of a flashlight.

The British have developed a similar system except that the transmitter is larger and is held against the cheek. As used in the Vickers Armstrongs six ton tank (British), the head sets for the crew are hooked up with the radio set. To reduce the drain upon the batteries in using the phone sets, all circuits are normally open. To close the circuit and use the system, a switch must be turned by the person desiring to speak.

Exterior Communication.

Semaphore apparatus has been used on U. S. and British tanks. This apparatus has been removed from U. S. tanks as the exterior semaphore arms were damaged by trees. Semaphore is not considered a satisfactory means of communication. If used, the metal arms by which the signals are transmitted must be so mounted that they can be turned to present their broad surface toward the person for whom the message is intended. However, this or any other visual means will be of limited value in action on account of obstructions to vision such as rough terrain features, brush, trees, buildings, fog, smoke, dust, haze and darkness. These obstructions will frequently prevent anything like concerted action as a result of orders transmitted by visual signals. Brief signal messages may, under favorable conditions, be sent to the rear by tank crews for short distances, but the prospect of signals being understood at any range of over 400 yards is poor.

Flags have been and are still being used by the tank crews. The number of flags used should be reduced to the minimum and they must be of the proper color. Orange is best as to general visibility and red is next. Although flags can be used only for short distances and their effectiveness is uncertain, they are considered preferable to semaphore equipment of the types used thus far.

A protected opening in the armor should be provided for giving flag signals when the tank is under fire. At the battle of Moreuil in July 1918, when British tanks supported French troops, English speaking French soldiers who were assigned, one to each tank, used flags from the rear doors of the tanks to give messages to the French infantry. While some good was accomplished by the use of the flags, they attracted the attention of the German machine gunners and a great many of these French liaison agents were killed early in the action.

The signal flags used with tanks of various countries range from twelve differently colored flags and a complicated combination of colors and positions, in one army, to the single orange colored flag which is to be used in different positions and with different motions for signal purposes by U. S. tanks. This flag replaces the former standard set of three flags which was cumbersome and difficult to use effectively.

Pyrotechnics may be of value under favorable conditions for sending messages from tanks, but if other troops are also using pyrotechnic signals, or if the atmospheric conditions are not right, these signals will not be of great value.

Special lights, which require protection against hostile fire when not in use, offer a possible means for visual signalling. It might be possible to protect signal lights while in use if operated in conjunction with a periscopic device. Like any other form of visual signals, lights are effective if observed by the persons for whom the message is intended, and by them only. But, when the light must be seen for several hundred yards, its effectiveness is uncertain due to one or more of the conditions previously mentioned. Moreover it is likely to be damaged by bullets and it may be seen by the enemy. A dim light installed or carried at the rear of the tank is useful, during the approach march, in guiding the following tank.

Flashlights, carried by hand and dimmed, enable the guide to transmit messages to the tank driver and are very useful on approach marches.

Audible signals have been advocated and, while they may be found to have little value under certain battle conditions, there is a possibility that some type of siren might be useful in attracting attention and thus assist in maintaining control within small tank units. The signal should be loud enough to be plainly heard by the intended tank crews. It might also perhaps be used to send a simple message to the infantry troops.

There is an instance on record of German infantry having been advised of the capture of a position by means of a bugle call sounded by a member of a German tank crew. A deterimental feature connected with special sirens is the space required. Most tanks are already overloaded with special equipment.

Pigeons have been used to send messages from tanks to the rear and probably will be so used in the future. A limit on their use is the necessity of having the pigeon loft established for about two weeks, before the pigeons will return to it.

Radio telegraphy and **telephony** seem to offer the best means of tank intercommunication thus far tried. From now on tank radio sets are expected to be a part of the tank equipment and space must be allotted for them.

Armament.

Weapons and sights are considered in Chapter XXV.

Shell bags are used in some tanks. They are attached to the machine gun in such a manner that the empty shells will be discharged into the bag instead of upon the floor of the tank where they are a hindrance to the crew and may interfere with the operation of the controls. The British have developed a means for ejecting empty machine gun shells to the outside of the tank.

Hand operated belt loading machines have been carried as tank equipment in some tanks. Such a machine is required if machine gun ammunition is not delivered loaded in tank machine gun belts. The Mark VIII carries 300 fifty-cartridge belts. It is preferable that tank machine gun ammunition be delivered to tank companies already loaded in tank machine gun belts.

Oil cans, cleaning rods, gun patches, gun tools, spare parts, etc., are provided for each weapon mounted, and the spare parts usually include a spare barrel for each machine gun mounted. Some tanks carry extra machine guns. The equipment of the Mark VIII includes five such guns in addition to the five mounted. This and other tanks also carry small tripods for mounting the machine guns upon the ground. The gunners are provided with asbestos mittens so that the machine guns may be handled safely even when they are hot.

Ammunition racks have been made of sheet iron, aluminum, and wood, for carrying machine gun and 37 mm ammunition, and of pipe of suitable size for shells of larger calibre. Machine gun ammunition, in belts, has also in some cases been stored in the tank in the regular machine gun ammunition chest container. Space in the tank being at a premium, a great deal of ingenuity is required to use the space available to the best advantage and provide means for carrying an adequate amount of ammunition together with other necessary equipment. Regardless of the method used to carry ammunition, the belts and shells must be accessible and so placed that they cannot fall to the floor when the tank operates at a sharp inclination due to rough terrain. It is customary to mount individual containers for the larger shells, to place the containers in the horizontal position and provide a movable shell stop to hold the shells in place.

Target designation equipment has been installed in some of the medium and heavy tanks. Such equipment is useful when the tank commander is placed at a higher or more favorable position than the gunners, where he will have better opportunities for observing. This equipment usually consists of a pointer which is located at the gunner's position and which may be moved through the arc represented by the complete traverse of the gun, by means of small control rods or flexible cables connected to a similar pointer at the commander station. The pointer is moved to the position which indicates the direction of the target observed by the tank commander. Range may be indicated by a similar method. Such a system, if reduced in size so as to occupy little space, is valuable even in tanks equipped with telephone communication because it forms an alternative method of communication in case the other method fails.

Protection.

Since the tank is subjected to fire from all classes of enemy weapons, experience has shown the necessity for giving the crew and machinery the fullest possible protection in order that the vehicle may continue in operation long enough to complete its mission. Some agencies aid in protecting from enemy weapons; others are necessary to protect the crew from injury caused by jolts; others protect against heat, and gas fumes generated within the tank.

Bullet splash will find its way into the tank through any opening however small and special arrangements are necessary to prevent the entry of bullet splash at all doors, gun mounts, vision devices, ventilation openings, pistol ports, signal openings and armor joints. The necessity for excluding bullets and bullet splash at points where vision devices are located has resulted in much research and the production of several types of equipment for this purpose. Probably the best form of eye slit protection developed during the war was that used on British tanks and installed on the Mark VIII. It can be closed when not in use and provides a large opening for good vision or smaller openings with a certain amount of protection. On other World War tanks open slits were used, the opening being about $\frac{1}{8}''$ by $4''$ and the armor at the inside edge of the opening being beveled to permit a view of the ground at various angles of elevation and depression.

The open slit is very dangerous as splash from bullets striking within two inches above or below the opening on vertical plates may enter the slit. In addition, the inner bevel mentioned so reduces the thickness of armor plate that the whole bullet may enter when it strikes within a quarter of an inch of the slit.

Many experiments have been conducted with nonshatterable laminated glass mounted between the eye and the open slit. The tests have indicated that, when the assembled glass is of **sufficient thickness and** well made, worth-while protection is afforded. The glass will be cracked by the bullets which strike it, but is not apt to shatter and cause splinters to damage the eyes as was the case with the glass prism periscopes used on the British Mark I. Arrangements are made to permit these protective glass shields to be replaced quickly when they are cracked or smeared sufficiently to interfere with vision. In one experiment in Vienna, with five layers of glass, 9, 10, 14, 10 and 9 mm in thickness and with four celluloid strips between the glass plates, the total thickness being 2.2 inches, the assembled plate withstood direct impact

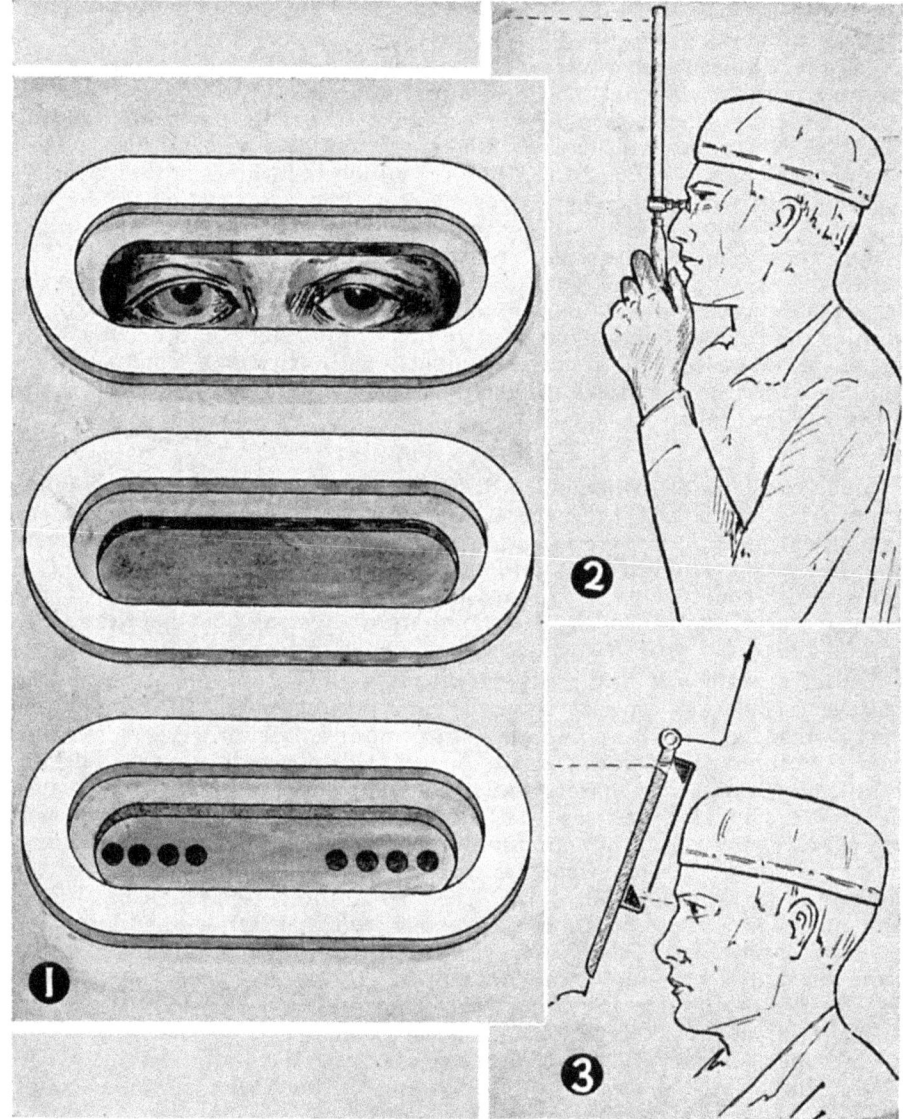

Plate 51.
1. Eye Slits of Mark VIII Tank.
2. Periscope of Type Used with Mark VIII Tank.
3. Proposed Form of Periscope for Tank Driver.

of steel jacketed bullets fired from the Austrian rifle at a distance of about 100 feet. Behind a narrow eye slit, a less thickness of high grade laminated glass may suffice. The eye slits of the U. S. Christie and the Medium T2 are protected with laminated glass approximately 1 inch in thickness.

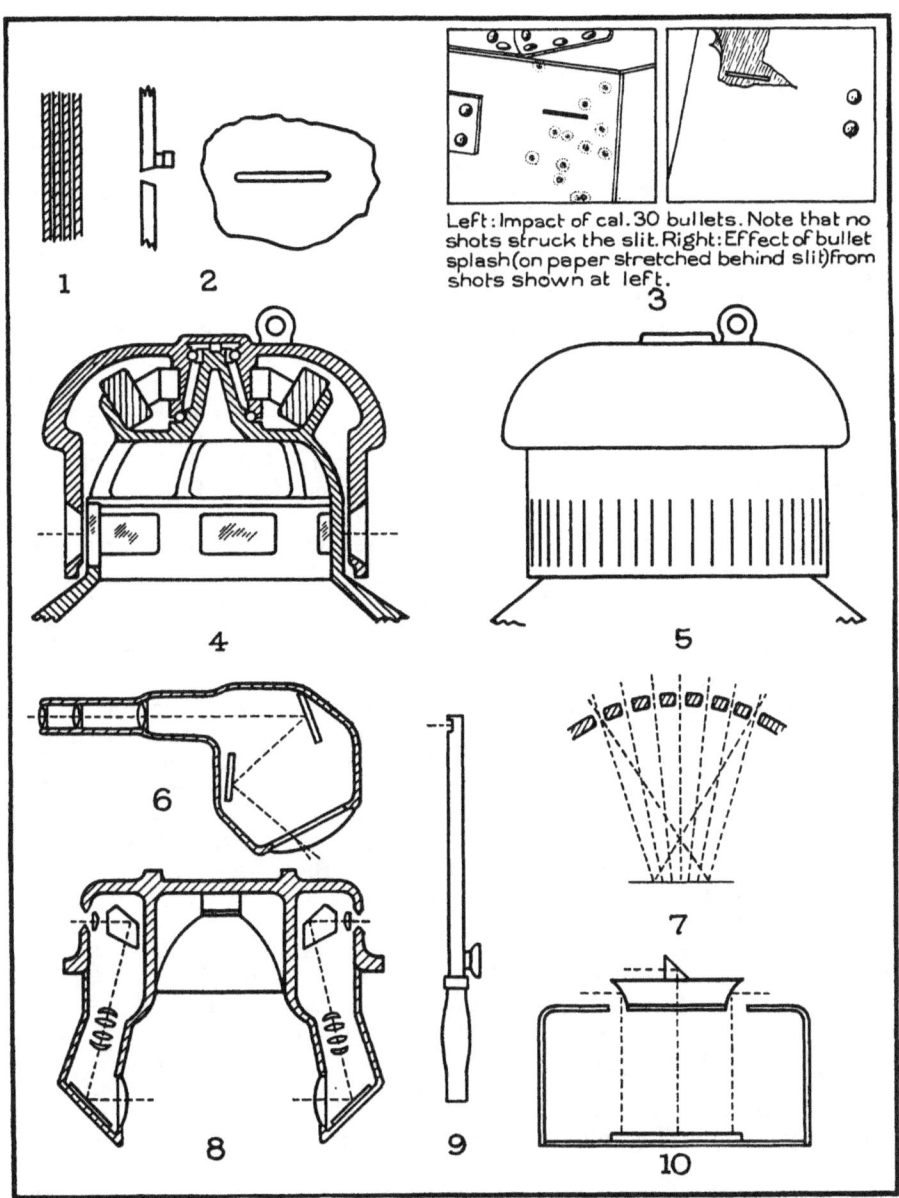

Plate 52.
Tank Vision Devices.

1. Bullet-proof triple glass. 2. Ordinary eye-slit. 3. Eye-slit test. 4. Interior arrangement of French drum stroboscope. 5. Exterior of stroboscope. 6. French geoscope. 7. Slit arrangement of new French stroboscope. 8. Goerz tank periscope. 9. Mark VIII hand periscope. 10. Karnes omniscope.

Periscopes provide the most safety of all vision devices. The Mark VIII equipment includes six small periscopes. Protected holes have been placed in the armor and, when the device used to close these holes has been moved aside, the periscope may be thrust upward through the armor and fairly good vision obtained. The field of vision is limited but the periscope can be turned in any direction. Some tanks carry periscopes mounted in position.

A proposed adaptation of this method to the driver's front vision slit on the U. S. Six Ton tank is shown in Fig. 3, Plate 51. The slit, which is above the driver's head, is covered with a beveled reflector which transmits the view to another reflector mounted on a level with the driver's eyes and, in rear of the opening a piece of armor which forms the frame of the device prevents bullets and bullet splash from entering the tank. The test of this device indicated its practicability provided extra reflectors were carried to replace the upper ones when they became broken It was also found that the reflectors collected dust and had to be removed and wiped off occasionally.

Since eye slits become the aiming points for hostile small arms fire, an effective form of protection, used toward the latter part of the war was the painting of dummy eye slits on the turrets and hull or the use of a checkerboard camouflage over the outer surface of turret for the purpose of preventing the locations of the real eye slits.

A new French vision device is called the **geoscope.** It is presumed to consist of a small lens system with a magnification of about 1½, reflecting an image to a double mirror system and thence to a large reflecting lens which is in no danger of breakage by fire and which presents an image that can be observed without blurring at any convenient distance from the eye.

An ingenious Austrian invention is the **Goerz tank periscope** for all-around vision. This consists of twenty-four prism systems arranged around a dome, twelve systems casting images on a large lens. The eyes must be kept about 10 inches away in observing. In case several prisms are destroyed by fire the whole system can be moved through 15 degrees to bring the other twelve into use and those injured can be quickly removed and replaced. In this case there is no danger to the observer but the system is complicated and very expensive, and the images are quite dark.

The French have conducted many experiments with the **stroboscope,** a device which provides all-around vision. It consists of a circular turret with strips of protective triplex glass mounted therein. Around this turret, an outer cylinder is caused to revolve at a speed of about 450 rpm. The outer cylinder is perforated with a number of vertical slits. When in operation, these slits appear as one opening around the entire device, in much the same way as the great number of pictures shown by the moving picture film appear as one picture—simply because the human eye cannot separate one picture from the other due to the speed with which they are successively flashed. The resulting vision from the stroboscope tested on the Mark VIII tank was cloudy and indistinct and, when subjected to a firing test, the protective glass plates in the inner turret were badly damaged. The same fault was found with the French stroboscope and objects viewed through it did not stand out in relief. The French have continued their experiments and have overcome the fault of depth of vision by cutting the slits in the outer cylinder not only at right angles to the surface of the outer cylinder but also at other angles. The latest form, used on the Chars 2C and 3C, have slits in nine groups of five each, the slits converging on the point of observation. The inner openings are protected by means of removable plates of triplex glass.

Other countries have utilized the stroboscope principle in somewhat different form. The Italian Ansaldo wheel tank has a conical stroboscope with slits cut radially. The shape exposed is such as to cause fire to glance off, reducing splash slightly. In addition vision is possible at several angles from horizontal depending on the pitch of the cone. Another variation is a revolving disc with radially cut slits and the inner opening protected by triplex glass.

Dust settles very rapidly on all outer glass in any vision device, necessitating that provision be made to wipe these surfaces clean quickly and effectively.

Goggles, properly designed, afford in most cases a very desirable protection to the eyes from bullet splash and from dust which comes into the tank through eye slits. In some tanks and especially those in which engine cooling air is drawn from the crew compartment, an excessive amount of dust and large particles of sand are drawn in through the eye slits and all other openings in the armor. In such tanks even well made goggles are not wholly satisfactory because they are quickly covered with dust and require frequent cleaning.

Goggles should be made of the nonshatterable type of glass, should be well ventilated (but without the ventilation currents passing in front of the eye), and comfortable. The material around the sides of the lens should fit the face closely at all points. If not properly ventilated, goggles are more likely to become clouded with vapor on the inner surface and have to be removed frequently for cleaning.

During the war the French used a mask of metal, lined with leather, the eyes being partially protected by metal discs in which slits were cut. Protection for the chin and throat was provided by a chainmail shield which was attached to the lower part of the mask. These masks are said to have been uncomfortable and were little used by the tank personnel.

Head protection for the members of the crew is necessary. This has been provided by leather covered cushion pads mounted above the eye slits to protect the forehead when observing and by a padded leather helmet for each member of the crew.

A certain amount of protection from observation is secured by means of **camouflage nets** and paulins which are a part of the tank equipment. Some tanks are provided with brackets to hold uprights for use with the nets or paulins This subject of camouflage is more fully discussed in Chapter XXXI.

Fire fighting equipment has been necessary in all gasoline propelled tanks and has taken the form generally of the liquid type of fire extinguisher. Our six ton tank carries two of this type, and the Mark VIII carries seven. Both tanks carry one refill for each extinguisher. In addition, the Mark VIII carries one foam type extinguisher. The Christie tank carries a powder type of fire extinguisher. Great improvements are being made in fire extinguishers suitable for tanks and other motor vehicles.

Protection from poisonous gas has usually taken the form of gas masks for each member of the crew, but a considerable amount of experimentation has been conducted by several countries to develop better methods on account of the reduction in the efficiency of the crew when forced to wear the masks.

Some foreign tanks use an air filter through which a sufficient quantity of air is drawn into the crew compartment by a fan to raise the air pressure slightly above atmospheric pressure, and thus prevent the entry of gas laden air through the unavoidable openings. Through these openings, the excess air is discharged. This method has another advantage, aside from the protection afforded from poisonous gases, in that it helps to ventilate the crew compartment and protects against gasoline, oil, and powder fumes.

The Russians are said to be using rubber devices to close the openings in the crew compartment and to include, as a part of their equipment, oxygen containers which supply oxygen to the crew when passing through gassed areas.

Although they do not affect all metals to the same degree, each of the following named commonly used chemical agents does have a considerable corrosive effect upon instruments and other metal parts: phosgene (CG), chlorine (CL), brombenzyl cyanide (CA), adamsite (DM) (a toxic smoke), titanium tetrachloride (FM) (a screening smoke), and hexachlorethane mixture (HC) (a screening smoke). Lewisite (M-1) and mustard (HS) do not corrode metal but do form a film on the surface which is very dangerous to personnel. Provision may be necessary in the future for carrying the solutions needed for cleaning the metal parts after a gas attack. Experiments have not yet shown definitely the effect of all gases upon the engine and its operation, but it may be that some protective device upon the carburetor will be found to be necessary as chemical agents are further developed.

Smoke projection equipment of various kinds has been tried with tanks as a protective measure for the tanks themselves as well as to provide a smoke screen for other troops. Experiments were conducted during the war with rifle smoke grenades and an efficient screen was produced by this means, the rifles being thrust through various openings in the tanks for firing. Another method was to permit oleum or titanium tetrachloride in liquid form to enter the hot exhaust pipe where clouds of nontoxic smoke are then generated and throw out. The British wartime Medium B was the first tank so equipped but, thus far, it is not known to be standard equipment on any other tank in any country.

Some years ago in Poland, a Renault tank was converted to a smoke tank by the installation of two large drums carried over the tracks with two blowers arranged above the hull for directing the smoke under pressure. For this purpose it was necessary to remove the turret. The experiment did not prove entirely successful. A test of smoke projection equipment on the Six Ton tank was made in this country. Two cylinders were mounted near the rear of the hull, one on each side. The cylinders each contained two chemicals which were carried in separate compartments of the cylinder and, by opening valves above the compartments by means of control rods operated from the crew compartment, the chemicals were permitted to mix. Another valve on each container, controlled in the same manner, liberated the resulting smoke which formed the screen. It was intended to use the container on the side which, depending upon the direction of the wind, would best permit the smoke to be carried away from the tank. This experiment was not entirely successful as it was found that, while an efficient smoke screen was produced, the smoke was also drawn into the tank and the crew was severely handicapped thereby. (See Figs. 3 and 4, Plate 53.)

The British experimented in 1928 with a smoke laying tank which produced a very good screen, using compressed air to discharge chlorsulphonic acid. They were not favorably impressed with the idea since they found that the smoke disclosed the general location of the tanks and the smoke tank became a fine target.

Thus, experience to date has shown that it is difficult to provide a satisfactory smoke screen by equipment attached to the tank and, in all experiments, the smoke screen served to outline the tank, making it an excellent target, with the probabilities all in favor of attracting so much

Plate 53.
1. **Mark V Double Star, with Mine Sweeping Equipment.**
(From *In the Wake of the Tank* by Martel.)
2. **Unditching Beam in Use.**
3. **Polish Smoke Tank.**
(From *Taschenbuch der Tanks* by Heigl.)
4. **Experimental Smoke Equipment on U.S. Six Ton Tank.**
(Photo by Ordnance Dept. U.S.A.)

attention to the tank laying the screen that it would not survive long enough to accomplish much.

Mines are one means of combatting tanks and the British brought out one piece of equipment intended to protect the tank from them. The Mark V Double Star tank was equipped with a heavy iron roller attached to a chain and spar at the front of the tank for the purpose of detonating the mines in the path of the tank. Such mine sweeping equipment may be of considerable value where mines are used extensively. (See Fig. 1, Plate 53.)

First aid equipment, for protection of the crew from the results of wounds and burns, and for reviving members of the crew who have been overcome by gasoline fumes in the tank, is generally carried. This equipment should include bandages, iodine, swabs, cotton, ampules of ammonia, adhesive tape, etc.

Equipment Pertaining to Tank Operation.

Speedometers to show the speed of the tank and odometers to show total distance and trip distance have been provided for various tanks, and are deemed necessary.

Map boards, located close to the tank commander's position, are supplied in some tanks and provide a convenient method for carrying the map in such a manner that it can be referred to easily.

Direction indicators are carried in some tanks. Magnetic compasses were used during the war. While they were not entirely satisfactory since they were easily thrown out of adjustment and were affected by the approach of other tanks, they were a big help in some operations.

The Sperry Gyroscope Company based an electrically operated tank **gyroscopic compass** on their experience and had it tested in the Mark VIII tank in 1925, but the compass and the large batteries required for its operation occupied too much room in quarters already cramped. The project was modified to reduce its size and employ vacuum operation but was later dropped. Subsequently the Sperry Company produced a very small vacuum driven gyroscopic direction indicator (not north-seeking), and more recently produced an electrically driven gyroscopic direction indicator (not north-seeking) of very moderate size. The latter is installed in the U. S. Medium T-2 tank and gives fairly good results so long dust or other dirt in material quantities can be kept from the mechanism. The new British 16 ton tank uses the Vickers-Schilovsky compass and courselayer which is reported as a satisfactory substitute for the magnetic compass. This gyroscopic compass operates by compressed air from a compressor on the tank engine. The housing for the device is only ten inches in diameter and the whole weighs about twenty-five pounds. It is reported to be very accurate.

Interior lights mounted in position so that tachometers, oil and temperature gauges may be seen and also trouble lights, are furnished on some tanks. It is very necessary that some kind of lights be supplied. If no other lights are available, the crew should be supplied with good flashlights.

Exterior lights, such as headlights and taillights, are convenient for night operations in rear areas. Where headlights are used, provision should be made to remove them before combat or before entering woods, as removal is less expensive than armoring the lights or replacing them after they have been shot away or damaged by trees.

Water expelling or bilge pumps have been used to force water from tanks during stream crossings. These are generally operated by the engine. A water tight hull is difficult to obtain, as openings must be left for oil drainage. Moreover, entrance doors, pistol ports and observation openings are not easily made water tight. A bilge pump, if available, will lower the water level in the engine compartment and so prevent it from reaching the flywheel or fan by which it might be thrown upon the ignition system. An efficient pump will materially increase the stream crossing ability of a tank.

Grousers are cleats that form part of or are attached to track plates. Sometimes, especially in the detachable form, they are relatively deep and thin. They greatly increase the ability of a tank to pass through mud, swamps, and up steep slopes. Most tank tracks have integral grousers. Some tanks are provided with detachable grousers which can be clamped quickly to track plates.

Unditching gear, in the form of small logs or ties which are fastened to the track plates by chains and thus pulled down under the tank, have been used on some British tanks to aid in moving a tank when it has become stuck. The first unditching gear consisted of a torpedo shaped spar for each track connected by chains and a clamp to a track plate. The spars were superseded by a single hardwood beam ten feet long by one foot square attached by chains to the tracks. Two rails on top of the tank carried the beam when not in use and enabled it to clear the cab when in use. (See Fig. 2, Plate 53.) In 1917 an attempt was made to make the beam a permanent fixture by providing for attachment by means of a clutch, without exposing the crew, but this was not adopted. The single beam was considered by some to be inferior to the first form, a spar for each track, inasmuch as it allowed for no difference in traction and restricted turning. Occasional use was also made during the war of detachable wooden and steel grousers. Detachable wooden grousers may be seen attached to the side of the Medium A tank in Fig. 1, Plate 4. It is believed that the detachable steel grouser is the result of these various experiments.

Large fascines (cylindrical bundles of brushwood) were used in the war to good advantage in crossing wide trenches They were made 10 feet 6 inches in length, were compressed by the use of chains and two tanks pulling in tug-of-war fashion. They were secured by chains and were carried on top of the tanks. A releasing device permitted them to be dropped into trenches without danger to the crew. Sometime later, the fascines were replaced by large cylindrical steel skeleton frames which weighed but 1200 pounds as compared to 3000 pounds for the fascines.

Engine Accessories.

Priming devices have been found to be of assistance in starting tank engines. The most efficient primer thus far tested is the type that injects a small quantity of gasoline, in the form of a spray, into the intake manifold. This device is better than a choke and makes the latter unnecessary.

Engine starting, where large engines are used, is usually difficult, especially in cold weather. At least two kinds of starting systems, the ordinary electrical starter and the hand crank, are provided for most tanks and usually the engine may be started from the outside or the inside of the tank. On one tank, the German A-7-V a number of methods were used, such as a priming pump, an electrical starting motor, a Bosch atomizer, acetylene, and a three-man hand crank. Two engines were used on this tank and, if one engine was started, the tank was moved forward and the other engine could then be started by engaging its clutch.

Even large engines, such as the 12 cylinder Liberty, can be started by hand, however, when a proper gear ratio between the crank and the engine is provided. The Liberty engine in the Christie tank, which is started quite easily by hand, uses a ratio of 12.25 to 1. An efficient primer is also used with this engine. The British Independent tank has such a large engine that a smaller gasoline engine is used to start it.

Inertia starters, both the manual and the electrically driven varieties have been used to start tank engines. They are generally efficacious, but expensive.

A **tachometer** is a desirable accessory as it enables the driver to watch the engine speed under all conditions and provides the only means for determining whether or not the governor is functioning effectively.

The **governor** is a very desirable engine accessory. Experience has shown that tank engines do not cool effectively or receive adequate lubrication when operated at or near their highest speed. For these reasons there will be more depreciation in a nongoverned engine than in one which is prevented by a governor from operating beyond the safe cooling and lubrication limits. It is better to accept a slight reduction in horsepower, at the upper end of the curve, which can rarely be used with safety, than to cause excessive depreciation due to incomplete lubrication by attempting to make available the last fraction of power and to have to accept the subsequent reduction in power as a result of overheating.

Air cleaners, if effective, will reduce engine depreciation. They are needed, due to the fact that the tank engine operates in a cloud of dust most of the time.

An **oil cooler** is an important tank engine accessory. Lubricating oil reaches a high temperature in a tank engine due to the fact that heat accumulates in a tank engine compartment to a greater degree than in the case of an ordinary motor vehicle. The tank engine, being surrounded by armor plate, receives no assistance in cooling from the rush of air due to the movement of the vehicle, as in the case of the truck, car or airplane and, unless the engine compartment is ventilated to remove the accumulating heat, the oil in the crankcase will become unduly hot and lose, in a considerable measure, its lubricating value. Oil coolers have been found effective for reducing the temperature of the oil and they thus insure better lubrication of the bearings and other working parts. By having an oil of lower temperature flow past the bearings, these parts are cooled to an appreciable degree and deterioration of the bearing metal is reduced.

Oil pressure gauges are beneficial as they give an indication of the circulation of the oil and the approximate pressures. A sudden drop in pressure, indicating a break in an oil line, can be detected promptly by the driver only through the use of a pressure gauge.

Oil and fuel quantity gauges are also of value as knowing the quantity of oil and fuel on hand at any time may prevent immobilization of the tank due to lack of fuel, or damage due to the use of excessive speeds when the oil level is low.

Water temperature gauges are of value on a water cooled engine because the tank may, upon some occasions, be operated at a reduced speed to reduce the temperature if the gauge shows that it is reaching the danger point, thus saving water and preventing damage to the engine.

Vehicular Maintenance.

Certain **tools** are necessary for maintenance operations and, while the number carried should be reduced as much as possible to conserve space, the crew should not be hampered, in their job of keeping the tank in an operating condition, by lack of tools to do those things which the crew can and should do. The number of tools actually needed may be reduced if the designer considers the tool question when specifying sizes and shapes of bolts and nuts to be used and gives consideration to the matter of *accessibility,* when the tank is designed, so that the need for special tools may be reduced to a minimum.

Among the maintenance operations to be carried out by the tank crew, for which tools may be required, are the following: Complete lubrication of engine, transmission, power train, suspension system, turret races, etc.; filling and draining of the cooling system, where a water-cooled engine is used; tightening engine, transmission and other support bolts; adjusting valve tappets and ignition equipment; adjusting controls; replacing broken fan belts, on water cooled tanks; adjusting fan belts to take up slack, on water cooled tanks; adjusting main and steering clutches and brakes; adjusting the track; replacing a broken track plate or a track pin.

Track replacement and adjustment require a jack, a sledge and, usually certain special tools for releasing the tension on the track. In general, maintenance work requires certain wrenches, hammers, drifts, screwdrivers, pliers, oil guns, etc. The same wrench may be used on several of these operations if thought is devoted to the subject when the vehicle is designed.

Spare parts are necessary but only such parts as can actually be replaced by the crew should be carried. These should include: Two or more track plates, an adequate number of track pins and retaining pins; an extra fan belt, on water cooled tanks; and an extra silent chain if these are used in accessible places; control linkage pins and cotter pins; extra gaskets, hose connections, hose clamps, and clamp bolts, for water cooled tanks; extra fuses for electrical circuits; extra glass plates for vision devices.

Trouble lights are very necessary accessories because maintenance work frequently must be done at night, or under conditions where the light is poor such as when the tanks are parked in dense woods, or in buildings. These lights must be shielded or dimmed and protected from damage. Flashlights will serve as trouble lights, but, on some tanks which are equipped with battery ignition, a 32 candle power lamp with a cord of the desired length may be used to advantage.

Miscellaneous Equipment.

Space being very limited, the storing of the very considerable amount of equipment and supplies requires that advantage be taken of all suitable space available inside the tank, and it is usually necessary that certain equipment be carried on the outside.

Among the miscellaneous equipment for which room must be made inside the tank is the following: Personal equipment of the crew and racks therefor, rations, drinking water, log book, instruction book, canvas water buckets, funnels for water, oil and gas, canvas engine covers, replacement glass for vision devices, and pigeon baskets.

It has been the practice in the past to carry the towing cable, the jack, the extra track plates, the axe, the pick and the shovel on the outside of the tank. In some cases grapnels may become tank equipment where the tank is required to tear out wire entanglements so that cavalry may pass. These hooks, with the chains or cables by which they are attached to the tank, will be carried on the outside of the tank.

CHAPTER XXII.

MISCELLANEOUS VEHICLES.

COMMAND AND SIGNAL TANKS.

As a result of Colonel Swinton's original *Notes on the Employment of Tanks,* written in February 1916, the British constructed special radio telegraph sets for use in tanks and trained the tank personnel in their use. But it was soon decided not to use them and reliance was placed instead upon visual signals of various kinds. However, following the battle of Arras in April 1917, tank signal companies were formed, which were equipped with means for visual and telephonic communication and which also had Mark IV tanks equipped with radio telegraph sets. In May 1917, the first experiments were made with these Mark IV tanks. Various types of antennae were tried. Six tanks thus converted were used to a limited extent during the Third Battle of Ypres in August 1917. Better use was made of them at Cambrai where they proved very valuable as relay points and information centers. The original radio equipment was of the spark type but early in 1918 this was replaced by the continuous wave.

The French radio telegraph tank was a modified Renault, the superstructure of which was enlarged, giving it a conspicuously characteristic appearance and causing it to be recognized instantly for what it was. The U. S. signal tank of the war period was copied from this.

Both the French and British assigned one signal tank to each tank company. In each case the purpose seems to have been to use them in the axis of signal communication. They were seldom nearer the front than 600 yards and were often at greater distances. They were therefore, in reality, signal tanks, not command tanks.

The present trend in construction and employment differs much from the practice of the war period. Radio equipment (usually telephone), for reception at least, is regarded as desirable for each tank. Radio equipment in tanks generally is not permitted to interfere with the use of normal tank armament. The demand for command tanks is, however, growing. In tank units, they are for the use of company commanders, higher commanders, and perhaps staff officers. It seems that such vehicles might also, under some circumstances, be of much value for the use of commanders or staff officers of large infantry units in some kinds of reconnaissance and inspection activities. According to the U. S. view, command tanks should differ in no way in external appearance from the combat tanks.

British. (Carden Loyd) Light Tank Mark II.

Produced in 1931 by Vickers Armstrongs, Ltd. Total production about 12.

Except for the different shape of turret, this tank is substantially the same as the (Carden Loyd) Light Tank Mark 1A. Tanks of this type are used mainly by commanders and liaison officers.

In the United States, although radio telephone equipment has been developed suitable for use in combat tanks, no special command tanks have as yet been constructed. Great Britain is using radio telephony to a great extent in her medium and six-ton tanks and has a few command tanks. The Czechoslovakian KH-50, when used as a command tank, carries three men,

Plate 54.
1. (Carden Loyd) Light Tank Mark II. (A command tank.)
(From *London Times*.)
2. Mark IV Tank with Early British Type of Antenna.
(From *Royal Tank Corps Journal*.)
3. U.S. Signal Tank. (Copied from a modified French Renault.)
(Photo by Tank School, U.S.A.)
4. Modern British Command Tank of Vickers Medium Type. (The one leading.)
(Wide World Photos.)

and both transmitting and receiving equipment, as does the Swedish M-21. In these latter cases the combat tanks carry receiving sets only.

The Marconi 50-Watt Tank Set Type S.B.1a is used in the Vickers Armstrongs 6 ton tanks. Telephonic communication can be carried on over a distance of 1½ miles between two moving tanks and a distance of 5 miles when both tanks are stationary. The antenna is a twelve foot vertical jointed steel rod, and the set is very compact, operating on a wave length of 7 to 8 meters. The transmitter is a master oscillator-power amplifier type while

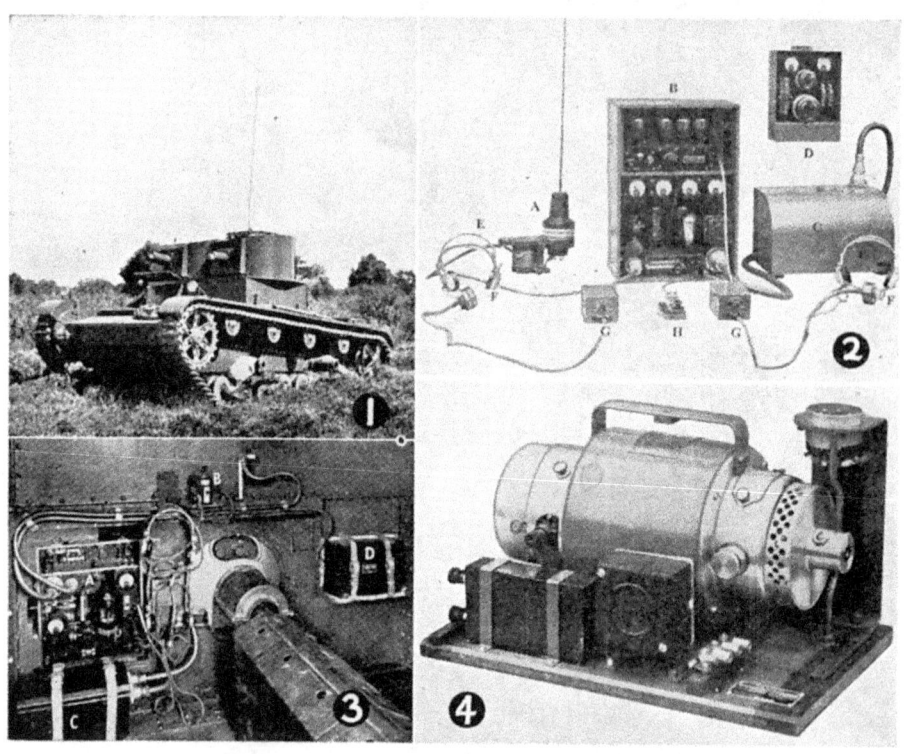

Plate 55.
1. **Vickers Armstrongs Six Ton Equipped with Radio.**
(Photo by Vickers Armstrongs, Ltd.)
2. **Type SB 4a Equipment with Intercommunication Gear.**
(From *Royal Tank Corps Journal*.)

A. Aerial and coupling coil	D. Switchboard	G. Intercommunication units
B. Instrument box	E. Telephones	H. Manipulating key
C. Rotary transformer	F. Microphones	

3. **Type SB 1a Radio Equipment as Installed in Vickers Armstrongs Tank.**
(Photo by Vickers Armstrongs, Ltd.)

A. Transmitter and receiver
B. Manipulating key and intercommunication unit
C. Case containing rotary transformer
D. Case containing H. T. battery

4. **Rotor Transformer and Relay. Cover removed.**
(From *Royal Tank Corps Journal*.)

the receiver has a detector and three stages of amplification. Either laryngaphone or ordinary microphone is used for telephony and a key is also cut into the circuit. The set itself is supported in the tank on thick rubber cushions. An improvement over this S.B.1a set is the recent Marconi 75 Watt Short Wave Tank Set Type, S.B.4a In the *Royal Tank Corps Journal* the improvements in the latter set are set forth as follows:

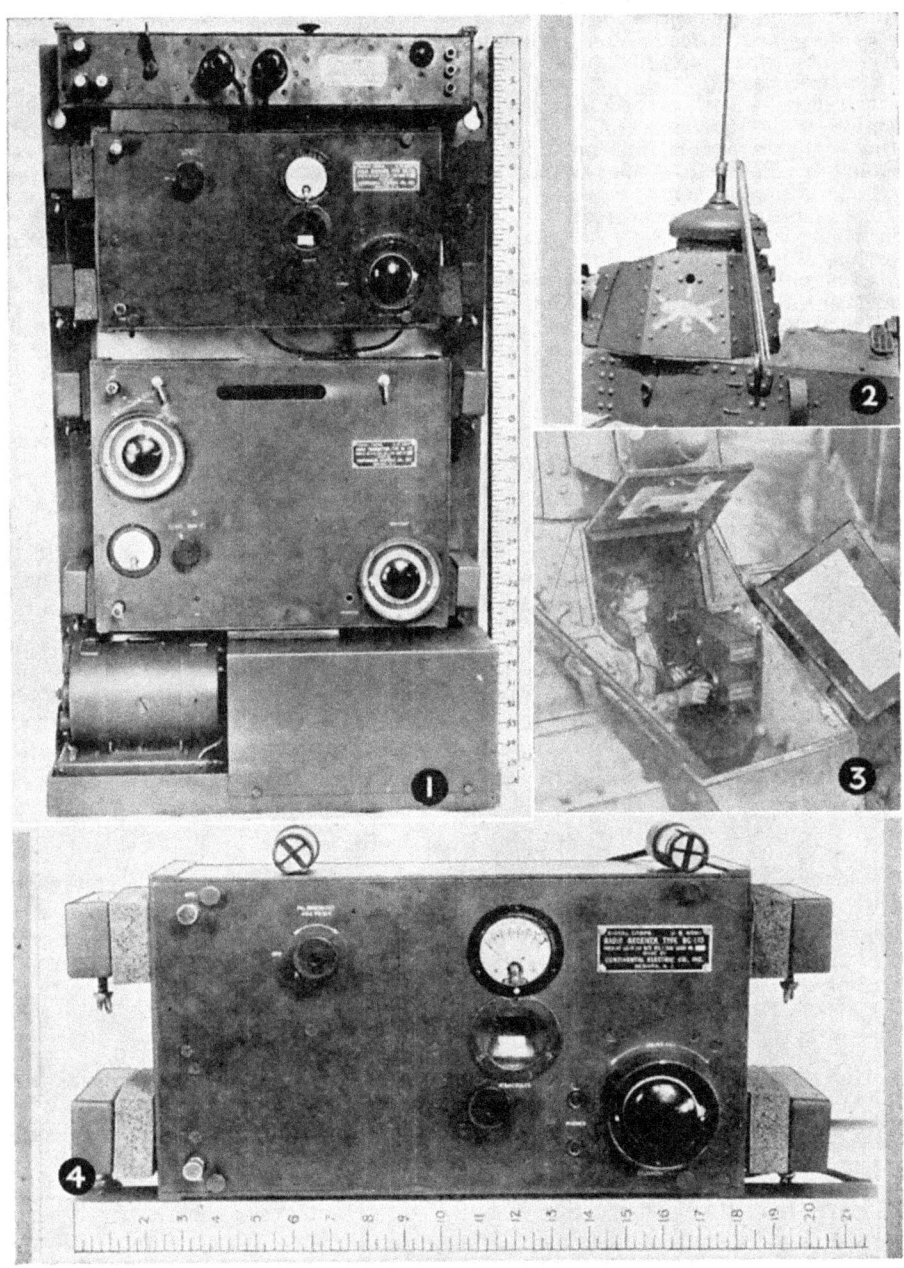

Plate 56.
1. U.S. Radio Set, SCR 189 T2-E1. Sending and Receiving.
   (Photo by Signal Corps, U.S.A.)
2. Antenna for U.S. Six Ton Tank.
3. Location of Radio Set in U.S. Six Ton Tank.
4. U.S. Radio Set, SCR 190 T2-E1. Receiving.
   (Photo by Signal Corps, U.S.A.)

Among these improvements may be briefly mentioned the following: (1) Increased power of the transmitter. (2) The use of three instead of four valves (tubes) in the transmitter. (3) Four valves of one type are used in the receiver. (4) The high tension battery is mounted in the instrument box. (5) The provision of a small switchboard to enable the filament battery to be charged from the power battery.

The Type S.B.4a set provides for the transmission and reception of telegraph and telephone signals and also for intercommunication and dual control if desired. The complete equipment, including its aerial system, can be permanently installed in the vehicle and enable communication to be carried out while the vehicle is on the move.

The outstanding features of the set, which render it particularly suitable for such purposes, may briefly be stated as follows: (1) The almost invisible aerial system. (2) The secrecy with which communication can be conducted. (3) Communication can be carried out while on the move. (4) Simplicity and ease of operation. (5) Compactness and portability.

When the set is installed in a tank or other vehicle the provision of the dual control and intercommunication system is particularly advantageous in that it provides a method whereby either of two persons can use the wireless set or they can communicate with each other without using the wireless.

The feature mentioned in the foregoing paragraph is regarded as particularly interesting and conducive to efficiency in communication.

In the U. S. Army, a Signal Corps radio telephone and telegraph set for tanks, as illustrated, has been tested and found satisfactory for short range work. It is in use in limited numbers in light tanks. This set is identified as follows: sending and receiving set complete, SCR 189 T2-E1; receiving set only, SCR 190 T2-E1; transmitter box only, BC-176; receiver box only, BC-175.

The tubular antenna is 15 feet tall and in five sections. It is easily depressed 90 degrees or more in any direction. It rights itself as soon as released. For telephone use, these sets operate very well over a range of two miles with the tanks in motion and over a greater range if the tanks are stationary. A still greater range is possible by using the telegraphic key.

Another set with greater range is being developed.

Command Post Vehicles.

The desirability should not be overlooked of providing a mobile command post vehicle with cross-country ability, for combat units generally. One such vehicle could serve for a battalion, and vehicles in larger numbers for the higher units as required. The equipment should be removable for transporation by hand in an emergency.

HALF-TRACK COMBAT VEHICLES.

Half-track combat vehicles constitute plainly a hybrid that is intermediate between the tank and the wheel-type armored car. Such a vehicle may be expected to have more cross-country ability than the armored car, and necessarily, less than the tank. This sort of vehicle has found considerable favor in Europe but it has never aroused much interest in the United States. It is believed that this construction has the disadvantages of both the armored car and the tank and the advantages of neither. Since it must always operate its short length track, it can never have a road speed comparable with that of a fast armored car. On the other hand, since it rests a considerable fraction of its weight upon a simple pair of wheels (normally at the front), it cannot possibly have cross-country ability equivalent to that of the tank.

It should be understood that the foregoing comment refers only to the construction wherein the track is not removable for travel upon roads. If such vehicles were built upon the wheel or track idea, the merits of the plan might be much greater.

Half-Track Combat Vehicles

British. Morris-Martel.

Produced in 1925-1926 by Morris Commercial Car Company. Total production, 16.

Crew: First tanks had crew of one, later model carried two.
Armament: One machine gun.
Armor: 0.4 in.
Maximum speed: 20 mph.
Suspension: Wheels spring suspended and half tracks with sprung rollers.
Tracks: Morris commercial plates, later modified to laterally flexible and finally rubber jointed track plates, width 10 in.
General arrangement: Engine in front; driver in center. In two man model, driver on right.
Dimensions: Length 9 ft. 10 in.; width, one man model, 3 ft. 9 in.; two man model, 4 ft. 7 in.; height, one man model, 4 ft. 10 in.; two man model, 5 ft. 6 in.
Weight: One man model 2.2 tons; two man model, 2.75 tons.
Engine: Morris, 14 HP.
Horsepower per ton: 6.4 and 5.1.
Transmission: Sliding gear, 3 speeds forward, 1 reverse, with additional planetary low gear.
Obstacle ability: Trench 3 ft. 6 in.; stream 1 ft. 10 in.; slope 40 degrees; vertical wall 18 in.; tree 7 in.

Plate 57.

1. Morris Martel.
2. Schneider Half Track.
3. Crossley Martel.
(From *Taschenbuch der Tanks* by Heigl.)
4. Autochenille, Model 1928.
(From *Royal Tank Corps Journal*.)
5. Austin Half Track.
(From *Taschenbuch der Tanks* by Heigl.)

Fuel distance: 62 miles. **Fuel capacity:** 11 gallons.
Special features: Tail wheels kept in contact with ground by heavy leaf springs. Wobbled badly at speeds over 15 mph. First British scout tank. Driver's seat could be lowered six inches so as to bring the driver's head within the turret. Tracks removable. It was hoped to place this vehicle in industrial use in large numbers, as a war reserve.

British. Crossley-Martel.

Produced in 1927 by Crossley Motor Company. Total production, 1.

Crew: 1.
Armament: One machine gun.
Armor: 0.4 in.
Maximum speed: 18.6 mph.
Suspension: Leaf springs, bogies and rollers.
Tracks: Kegresse rubber friction driven, width 5½ in.
General arrangement: Driver in front; engine in rear.
Dimensions: Length 10 ft.; width 4 ft.; 9 in.; height 5 ft. 4 in.
Weight: 1.5 tons.
Engine: Crossley, 14 HP.
Horsepower per ton: 9.3
Transmission: Sliding gear, 3 speeds forward, 1 reverse.
Obstacle ability: Trench 3 ft. 11 in.; stream 2 ft.; slope 35 degrees; vertical wall 1 ft. 4 in.
Fuel distance: 124 miles. **Fuel capacity:** 14 gallons.

French. Schneider Half-Track.

Produced in 1922. Total production, 100.

Crew: 3.
Armament: One 37 mm (1.46 in.) semiautomatic gun and one machine gun.
Armor: 0.236 in.
Maximum speed: 16 mph.
Suspension: Kegresse.
Tracks: Rubber track 9½ in. wide.
General arrangement: Engine in front; driver in center; gunners in rear.
Dimensions: Length 11 ft. 2 in.; width 4 ft. 7 in.; height 7 ft. 6½ in.
Weight: 2.4 tons.
Engine: Citroen, 18 HP, water cooled.
Horsepower per ton: 7.5
Obstacle ability: Slope 30 degrees.

French. Autochenille 1928-1929 (Half-Track).

One type produced in 1928, another in 1929, by Schneider. Total production, several.

Crew: 2 or 3.
Armament: 1928, one 20 mm (.79 in.) machine gun forward and one rifle caliber machine gun rear. 1929, one 37 mm (1.46 in.) gun and one machine gun in one mount and one additional machine gun.
Armor: 0.5 in.
Maximum speed: 35 mph on roads.
Suspension: Kegresse.
Tracks: Kegresse, half metal.
Dimensions: Length 15 ft. 9 in.; width 5 ft. 10 in.; height 8 ft. 2 in.
Weight: 6.6 tons (1928 model).

Russian. Austin Half-Track.

Produced in 1916-1917 by Austin Motor Company. Modified in Russia.

Crew: 5.
Armament: Two machine guns.
Armor: 0.3 in.
Maximum speed: 16 mph.
Suspension: Early Kegresse.
Tracks: Rubber, 11.8 in. wide.
Dimensions: Length 20 ft. 8 in.; width 6 ft. 2 in.; height 7 ft. 10 in.
Weight: 6.4 tons.
Engine: 4 cylinder, 50 HP, water cooled.
Horsepower per ton: 7.8.
Transmission: Sliding gear.
Obstacle ability: Trench 5 ft. 10 in.; stream 2 ft.; slope 30 degrees; vertical wall 15 in.
Fuel distance: 50 miles.
Special features: Two models; one with turrets abreast, one with turrets echeloned. Turrets have 270 degree traverse. Converted from armored cars. Extension rollers were added to increase obstacle crossing ability.

ARMORED CARS.

A military armored car occupies an intermediate position between an automobile and a modern fast tank. It partakes of as many of the desirable qualities of the tank as practicable without losing its likeness to the automobile. Its combat value is higher than that of an automobile bearing men and weapons, but materially less than that of a fast tank. For the efficient employment of armored cars a reasonably good network of roads is necessary.

The value of armored cars in warfare depends upon the general situation. They are most useful in open warfare when the field of operation is very extensive. If the situation is compact and more or less stabilized, the roads cannot be used advantageously to any considerable degree, the armored cars cannot employ their mobility, and, since their ability off roads is limited and uncertain, their tactical value is very much less than in the former situation.

Armored cars are suitable for the following tactical purposes: (1) Long distance reconnaissance. (2) Screening. (3) Pursuit of disorganized and demoralized hostile forces. (4) Delaying action. (5) In a sufficiently open situation, for raiding.

Armored cars will probably be employed, for the most part, in collaboration with cavalry or fast tanks, rather than independently.

The main questions involved in the determination of a suitable type of armored car are the following:

(1) Is a standard chassis of an automobile or fast motor truck from the commercial world to be used, or is a special chassis to be constructed? (The former is very advantageous from the standpoints of economy and rapid production in large numbers. The latter will probably be more efficient.)

(2) Is it necessary to have the desirable qualities of efficient steering and full speed when traveling to the rear?

(3) How many wheels should there be?
(4) How many wheels should drive?
(5) What kind of tires are best?
(6) What speed is desired?
(7) What should the armament be?
(8) What armor protection is necessary?
(9) Should tracks be carried for cross-country travel?

Some of the desirable qualities in an armored car are in conflict. They cannot all be incorporated in one type. If we are to have the advantages of a commercial chassis, we cannot have efficient steering and high speed when moving to the rear, nor can we have six wheel drive unless we find a chassis of that unusual type that is otherwise satisfactory. If we are to have the speed and ability to negotiate difficult ground that pneumatic tires will give, we must expect our tires to be more vulnerable to hostile fire than are solid tires.

Since the value of armored cars lies, to a considerable degree, in their availability in large numbers, it is thought that the use of a commercial chassis may, at the beginning of a war, be regarded a necessity. All things considered, then, it is believed that initially a very satisfactory armored car would be such as might be constructed in accordance with the following general specifications:

(1) Built upon a standard commercial chassis.
(2) Not suitable for rapid travel to the rear (necessity).
(3) Six wheels.
(4) Four or six wheel drive (preferably six).
(5) Tires to be pneumatic or some kind of a relatively soft, puncture proof, cushion type.

(6) Maximum speed of at least 50 miles per hour.

(7) Armament to include one antitank weapon at least as effective as the cal. .50 machine gun; also one or two ordinary (cal. .30) machine guns.

(8) Armor to be sufficient generally to stop the cal. .30 service ammunition and over the vital parts to stop the cal. .30 armor piercing ammunition.

Due to the poor cross-country ability of the ordinary wheeled vehicle, and because the commercial truck or car chassis is not built for the rough cross-country work to be done by the armored car, it is desirable that there also be a vehicle that is designed especially for this work. It should embody the wheel and track principle, or a device similar to the Hipkins traction device should be carried, because the armored car equipped with wheels only will be, especially in wet weather, practically confined to the roads and it will, on this account, be severely handicapped in many situations.

British. Lanchester Six Wheel.

Produced in 1929 by Lanchester Motor Company.

Crew: 3 or 4.
Armament: Two machine guns (one cal. .50 and one cal. .303).
Armor: 0.31 in.
Maximum speed: 45 mph.
Wheels: 6 wheels, 4 wheel drive.
Tires: Pneumatic, all rear wheels dual.
Length: 17 ft.
Weight: 7.44 tons.
Engine: Lanchester, 45 HP.
Horsepower per ton: 6.
Transmission: Planetary, 3 speeds forward, 1 reverse.
Fuel distance: 200 miles.
Special features: Will ford a depth of 2½ ft. One model with a slightly modified turret has a double machine gun mount therein.

British. Crossley-Vickers Armstrongs Heavy.

Produced by Vickers Armstrongs, Ltd.

Crew: 4.
Armament: Two machine guns, with armored water jackets.
Armor: Covering vital parts protects against rifle fire.
Maximum speed: 50 mph.
Wheels: 6 wheels, 4 wheel drive.
Tires: Pneumatic, all dual at rear.
Weight: 8.28 tons.
Fuel distance: 150 miles. **Fuel capacity:** 30 gallons.
Special features: Four machine gun mountings permitting very quick shifting of machine guns. The dummy axles carrying the spare tires enable the car to cross a trench 3 ft. 6 in. wide. Body and turret are lined with asbestos. Two ditching boards are carried that enable the car to pass over gullies or wide holes. Light tracks are provided for the two pairs of rear wheels for use over soft ground.

British. Crossley-Vickers Armstrongs Light.

Produced by Vickers Armstrongs, Ltd.

Crew: 3.
Armament: Two machine guns with armored water jackets.
Armor: Vertical plates, 0.276 in.
Maximum speed: 40 mph.
Wheels: 6 wheels, 4 wheel drive.
Tires: Pneumatic, all single.
Dimensions: Length 17 ft. 3½ in.; width 6 ft. 4 in.; height 8 ft.
Weight: 5 tons.
Engine: 50 HP.
Horsepower per ton: 10.
Fuel capacity: 24 gallons.
Special features: Light tracks are provided for the two pairs of rear wheels for use over soft ground. Two ditching boards are carried that enable the car to pass over gullies or wide holes.

Swedish. M 1926.

Produced in 1926 by Artillery Department.

Crew: 5.
Armament: One 37 mm (1.46 in.) gun and one machine gun, both in rotating turret.

Maximum speed: 35 mph forward, 32 mph backward.
Wheels: 4 wheels, 4 wheel drive.
Tires: Single, hard.
Height: 8 ft. 6 in.
Wheel base: 11 ft. 6 in.
Weight: 5.5 tons.
Engine: Tidaholms, 4 cylinder, 45 HP.
Horsepower per ton: 8.2.
Transmission: Same number of speeds forward and backward.
Special features: Steers at both front and rear.

Plate 58.

1. **Lanchester Six Wheel, British.**
   (From *Royal Tank Corps Journal.*)
2. **Crossley Vickers Armstrongs Heavy, British.**
   (Photo by Vickers Armstrongs, Ltd.)
3. **Model 1926, Swedish.**
   (From *Army Ordnance*, U.S.A.)
4. **Crossley Vickers Armstrongs Light, British.**
   (Photo by Vickers Armstrongs, Ltd.)
5. **PA II 1923, Czechoslovakian.**
   (From *Taschenbuch der Tanks* by Heigl.)

Czechoslovakian. PA 2 (M 1923).

Produced in 1923 by Skoda Company. Total production, 30.

Crew: 5.
Armament: Four machine guns.
Armor: 0.315 in. on front and sides.
Maximum speed: 38 mph.
Wheels: 4 wheels, 4 wheel drive.
Tires: Single, pneumatic.
Dimensions: Length 20 ft. 2 in.; width 7 ft. 2 in.; height 8 ft. 5 in.
Weight: 7.7 tons.
Engine: 84 HP, water cooled.
Horsepower per ton: 11.
Transmission: Sliding gear, 4 speeds forward, 4 reverse.
Steering: Double steering controls forward and backward.
Special features: Armored headlights. Difficult to determine front or rear. Each gun has 90 degrees traverse. Excellent form of armor but very expensive; these cars are said to have cost $60,000 each.

U. S. Light Armored Car (Chevrolet).

Produced in 1930 by Holabird Quartermaster Depot. Total production, 5 (3 Whippet chassis, 1 Plymouth, and 1 Chevrolet).

Crew: 3.
Armament: One cal. .30 machine gun, air cooled and one Thompson submachine gun.
Armor: ⅛ in.
Maximum speed: 60 mph.
Wheels: 4 wheels, 2-wheel drive.
Tires: All single and pneumatic.
Dimensions: Length 13 ft.; width 5 ft. 8 in.; height 5 ft. 9 in.
Wheel base: Varies with chassis used.
Weight: About 2 tons.
Engine: 3 Whippet, 4 cylinder; 1 Plymouth, 4 cylinder; 1 Chevrolet. 6 cylinder; forced water cooling; 19 to 23 HP, depending upon make.
Horsepower per ton: Approximately 10.
Transmission: Sliding gear; 3 speeds forward, 1 reverse.
Steering: Steers front wheels only and only from the front.
Fuel distance: 250 miles. **Fuel capacity:** 19 gallons.

U. S. Four Wheel Drive Armored Car (Franklin).

Produced in 1930 by Holabird Quartermaster Depot. Total production, 6.

Crew: 4.
Armament: One cal. .50 water cooled machine gun, two cal. .30 air cooled machine guns, and one Thompson submachine gun.
Armor: .25 in.
Maximum speed: 65 mph.
Wheels: 4 wheels, 4 wheel drive.
Tires: Single in front, dual in rear, all pneumatic.
Dimensions: Length 13 ft. 10 in.; width 5 ft. 9 in.; height 7 ft. 7½ in.
Wheel base: 106 in.
Weight: 3.6 tons, complete.
Engine: Franklin, 6 cylinder, 95 HP, air cooled.
Horsepower per ton: 26.4.
Transmission: Sliding gear, 8 speeds forward, 2 reverse.
Steering: Steers front wheels only and only from the front.
Fuel distance: 300 miles. **Fuel capacity:** 30 gallons.

U. S. Armored Car T4.

Produced in 1931 by James Cunningham Sons, and Company. Total production, 2.

Crew: 4.
Armament: One cal. .50 machine gun and one cal. .30 machine gun in combined mount; one cal. .30 machine gun, AA.
Armor: 5/32 in. to ⅜ in.
Maximum speed: 55 mph.
Wheels: 6 wheels; 4 wheel drive.
Tires: All single pneumatic.
Dimensions: Length 15 ft.; width 6 ft.; height 6 ft. 11 in.
Wheel base: 99 in.; 42 in. (axle to axle, 3 axles).
Weight: 4¼ tons.
Engine: Cunningham, V type, 8 cylinder, 135 HP, forced water cooling.
Horsepower per ton: 31.8.
Transmission: Sliding gear, 4 speeds forward, 1 reverse.
Steering: Steers front wheels only and only from the front.
Fuel distance: 200 miles. **Fuel capacity:** 30 gallons.
Special features: The two spare wheels are so mounted that, in some circum-

stances, they may make contact with the ground, rotate, protect the belly of the vehicle, and contribute slightly to cross-country ability. The car has six-wheel mechanical brakes. Ten more of these cars are being built. The fact that the two front wheels do not drive reduces materially the cross-country mobility of this vehicle.

Plate 59.
1. **Light Armored Car, Chevrolet.**
(Photo by Signal Corps, U.S.A.)
2. **Four Wheel Drive Armored Car, Franklin.**
(Photo by Signal Corps, U.S.A.)
3. **Armored Car T-4.**
(Photo by Ordnance Dept., U.S.A.)

CHAPTER XXIII.

MISCELLANEOUS VEHICLES (Continued).

SELF-PROPELLED ARTILLERY.

The more than casual interest of tank personnel in self-propelled artillery is due to the fact that some such type of artillery, if sufficiently mobile, will be suitable for accompanying a tank force when the latter is out of touch with the normal artillery which supports infantry. When an armored force is acting independently and its assault tanks must move against hostile fire, if the supporting fire from its 2d echelon tanks is inadequate, it would seem likely that the larger guns of self-propelled artillery might be the solution of the problem. There is the possibility, however, that all-around armor protection for such vehicles might be found necessary, in which case the self-propelled artillery to accompany tanks would be nothing more or less than supporting tanks or, as the British would call them, close support tanks.

There is a marked tendency among tank personnel to regard tractor drawn artillery as unsuitable for the purpose of accompanying tanks. In the first place, if the tanks are slow, they will operate close to and in close cooperation with the foot troops. In such a case the normal artillery is expected to support them, although there is doubt as to its ability to do so adequately. On the other hand, if the tanks are fast and are used on relatively independent missions, artillery which accompanies them must have substantially equal mobility. Judging by past experience tractor-drawn artillery is quite slow and hence unsuitable. Even if it can be adequately speeded up, there still remains the limbering and unlimbering feature. It is believed that the supporting fire for fast tanks on independent missions must be very close. Otherwise the fire may not be put down with sufficient promptness or it may be too inaccurate, and there is a gravely increased danger that the tanks and their supporting artillery may become lost from each other. In this very close support, which seems to be necessary, it appears that the delay and exposure incident to unlimbering and limbering is very undesirable and should be avoided if possible. These various considerations cause the preference for the self-propelled type.

In connection with supporting fire for assault tanks, in addition to close support tanks, there is a widely recognized need for a type of highly mobile vehicle that can put down smoke effectively. Long range and great accuracy not being required, a mortar seems to be indicated. Such a vehicle, in addition to being an important tank auxiliary, would perhaps have a high value in support of foot troops. It seems altogether probable that a vehicle of this general nature will constitute an important element in the armored forces of the future.

British. Gun Carrier Tank.

Produced in 1917 by William Foster and Company, Ltd. Total production, 48.

Crew: 3 exclusive of gun crew.
Armament: 60 pounder (5 in.) rifle or 6 in. howitzer.
Armor: 0.3 in.
Maximum speed: 3 mph.
Suspension: Rigid, with rollers.
Tracks: Flat steel plates, width 17 in.

Self-Propelled Artillery

Plate 60.
1. **Gun Carrier Tank.**
(Photo by Wm. Foster and Co., Ltd.)
2. **Self-Propelled Mount for 18-Pounder, Model 1926.**
(From *Royal Tank Corps Journal.*)
3. **Self-Propelled Mount for 18-Pounder, Model 1927. Birch Gun.**
(Fox Photos, London.)

General arrangement: Driver and brakeman in separate cabs over tracks, left and right, respectively; engine and final drive in rear.
Dimensions: Length 30 ft.; with gun 42 ft. 10 in.; width 11 ft.; height 9 ft. 4 in.
Weight: 30.2 tons empty; 38 tons loaded.
Engine: Daimler, 6 cylinder, 105 HP, forced water cooling.
Horsepower per ton: 2.76 loaded.
Transmission: Same as Mark IV but modified to accomodate new position of engine.
Obstacle ability: Trench 11 ft. 6 in.; stream 2 ft. 7 in.; slope 35 degrees; vertical wall 2 ft. 5 in.
Fuel capacity: 70 gallons.
Special features: Winch drums inside hull. Gun carried on frames, one of which was used as a loading ramp. Gun could be fired from tank or dismounted, usually the latter. Used mainly as supply tanks. Two modified as salvage tanks with 3 ton hand operated jib cranes and no front cabs. Bracing legs used to increase lifting capacity. Driver in turret in front of engine. A Gun Carrier Mark II and Salvage Mark II were built, based on Mark V in both appearance and construction.

British. Self-Propelled Mount for 18-Pounder, M 1926.

Produced in 1926 by Vickers Armstrongs, Ltd. Total production, several.

Crew: 5.
Armament: One 18-pounder (75 mm-2.95 in.) gun. Could be elevated to 90 degrees for antiaircraft fire.
Maximum speed: 15 mph.
Suspension: Same as Vickers Medium tank, Mark I.
Tracks: Same as Vickers Medium tank, Mark I.
General arrangement: Driver right front; engine left front; gun crew in rear.
Engine: Armstrong-Siddeley, 90 HP, air cooled.
Special features: Built on Vickers tank chassis. Crew not protected.

British. Self-Propelled Mount for 18-Pounder, M 1927 (Birch Gun).

Produced in 1927 by Vickers Armstrongs, Ltd. Total production, several.

Crew: 5.
Armament: One 18 pounder (75 mm-2.95 in.) gun.
Maximum speed: 15 mph.
Suspension: Same as Vickers Medium tank Mark II.
Tracks: Same as Vickers Medium tank Mark II.
General arrangement: Driver right front; engine left front; gun crew in rear.
Special features: Driver unprotected. Gun crew partly protected by shield. Built on Vickers tank chassis.

British. Self-Propelled Mount for 18-Pounder, M 1929.

Produced in 1929 by Vickers Armstrongs, Ltd.

Crew: 3.
Armament: One 18-pounder (75 mm-2.95 in.) gun.
Armor: 0.4 in.
Maximum speed: 12 mph.
Suspension: Same as Vickers Mark II.
Tracks: Same as Vickers Mark II.
General arrangement: Engine in front; crew in center; final driver in rear.
Dimensions: Length 19 ft.; width 7 ft. 10 in.; height 7 ft. 6 in.
Weight: 13.4 tons.
Engine: Armstrong-Siddeley, 8 cylinder, 90 HP, air cooled.
Horsepower per ton: 6.7.
Transmission: Same as Vickers Mark II tank.
Obstacle ability: Trench 6 ft. 9 in.

British. Carden Loyd With Stokes Mortar.

Produced in 1930 by Vickers Armstrongs, Ltd.

Armament: One Stokes mortar.
Special features: All features same as Carden Loyd Mark VI. Stokes mortar mount on left of front deck. Can be fired from vehicle but is usually dismounted for firing.

British. Carden Loyd With 47 mm Gun.

Produced in 1930 by Vickers Armstrongs, Ltd.

Armament: One Vickers short 47 mm (1.85 in.) gun.
Special features:. All features same as Carden Loyd Mark VI except armament.

SELF-PROPELLED ARTILLERY

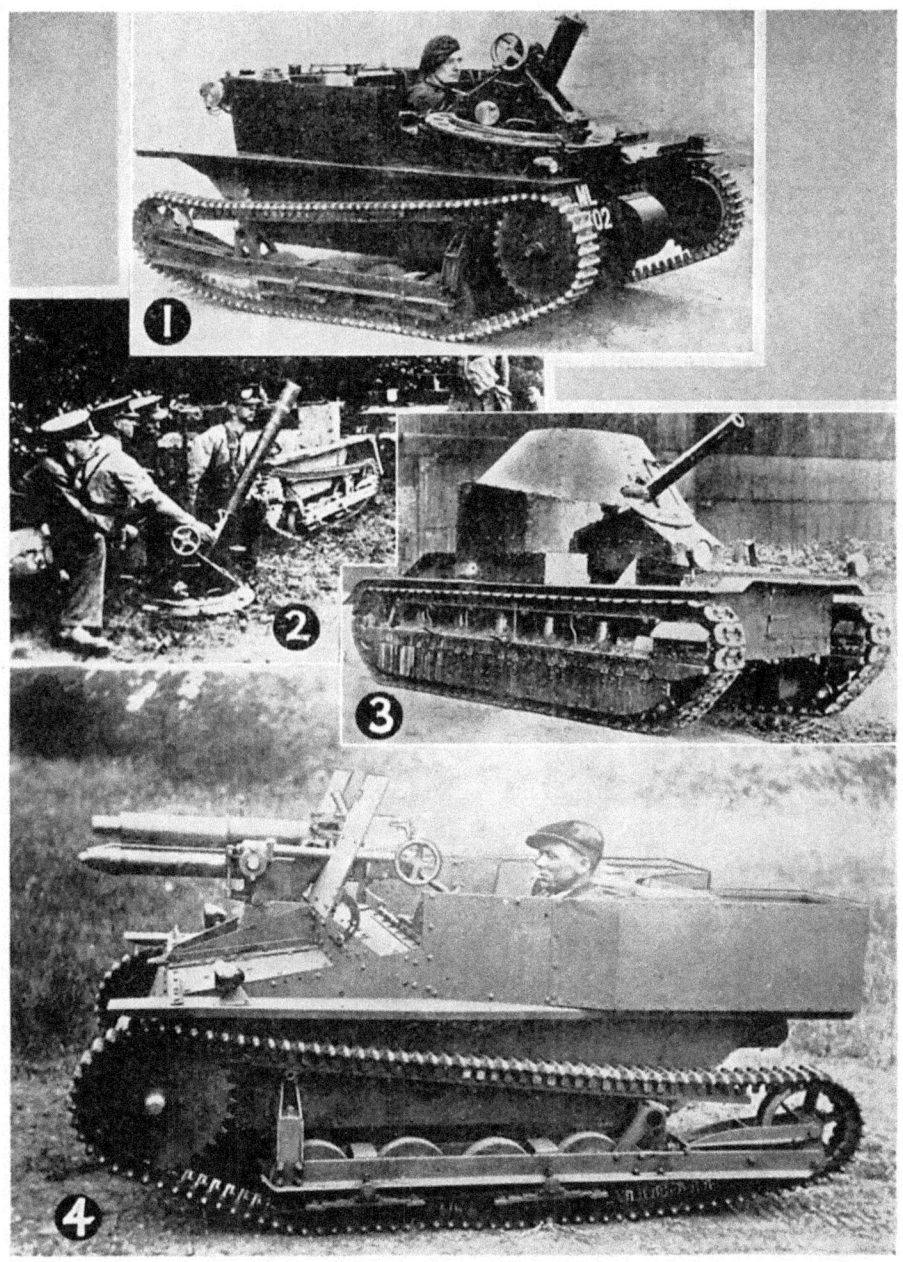

Plate 61.
1. Carden Loyd Stokes Mortar.
(From *Royal Tank Corps Journal*.)
2. Mortar Dismounted from Carden Loyd Carriage.
3. Self-Propelled Mount for 18-Pounder, Model 1929.
(From *Royal Tank Corps Journal*.)
4. Carden Loyd with 47 mm Gun.
(Photo by Vickers Armstrongs, Ltd.)

U. S. Christie Motor Carriage for 8 Inch Howitzer.

Produced in 1918 by Front Drive Motor Company. Total production, 1.

Armament: 8 inch howitzer, Mark VI.
Maximum speed: 16 mph.
Suspension: Rubber tired wheels; center wheels sprung; no springs at front or rear.
Tracks: Removable; width 15 inches.
Dimensions: Length 16 ft. 5 in.; width 9 ft. 7 in.; height 6 ft. 7 in.
Weight: 17 tons.
Engine: Christie, 6 cylinder, 120 HP, forced water cooling.
Horsepower per ton: 7.1.

U. S. Christie Motor Carriage for 155 mm Gun.

Produced in 1920 by Front Drive Motor Company. Total production, 4.

Armament: One 155 mm (6.1 in.) gun.
Maximum speed: 11.5 mph.
Suspension: Front and rear wheels rigid; center wheels sprung.
Tracks: Width 22 in.; pitch 9.75 in.
Dimensions: Length 19 ft. 8 in.; width 9 ft. 8 in.; height 6 ft. 8 in.
Weight: 19.75 tons.
Engine: Christie, 6 cylinder, 120 HP, forced water cooling.
Horsepower per ton: 6.1.
Transmission: Sliding gear; 4 speeds forward and 4 reverse.

U. S. Christie Motor Carriage for 75 mm Gun or 105 mm Howitzer.

Produced in 1920 by Front Drive Motor Company. Total production, 2.

Armament: One 75 mm (2.95 in.) gun, or one 105 mm (4.1 in.) howitzer.
Maximum speed: 20 mph.
Suspension: Rubber tired wheels; center wheels adjustable.
Tracks: Width 10 in., pitch 7¾ in.
Dimensions: Length 12 ft.; width 7 ft. 6 in.; height 6 ft. 4 in.
Weight: 8 tons.
Engine: Christie, 6 cylinder, 90 HP, forced water cooling.
Horsepower per ton: 11.25.
Transmission: Sliding gear, 4 speeds forward, 4 reverse.

U. S. Mark VII Motor Carriage.

Produced in 1919 by Holt Caterpillar Company. Total production, 2.

Crew: 4.
Armament: One 75 mm (2.95 in.) gun, M1916.
Armor: 0.25 in.
Maximum speed: 9½ mph.
Tracks: Width 8 in., pitch 5 in.
Dimensions: Length 11 ft. 3 in.; width 6 ft.; height 6 ft.
Weight: 5¼ tons.
Engine: Cadillac, V type, 8 cylinder, 70 HP, forced water cooling.
Horsepower per ton: 13.3.
Transmission: Sliding gear, 3 speeds forward, 1 reverse.
Obstacle ability: Stream 34 in., slope 35 degrees.
Fuel distance: 36 miles. **Fuel capacity:** 18 gallons.
Special features: Maximum elevation, 900 mils. Depression, none. Maximum traverse, 322 mils each side of center.

U. S. 75 mm Howitzer Motor Carriage, T1.

Produced in 1930 by James Cunningham Sons, and Company. (Ordnance design.) Total production, 1.

Crew: 4.
Armament: One 75 mm (2.95 in.) pack howitzer, M1.
Armor: 0.25 in. to 0.375 in.
Maximum speed: 21 mph.
Suspension: Bogies, rollers, and leaf springs.
Tracks: Open plates, width 10½ in., pitch 5 in.
General arrangement: Engine and howitzer in front; crew in rear.
Dimensions: Length 10 ft. 9 in.; width 6 ft. 1 in.; height 6 ft. 3 in.
Weight: 5.65 tons.
Engine: LaSalle, V type, 8 cylinder, 87 HP, forced water cooling.
Horsepower per ton: 15.4.
Transmission: Sliding gear, 3 speeds forward, 1 reverse.

Self-Propelled Artillery

Plate 62.
1. Christie Motor Carriage for 8 Inch Howitzer.
   (Photo by Ordnance Dept., U.S.A.)
2. Christie Motor Carriage for 155 mm Gun.
   (Photo by Ordnance Dept., U.S.A.)
3. Christie Motor Carriage for 75 mm Gun or 105 mm Howitzer.
   (Photo by Ordnance Dept., U.S.A.)

Plate 63.

1. Mark VII Motor Carriage.
 (Photo by Ordnance Dept., U.S.A.)
2. Motor Carriage for 75 mm Howitzer, T1.
 (Photo by Mechanized Force, U.S.A.)
3. Motor Carriage for 4.2 Inch Mortar, T1.
 (Photo by Ordnance Dept., U.S.A.)

Obstacle ability: Stream 22 in., slope 32 degrees.
Fuel distance: 60 miles. **Fuel capacity:** 30 gallons.
Special features: Maximum elevation of howitzer, 375 mils. Maximum depression. 120 mils. Maximum traverse, 270 mils each side of center. Hand brake locks both tracks when firing.

U. S. 4.2 Inch Mortar Motor Carriage, T1.

Produced in 1928 by James Cunningham Sons, and Company. (Ordnance design.) Total production, 1.

Crew: 4.
Armament: One 4.2 in. chemical mortar.
Maximum speed: 21.9 mph.
Dimensions: Length 13 ft. 5 in.; width 8 ft.; height 5 ft. 4 in.
Weight: 7.4 tons.
Engine: Cunningham, V type, 8 cylinder, 110 HP, forced water cooling.
Horsepower per ton: 15.4.
Transmission: Sliding gear, 3 speeds forward, 1 reverse.
Fuel distance: 50 miles. **Fuel capacity:** 50 gallons.
Special features: Chassis same as T1-E1 tank. Carries 75 rounds of ammunition. Crew unprotected.

MILITARY TRACTORS.

British. Dragon, Mark I (Artillery Tractor).

Armor: 0.25 in.
General arrangement: Engine in left front; driver in right front; final drive in rear.
Dimensions: Length 15 ft. 9½ in.; width 9 ft. 3½ in.
Weight: 7½ tons.
Fuel distance: 70 miles.
Special features: Will carry 12 passengers. Inadequately powered.

British. Dragon, Mark II (Artillery Tractor).

Armor: 0.25 in.
Maximum speed: 15 mph.
General arrangement: Engine in left front; driver in right front; final drive in rear.
Dimensions: Length 15 ft. 9 in., width 7 ft. 9 in.
Weight: 5.25 tons.
Engine: Armstrong-Siddeley, 90 HP, air cooled.
Horsepower per ton: 17.2.
Special features: Will carry 12 passengers. A material improvement over the Mark I. Better maneuverability.

British. Dragon, Mark III (Artillery Tractor).

Produced in 1927.

General arrangement: Engine in left front; driver in right front; final drive in rear.
Special features: An improvement over the Mark II. Details not available.

British. Special Carden Loyd Light Tractor.

Produced by Vickers Armstrongs, Ltd.

Crew: 1.
Maximum speed: 25 mph.
Suspension:. Two sets of pivoted and sprung bogies on each side, large rubber tired wheels.
Tracks: Carden Loyd type.
General arrangement: Final drive in front.
Dimensions: Length 12 ft.; width 6 ft. 2 in.; height 4 ft. 3 in.
Weight: 5680 pounds empty.
Engine: 50 HP, water cooled.
Obstacle ability: Trench 4 ft.; stream 2 ft. 6 in.
Fuel capacity: 30 gallons.
Special features: Can tow 4680 pounds up an 18 degree slope, or 7¼ tons up a 9.5 degree slope.

Plate 64.
1. Dragon Mark I.
2. Dragon Mark III.
(From *Royal Tank Corps Journal*.)
3. Special Carden Loyd Light Tractor.
(Photo by Vickers Armstrongs, Ltd.)
4. Carden Loyd Machine Gun Carrier, Towing a Trailer.

British. Carden Loyd Machine Gun Carrier.

Tows a Trailer. Produced by Vickers Armstrongs, Ltd.

Maximum speed: 16 mph with 1000 pound load in trailer on good roads.
Dimensions of trailer: Length inside 50 in., width inside 38 in., depth inside 14 in.
Special features: These trailers are furnished for numerous purposes. Trailer capacity, 1500 pounds.

French. Citroen Kegresse Light Tractor.

Produced by Citroen and Co. Large numbers of these vehicles have been produced.

Crew: 2 (on front seat); room for 3 or 4 additional men on rear seat.
Maximum speed: About 20 mph.
Suspension: The weight of the rear of the vehicle is chiefly supported through leaf springs by a dead axle connecting the centers of the two track assemblies. The front wheels of track assemblies are elevated somewhat above the ground drive the tracks, and are sprung to a limited degree in their connection with the body. The steel rollers within the track assembly are pivotally mounted. The length of the track assembly is 73 inches.
Tracks: The early types of this vehicle had simple rubber tracks that were unsatisfactory as to durability, traction, and stretching. The present track is of rubber and fabric. It is 8¾ inches wide and has metal plates with rubber cleats superimposed upon them rivetted to the exterior surface. Upon the interior surface are three rows of rubber cleats. The center row (high cleats) serve only to guide the track with reference to the wheels and rollers. The outside rows (low cleats) mesh with the front (sprocket) wheels and so enable those wheels to drive the track.
Wheels: The front wheels of the vehicle have single pneumatic tires.
Dimensions: Length 14 ft.; width 5 ft. 2½ in.; height 6 ft. 6½ in.
Weight: 4130 pounds.
Engine: Citroen, 4 cylinder, 30 HP, forced water cooling.
Horsepower per ton: 14.4 empty.
Transmission: Sliding gear, 5 speeds forward, 2 in reverse.
Fuel capacity: 10 gallons.
Special features: This light tractor has a number of uses, but its principal purpose is the towing of the 75 mm gun. For this, the French use it to a large and increasing extent. When so employed, the entire weapon and mount is supported upon two spring-suspended bogies, each of which has two small wheels with rubber tires. (Shown in Fig. 1, Plate 65). This tractor is very highly thought of in France. The French regard the tractor as suitable, also, for the following purposes: To carry miscellaneous cargo; to tow trailers carrying infantry howitzers or other accompanying weapons; to serve as an accompanying vehicle for machine gun equipment; to be adapted for use as a reconnaissance vehicle, armored car or ambulance.

French. Light Infantry Supply Tank, Type U E.

Produced in 1931 by The Renault Mills.

Crew: 2.
Armor: 0.157 in. to 0.276 in. chromium steel.
Maximum speed: 8.75 mph.
Suspension: Leaf springs.
Tracks: Metal plates, width 6.69 in.; or rubber tracks, width 9.05 in., as alternative.
General arrangement: Engine, final drive and crew in front, cargo space in rear.
Dimensions: Length 8 ft. 10 in.; width 5 ft. 7 in.; height 3 ft. 10 in.
Weight: Empty, 4290 lbs.; loaded, 5555 lbs.
Engine: 4 cylinder, 35 HP.
Horsepower per ton: 12.6 loaded.
Transmission: Tandem sliding gear, 6 speeds forward, 2 speeds in reverse.
Obstacle ability: Trench 4 ft., slope 39 degrees.
Fuel capacity: 21 gallons.
Special features: Ammunition carrier for front line troops. This vehicle also tows a cargo trailer, the body of which is 4 ft. 1 in. long, 2 ft. 9 in. wide, and 1 ft. 2 in. deep. The trailer is also a track vehicle with leaf spring suspension. This whole assembly of cargo vehicle and trailer is transported on the road by means of a larger trailer towed by a truck.

Plate 65.
1. Citroen Kegresse Light Tractor, Towing a Field Piece Mounted upon Sprung Rubber Tired Bogies.
   (Photo by Tank School, U.S.A.)
2. Track and Suspension of Citroen Kegresse Light Tractor.
   (Photo by Ordnance Dept., U.S.A.)
3. Light Infantry Supply Tank, U.E., with Trailer.

CHAPTER XXIV.

MISCELLANEOUS VEHICLES (Continued).

CROSS-COUNTRY TRACTION AIDS.

Removable Light Track for Rear Wheels of Six-Wheeled Vehicles.

This device was produced in experimental numbers by U. S. Ordnance Department. Among the vehicles for which it is especially intended are the antiaircraft six-wheeled mounts. The British and others use removable tracks of this general nature with six-wheeled armored cars. Devices of this character, if strong, durable, easily-adjustable, and not too heavy, will add very much to the value of vehicles for various military purposes by increasing greatly their cross-country ability.

Hipkins Traction Device.

In the form that is probably of greatest military value, this device, manufactured by O. F. Hipkins, Port Deposit, Md., is wrapped about a dual-tired truck wheel. In this form (for tires 32 by 6.00) the plates are 8 in. by 12 in. The chain is similar to that used in the final drive of Mack trucks. A complete device of this kind for one wheel weighs 125 lbs. It is of course normally used upon the two rear wheels.

The aid given to the traction of a two-wheel-drive truck by the device is remarkable. By its use an ordinary commercial truck is at once converted temporarily into a cross-country cargo carrier and tractor of surprising ability. One man can put on or remove a complete set for two wheels within a few minutes. The device on each wheel is secured in place by the application of one nut. The act of removal is simple, whereupon the vehicle at once regains its high road speed. Even with the Hipkins device installed, the Ford truck can, over favorable ground, be run at 30 miles per hour.

One might at first think that this device is merely a variation in tire chains. Although the purpose is the same, the effect is quite different. The Hipkins traction device improves the factor of flotation by more than 100%, and in mud gives a vehicle so equipped a minimum tractive superiority of 80% over one equipped with chains. Under some conditions this superiority is much greater. There are many places, impassable to a vehicle with chains, through which a vehicle with this device can pass with ease. When the traction device is used, the tires roll upon a smooth surface and are protected against punctures.

No half-track vehicle nor any specially equipped wheel vehicle can possibly have such excellent cross-country ability as a tank. There are situations where only the fully-tracked vehicle can get through. Yet the light tractor (Fig. 1, Plate 65) has good, though limited, cross-country ability. The French rely upon it as a very important war vehicle. The Hipkins traction device, however, installed upon a U. S. commercial truck, can out-do the light tractor in nearly all situations. In addition to other advantages it does not alter the vehicle permanently. The vehicle is quickly convertible for road use at higher speed. It is difficult to manufacture, in time of peace, special military tractors (like that in Fig. 1, Plate 65) in adequate numbers for use in war. But if a traction device can make ordinary trucks suitable for the purpose, the trucks will already be available and the chief manufacturing requirement will be eliminated. Herein, from the military viewpoint, lies a great advantage in this Hipkins traction device. This type of device, or something similar for the same purpose, should in time of war make quickly available the means for carrying or towing cargo across country on a large scale, and thus contribute greatly to the mobility of troops in the early part of the war.

A variation in the device (Fig. 4, Plate 66) is adapted to single-tired wheels. By its application to the ordinary light touring car, it is possible that a materially improved reconnaissance car might be obtained.

Plate 66.

1. **Hipkins Traction Device.** Showing how the plates are connected with chain segments.
   (Photo by Ordnance Dept., U.S.A.)
2. **Hipkins Traction Device Installed on a Ford Truck.** The wheel rests upon two plates much of the time. Each plate successively is laid flat upon the ground, immediately after which the wheel rolls upon it.
   (Photo by Ordnance Dept., U.S.A.)
3. **Removable Light Track for Rear Wheels of Six Wheeled Vehicles.**
   (From *Army Ordnance*, U. S. A.)
4. **Hipkins Traction Device, a Variation Adapted to Single Tired Wheels.**
   (Photo by Ordnance Dept., U.S.A.)
5. **Hipkins Traction Device.** This Ford truck is pulling a 155 mm gun, weighing 13½ tons through the mud, with a dynamometer between tractor and tow.
   (Photo by Ordnance Dept., U.S.A.)

CARGO CARRIERS, CONVERTED FROM TANKS.

British. Supply Tank, Mark IV.

Some of the Mark IV tanks were modified and used in the War as supply vehicles. As an instance of their value during combat the following quotation is of interest: "Sixteen Mark VI Tank Tenders carried the following: grenades, trench mortar ammunition, small arms ammunition, screw pickets, barbed wire, picks, shovels, etc. The total weight of the load was 120,184 pounds. The crew of 16 tanks numbered 64. If a carrying party had been used, each man carrying 56 pounds, the number of men required would have

Plate 67.
1. **Supply Tank, Mark IV.**
2. **Supply Tank, Modified Gun Carrier.**
(French official photo.)
3. **Supply Tank, St. Chamond.**
(Photo by Signal Corps, U.S.A.)

been about 2,146. In other words over two thousand men were saved for the firing line by 16 tanks of obsolete pattern." (*Royal Tank Corps Journal*, May, 1920.)

British. Supply Tank, Modified Gun Carrier.

Produced in 1917 by William Foster and Company, Ltd. Total production, several.

Crew: 3 (for supply use.)
Armor: 0.3 in.
Maximum speed: 3 mph.
Suspension: Rigid, with rollers.
Tracks: Flat steel plates, width 17 in.
General arrangement: Driver and brakeman over tracks in front; driver on left; engine and final drive in rear.
Dimensions: Length 30 ft.; width 11 ft.; height 9 ft. 4 in.
Weight: 30.2 tons empty; 40.2 loaded.
Engine: Daimler, 6 cylinder, 105 HP, forced water cooling.
Horsepower per ton: 3.5 empty; 2.6 loaded.
Transmission: Same as gun carrier tank.
Obstacle ability: Trench 11 ft. 6 in.; stream 2 ft. 7 in.; slope 35 degrees; vertical wall 2 ft. 5 in.
Fuel capacity: 70 gallons.
Special features: This was the gun carrier tank modified for use as a supply tank. As a rule, artillery ammunition was carried, with an average load per tank of about 8 tons.

French. Supply Tank, St. Chamond.

Some of the St. Chamond tanks were modified and used in the War as supply vehicles.

CROSS-COUNTRY CARGO CARRIERS.

Recognizing the need for a more efficient means with which to transport ammunition and other supplies across country to troops engaged in action, our Ordnance Department has built some experimental cross-country cargo carriers, and, previously, some power carts.

U. S. Light Cargo Carrier, T1.

Produced in 1928 by James Cunningham Sons, and Company. (Ordnance design). Total production, 1.

Crew: 1.
Maximum speed: 20 mph empty.
General arrangement: Engine in front; driver in center; cargo at rear and sides.
Dimensions: Length 12 ft. 11.5 in.; width 8 ft. 5 in.; height 5 ft. 2 in
Weight: 13,350 pounds
Cargo capacity: 2 tons.
Special features: The Light Tank, T1, was converted into this carrier. It has a wooden body. The suspension of this vehicle was later modified by substituting hydraulic and spring mechanism for the link type mechanism. This cargo carrier was later converted into an artillery motor reel.

U. S. Light Cargo Carrier, T1-E1.

Produced in 1929 by James Cunningham Sons, and Company. (Ordnance design). Total production, 2.

Crew: 1.
Maximum speed: 18 mph empty.
General arrangement: Engine in front; driver in center; cargo at rear and sides.
Dimensions: Length 12 ft. 10 in.; width 8 ft. 2 in.; height 5 ft. 2 in.
Weight: 7½ tons
Cargo capacity: 2 to 3 tons.
Special features: Same type chassis as Light Tank, T1-E1. One of these cargo carriers was later converted into a chemical mortar mount.

POWER CARTS.

Of the several power carts built, the two illustrated herein are most typical. When using such a vehicle, the cargo is placed in the body and the operator

Plate 68.
1. **Light Cargo Carrier, T1.**
(Photo by Ordnance Dept., U.S.A.)
2. **Light Cargo Carrier, T1-E1.**
(Photo by Ordnance Dept., U.S.A.)
3. **Small Power Cart, Model 1924.**
(Photo by Ordnance Dept., U.S.A.)
4. **Large Power Cart, Model 1922.**
(Photo by Ordnance Dept., U.S.A.)

walks or runs at the rear. The cart shown in Fig. 4, Plate 68, is the largest model built. The carts have given some creditable performances but none has been adopted for manufacture.

U. S. Large Power Cart, M 1922.

Produced in 1922 by Ordnance Department. Total production, 1.
Crew: 1.
Maximum speed: About 6 mph.
Suspension: Spring.
Tracks: Light in weight but insufficient in durability.
General arrangement: Power plant at the rear but within the cargo body.
Dimensions: Length 8 ft. 4 in.; width 3 ft. 8½ in.; height 3 ft. 8 in.
Weight: 895 pounds, no load
Cargo capacity: 450 pounds.
Engine: Harley-Davidson, 2 cylinder opposed, 7½ HP air cooled.

U. S. Small Power Cart, M 1924-E.

Produced in 1924 by Ordnance Department. Total production, 1.
Crew: 1.
Maximum speed: About 6 mph.
Suspension: Spring.
General arrangement: Power plant at the rear and outside the cargo body.
Dimensions: Length 6 ft. 8 in.; width 3 ft. 5½ in.; height 3 ft. 3 in.
Cargo capacity: 450 pounds
Engine: Harley-Davidson, V type, 2 cylinder, air cooled.

TANK CARRIERS.

U. S. Mack Tank Carrier.

Produced by International Motor Company.
Maximum speed: 10 mph loaded; 14 mph unloaded.
Dimensions: Length 21 ft. 7½ in.; width 7 ft. 6½ in.; height 8 ft. 3 in.; height of floor, loaded, 4 ft. 2 in
Weight: 5½ tons.
Engine: Mack, 4 cylinder, 42 HP, forced water cooling.
Transmission: Sliding gear, 3 speeds forward, 1 reverse.
Fuel distance: 90 miles. **Fuel capacity:** 30 gallons.
Special features: This is a commercial chain-drive Mack truck with added leaves in the rear springs to enable it to carry the heavy load. It has been used as a tank carrier ever since the World War.

U. S. TCSW (Tank Carrier, Six Wheel).

Produced in 1928 by Quartermaster Corps. Total production, 48.
Crew: 2.
Maximum speed: 20 mph, loaded; 25 mph, unloaded.
Dimensions: Length 25 ft. 1½ in.; width 8 ft. 2½ in.; height 8 ft. 9 in.; height of floor, loaded, 3 ft 2½ in
Weight: 8½ tons.
Engine: Continental 15-H, 6 cylinder, 105 HP, forced water cooling.
Transmission: Class B sliding gear, 4 speeds forward, 1 reverse.
Fuel distance: 90 miles. **Fuel capacity:** 60 gallons.
Special features: Four wheel drive. Tail gate and loading ramp are combined.

U. S. TCSW, Pneumatic Tired.

Produced in 1930 by Quartermaster Corps. Total production, 12.
Maximum speed: 30 mph loaded; 40 mph unloaded.
Height of floor, loaded: 3 ft. 9½ in.
Special features: Pneumatic tires. Sloping floor at rear reduces length of tail gate ramp. Otherwise substantially the same as the TCSW with hard tires. Most of the parts are identical with those in the hard tired model. Pneumatic tires are contemplated for any future tank carriers.

U. S. G. M. C. 10 Ton Truck, Towing Medium T2 Tank.

Produced by General Motors Corporation.
Maximum speed: 40 mph.
Dimensions of truck: Length 24 ft.; width 8 ft.; height 8 ft. 1 in.
Weight: 9 tons.
Cargo capacity: 8 tons.
Engine: 6 cylinder, 175 HP, forced water cooling.
Transmission: Sliding gear, 12 speeds forward.

TANK CARRIERS 247

Plate 69.
1. Mack Tank Carrier. Tank entrucking.
   (Photo by Tank School, U.S.A.)
2. T C S W (Tank Carrier Six Wheel.)
   (Photo by Tank School, U.S.A.)
3. T-C-S-W, Pneumatic Tires.
   (Photo by Tank School, U.S.A.)
4. G M C Ten Ton Truck Towing Medium T2 Tank.
   (Photo by Ordnance Dept., U.S.A.)

Special features: Four wheel drive, pneumatic tires. The trailer shown with this truck is a Holstein trailer with a capacity of 30 tons. Its length (with draw bar) is 35 ft. 11 in., width 7 ft. 10 in., floor height 22 in. The length of the truck and trailer together is 59 ft. 11 in. This combination was used to transport the Medium T2 tank on good roads.

SPECIAL MAINTENANCE VEHICLES.

U. S. Wrecking Truck (Mechanized Force).

Produced in 1930.

Maximum speed: 35 mph.
Wheels: 6 wheel, 4-wheel drive.
Tires: Dual pneumatic all around. The outside tire of each front wheel is not in contact with the ground on hard surfaces. All tires have puncture proof inner tubes.
Weight: 8 tons.
Engine: Hercules, 6 cylinder, 93 HP, forced water cooling.
Horsepower per ton: 11.6.
Transmission: Sliding gear, 8 speeds forward, 2 reverse.
Obstacle ability: 14 degree slope.
Special features: The crane carried has a 5000 pound capacity. Light detachable tracks are provided for the 4 wheels in rear.

Plate 70.

1. Wrecking Truck, Mechanized Force.
2. Ten Ton Tractor with Cranes.
(Photo by Ordnance Dept., U.S.A.)

CHAPTER XXV.

TANK ARMAMENT AND GUNNERY.

ARMAMENT.

It is the consensus of opinion that a fighting tank should carry more than one weapon, unless it is intended for some exceptional purpose for which one weapon is sufficient. In our service, the minimum armament for tanks is considered to be one antitank gun, one machine gun and, for each member of the crew, a hand weapon to be used in covering dead space about the tank and for personal protection in case the tank must be evacuated. World War experience indicated the necessity of providing the tank gunner with more than one type of weapon in order that the tank might be effective in attacking both personnel and materiel. It is generally considered uneconomical to spend ten or fifteen thousand dollars or more in building a vehicle that carries only one small caliber weapon and hence is capable of accomplishing only one of these purposes.

A number of tanks now carry a machine gun and a gun of larger caliber in one mount. Where only one gunner is provided this method conserves space and gives the gunner a choice of weapons, enabling him to select and use immediately the weapon best suited to the target of the moment.

In medium and heavy tanks space is usually available for one or more weapons of fairly large size such as the 3 pounder, 6 pounder, or larger gun. Many who have used such weapons are in favor of building tanks, even for the infantry accompanying mission, sufficiently large to accommodate these weapons. It is claimed that the large guns, in addition to the effect of their H. E. projectiles against materiel, are highly effective against personnel by the use of canister, and that they may even fire smoke to advantage under favorable conditions. A tank cannon should preferably be of the semiautomatic type.

Whatever the type or number of weapons carried, the efficiency of the tank is increased by providing the weapons with a maximum degree of traverse, elevation and depression. German tank combat history records an instance of the inability of the tank gunner to depress his machine gun sufficiently to hit British soldiers in plain sight and at close range. Many such instances actually occurred, and it is a well known fact that relatively few tank weapons are so mounted that they can be depressed sufficiently to strike targets close to the tank. The British 16 ton tank is a rare instance wherein the low construction of the rear part of the tank and the mounting of the gun permits a gun in the forward portion to be fired to the rear at a suitable angle of depression.

Mounts for antiaircraft machine guns have been installed in a number of tanks. The general practice has been to mount these weapons in ball mounts placed at or near the top of the turret in armor standing at an angle of about 45 degrees with the horizontal, thus giving the weapon favorable angles of elevation, while direction is dependent chiefly upon the rotation of the turret. The value of antiaircraft guns mounted in a tank or armored car has been questioned in our service, due to the interference by the armor with vision and the lack of free traverse. Moreover, these vehicles will often be in motion when the need for such a weapon occurs. If a gunner in his effort to hit a hostile airplane is under the triple handicap of greatly restricted vision, slow traverse, and a moving gun platform, it is believed that the effectiveness of his fire will be negligible. It is expected that our fighting vehicles will be protected against aircraft by weapons firing under more favorable conditions.

In the U. S. Mark VIII and a number of other tanks there are semi-cylindrical turrets located in sponsons, and which have a limited horizontal rotation. The Mark VIII has two of these with a cannon in each. In this tank is also exemplified the hemispherical turret, there being one in each side door and another high in the center of the tank, which supports the machine gun that fires to the rear. This turret is a hemispherical shell of armor, bulging outward and pivotally mounted in the armor wall by means of an upper and a lower trunnion. The purpose is the same as that of the semicylindrical turret—to increase the traverse. Each hemispherical turret in the Mark VIII contains a machine gun mounted to and within it by means of a ball mount. Most tank machine guns in the past, whether mounted to a wall of armor or some sort of turret, have used a ball mount. It consists of a major portion of a sphere to which is fixed the machine gun and which fits with a slight clearance in a corresponding shell, the latter being fixed to the armor. Such a mount gives to the weapon a limited traverse and permits it to be elevated and depressed. As a substitute for the ball mount, the double trunnion or gimbals mount has come into use to a considerable degree. In this, two trunnion bearings permit the limited traverse, two other trunnion bearings permit elevation and depression, and protection from bullets is secured by a more or less nearly hemispherical shell which projects to the front, which moves with the weapon, and fits closely to a corresponding interior projection in order to reduce bullet splash. The chief disadvantage of the ball mount is that it is impossible to keep sand and other dirt from the large surfaces that form the bearing. As a result the ball mount frequently binds or sticks, unless unduly loose, and can seldom be moved to the various positions with a uniform pressure upon the rear part of the weapon. The double trunnion mount, on the other hand, moves freely and smoothly, thus favoring accurate shooting, but, unless skilfully designed, admits more bullet splash than a well-adjusted ball mount. Cannons are more commonly mounted by means of double trunnions.

Flash hiders have been tested on machine guns and found to have value in reducing the flash and aiding the gunner to follow the course of tracer bullets.

Exact sighting is more important for tank cannons than for tank machine guns since fire from the latter can more readily be brought upon the target through observation of bullet strike. Telescopic sights promote exact sighting, and hence are favorably regarded for cannons. The telescopes of sights for tank use should magnify but little (probably not more than 2 times); otherwise the field of view will be unduly restricted and the dancing of the target and its surroundings, as viewed by the gunner, will be excessive. The sight should be adjustable so that the axis of its telescope can be brought into parallelism with the axis of the weapon both quickly and accurately. Since dust, oil mist, etc. frequently settle upon the object glass and thus make the sight temporarily useless, and since a gunner cannot during action go outside to wipe his object glass, the mechanical construction should be such that the sight may be quickly withdrawn to the interior of the tank, cleaned and replaced without disturbing to a material degree the adjustment as to parallelism. A good pair of open sights is more satisfactory than a defective telescopic sight, especially for machine guns. Either the front or the rear open sight must be adjustable as to elevation and deflection. It is somewhat difficult to design such a sight so that adjustments can be made quickly and yet will not be deranged by vibration caused by the motion of the tank and the firing. An adjustable open sight should be preferably so constructed that any usable range may be set accurately, and with added provisions for one or two battle sight settings.

GUNNERY.

Tank gunnery differs materially from other prior types of gunnery. The tank gunner must do most, if not all, of his firing while the tank is in motion, usually over a surface that is not smooth. Accordingly his fire is not like that of the rifleman, the infantry machine gunner, or the artillery gunner. The tank gunner's problem is that of the naval gunner, only still more difficult. The waves that rock the vessel are relatively smooth and regular. The bumps and pitching motions inflicted upon the tank when passing over moderately rough ground are very irregular. The tank gunner has no means of foretelling the direction, degree, severity, or proximity of the next sudden motion that will be imparted to the tank by the ground. Since any such motion will derange his aim one way or another, it is obvious that the motion of the tank places upon the gunner a handicap not encountered elsewhere in land or naval warfare.

And yet, in spite of this handicap, with a liberal amount of the right kind of training, gunners can develop a surprising degree of accuracy in their fire from tanks in motion. Hence, in justification of the money and effort put into the development of tank organizations, gunners must be trained so that they can fire quickly and accurately while in motion.

The chief purposes of tank gunnery training are these:

(1) To teach the gunner familiarity with his weapon so that he may operate it and care for it properly.

(2) To teach him to observe through the small eye slits of the tank and to locate indistinct targets while in his unsteady position within the moving tank.

(3) To teach him to fire effectively while stationary upon stationary and moving targets of various kinds.

(4) To teach him to fire effectively while in motion upon stationary and moving targets of various kinds.

(5) To teach him the capabilities of his weapon and what he needs to know regarding possible actual hostile targets (such as the vulnerable parts of hostile tanks, their degree of vulnerability, etc.)

(6) To teach him those additional special points of tank gunnery that pertain to combat and tactics.

The first material requirement for a tank gunner is a good gun platform. This means that the tank should have an efficient suspension—one that will protect him in a large measure from vibrations and jolts. His next requirement is a suitable weapon, with a satisfactory sight, and adequate vision. The interior arrangements should be such that he can assume a favorable posture for firing, one that will not place an excessive strain upon his muscles. He must also have conditions favorable for aiming; in order words, he needs wide limits for elevating, depressing, and traversing, a mechanism that will permit this to be done easily and accurately, and, if there is a rotating turret, it should be one that can be rotated rapidly and clamped in position, and not likely to get out of order.

The tank gunner operates, in most cases, either a cannon, a machine gun, or a cannon and machine gun in one mount. In controlling his weapon or weapons, he normally requires a shoulder piece so designed as to give him a maximum of assistance in elevating, depressing or traversing the weapon by a slight motion of the shoulder. However, when a member of a crew (as, for example, a platoon commander) is required to observe in various directions much of the time and is required to fire only a machine gun, there are advocates of the plan of his firing while in an erect posture without the use of sights, using his weapon like a garden hose. This method seems to have no great disadvantages when the conditions are such that the strike of the bullets can be seen.

Gunnery should teach economy of tank ammunition. Although a tank carries a certain ammunition supply very readily and without effort on the part of personnel, space requirements do not permit the liberality that would otherwise be desirable. Economy is therefore necessary, and the assurance of ammunition when needed is obviously important.

The considerations of economy, together with the fact that, except for the first shot or two, the fire delivered during a burst from the machine gun of a tank in motion is necessarily extremely inaccurate, seem to indicate that the bursts should habitually be very short. Numerous experienced tank gunners believe that they should. Some of these go so far as to state that the tank gun should be constructed so as to facilitate semiautomatic fire at will, and that such fire may be used to best advantage much of the time. On the other hand some officers advocate fairly long bursts habitually, due to the fact that such bursts permit seeing the bullets strike. There seems to be no doubt that fairly long bursts are needed under some conditions. How short the bursts should be under other conditions is probably a matter that will be better understood several years hence than it is now.

In the training of the tank gunner, after the preliminary training, the progress is from short range work (as 1,000 inches) to field ranges, from fire from a stationary tank to fire from a moving tank, from subcaliber fire to fire with service ammunition, and, finally, from instructional firing to firing under conditions approximating those of the battlefield.

The British have excelled all others in the thoroughness and effectiveness of their short range training. At their Tank Corps School of Fire at Lulworth a great many different types of firing have been embodied in the courses at the miniature range.[1] Much thought, effort and money have been devoted to the upbuilding of equipment for this range to meet all requirements in an interesting, practical and realistic way. After learning to aim properly and simulate fire at stationary targets, the student, from within a special tank turret, endeavors to aim his weapon at a target that merely rises and falls. The accuracy of his aiming is shown at once by means of electrical dotter equipment. Later, with other equipment he endeavors to aim his weapon at various types of moving targets before backgrounds of realistic appearance, all on a miniature basis. A miniature target in the form of a tank moves as though it were advancing over the hills and valleys of actual terrain. In another phase of the training, a spotlight, thrown upon a representation of terrain in miniature, represents a moving hostile tank. A tank commander directs his gunners, who operate air rifles representing machine guns, in the delivery of fire upon the spotlight. In due time the student aims a gun that is mounted upon a platform which, by motions imparted to it, simulates the rolling, pitching and bumping of a tank passing over moderately rough ground. From this, he practises aiming at both stationary and moving targets. These are but a few of the ingenious devices by means of which the training of the British tank gunner is made practical and thorough.

Before starting actual firing at field ranges, the student should be trained in estimating distance not only from the ground but from within a tank in motion.

In firing, both at miniature and field ranges, the targets should generally simulate the kinds of targets that tank gunners would normally fire upon

---

[1] This information has been obtained from articles written for publication by Captain B. H. Liddell-Hart, in the *Daily Telegraph* and *Royal Tank Corps Journal*.

in action; which are moving tanks, partly concealed guns, and groups of foot soldiers.

Before firing in combat practice, the crews of light tanks should be instructed and practised in the teamwork that will facilitate effective results in the shooting. Advantage should be taken of level spots on the ground to increase the accuracy of the fire. It is even desirable at times that the driver slow down momentarily for the same purpose. These and other associated matters require communication between the gunner and driver, also instruction and practice if satisfactory results are to be attained.

Gunnery training culminates in combat practice. In this, each man of the tank crew and each tank of the unit play their parts. The tanks advance under an assumed situation as though in battle. Targets of realistic types are placed or moved upon the ground as they might be in war. All concerned are on the alert for targets and the gunners make as many hits as they can.

As the interest in tanks has been less in the United States than in England and France, the gunnery training in our country has been less thorough. By proper instruction and practice the tank gunner can be trained to adapt his body and his weapon to the motion of the tank. He can learn to fire quickly and accurately from a tank in motion. The British have probably been most successful in this field. Their standard of proficiency in one type of 3 pounder fire is three hits out of five shots at 1000 yards from a tank moving 10 mph. against a target moving also at 10 mph. In many cases this is bettered. This standard is attained, not on level ground but on average ground and under average operating conditions. In England, firing from tanks is contemplated up to a range of 1500 yards, although ranges of from 500 to 1000 yards are expected to be most suitable for tank fire.

An example of the excellent results secured by the British previous to 1927 follows:[1]

Fifty yards to our front on the slope were two little clumps of figure targets each representing an infantry section, and frankly less visible and smaller than would real men have been. Behind us, at a range of 400 yards, the four tanks, running parallel with the target at a speed of ten mph, opened fire on one of these little groups. The first bullets fell just above, but a second later a myriad little flicks of dust were spurting up in amongst the targets, and this accurate rain of bullets was maintained. In war it would have been needless, for the infantry section could not have survived more than a few seconds, and inspection showed sufficient holes to put the question beyond doubt. Next the other little target group was opened on at a range of 900 yards, and, though the shooting was not as deadly, it was adequate. No more trying or practical conditions could have been conceived for a demonstration, for a wind of gale force, with driving rain, was blowing from the southwest. The demonstration against an infantry target was followed by one against the still more difficult one, a moving one—a six foot frame drawn on rails extending for several hundred yards.

The tanks, 500 yards away and moving parallel to it, opened fire with their 3 pounder guns and their machine guns, and though it was moving as fast as 10-20 mph, few of the shots were not close to the target. It is true that the human factor was apparent in the better shooting of one tank than its three companions, but in such weather conditions the margin of error was surprising not in its extent but in its narrowness. I gathered that at a previous demonstration one tank actually put four shells out of five through this target—far smaller than an enemy tank or lorry.

[1] By Captain B. H. Liddell-Hart, in *Royal Tank Corps Journal*

CHAPTER XXVI.

OBSTACLES AND DEFENSE AGAINST TANKS.

The principal means of defense against hostile tanks are natural obstacles, antitank weapons, and tanks.

Obstacles.

Obstacles, both natural and artificial, are important factors for consideration. Natural obstacles include: inland waters (streams, ponds and lakes), swamps and deep mud, woods, precipitous slopes and boulders. Artificial obstacles include: canals, stumps, trenches, large shell craters, walls, barricades and tank traps. So far as planning defense is concerned, artificial obstacles that are already in place and not made specially for defensive purposes, are best considered in the class with natural obstacles.

Water courses are the most serious obstacles. Ferrying or bridging is often necessary. Tanks can ford such depth as will not interfere with the functioning of their ignition systems or their carburetion. This depth varies according to the design of the tank. If narrow streams can be crossed at a fairly high speed, a greater depth of water can be negotiated as the water will have less time in which to leak into the tank. Tanks with hulls that are substantially water tight naturally have a material advantage in fording. The development and use of amphibious tanks may nullify the seriousness of the water-course obstacles.

Mud decreases the possible speed of tanks and deep mud is a very serious handicap. A tank has approximately the same ground pressure as a man. Therefore, if a man can cross a swamp without great difficulty, a tank can probably also cross it unless the depth of water is excessive. In crossing swamps tanks keep in motion and pursue a straight course to lessen the danger of becoming mired. Mud is most troublesome to tanks in low ground, trenches and ravines.

A tank is bellied when the floor of the tank is supported in such a way that both tracks slip whenever an attempt is made to move the tank. This may occur in attempting to pass over a stump or a boulder, or when traveling longitudinally over logs or over knocked down trees that are too large or too numerous, or when attempting to come out of a mire.

Woods constitute an obstacle which varies according to the density, size and strength of the trees. The trees may be so strong that the tank cannot knock them down and so close together that it cannot pass between them. Frequently, however, the difficulty is not the inability of the tank to knock down the trees individually but the inevitable straddling by the tank of trees and soil dug up by the roots, with consequent bellying. Trails through dense woods, if not of sufficient width to permit the passage of tanks, leave the woods nearly as great an obstacle as if there were no trails. A wooded area may conceal many obstacles in the form of impassable ravines, wooded slopes, swamps, rocks, tank traps and mines.

If trenches are to be specially constructed or adapted as tank obstacles, the type of tank against which they are to be used must be considered. Such trenches should have steep walls, a width somewhat greater than half the length of the tank they are intended to stop, and be sufficiently deep to be effective. A tank trap is such a pit or trench with a light camouflage covering.

Plate 71.
1. A Tank in Difficulty Crossing a 7 Foot Trench.
   (Photo by Tank School, U.S.A.)
2. A World War Tank Immobilized in a Trench. (Trenches should be crossed at right angles.)
   (Photo by Wm. Foster and Co., Ltd.)
3. A British Tank Crushing Wire in the World War.
   (Photo from British Imperial War Museum.)

Obstacles of different types in combination are sometimes very effective where one alone would be of negligible importance; for example, shell craters with mud or slopes with woods.

Antitank Weapons.

An antitank weapon is one that has sufficient caliber and muzzle velocity to insure penetration of tank armor at combat ranges and that is well suited for direct fire upon rapidly moving targets. Such a weapon should have a wide angle of fire, be very accurate, and, if not mounted upon a vehicle, be easily concealed. It should be possible to fire it rapidly and aim it very quickly at a target in a new direction. Antitank weapons include semiautomatic cannons and large caliber machine guns.

Other Means of Defense.

Additional means of defense against hostile tanks include artillery fire, chemical warfare concentrations, aerial bombing, small arms fire, special hand grenades and antitank mines.

Artillery Fire. Ordinary artillery pieces that cannot be manipulated with sufficient speed to qualify as antitank weapons are, in most cases, of little use for direct fire against fast moving tanks. It seems altogether probable that, in future construction, standard light artillery pieces will have their qualities in this respect materially improved. In most situations indirect artillery fire is relatively ineffective against tanks unless a very heavy concentration can be put down upon a restricted area through which tanks must pass. The reasons for such ineffectiveness are: first, the tanks move so rapidly that the artillery cannot readily predict where they will be at the time fire is actually put down; second, except for direct hits (and each tank is a small target), artillery fire has little or no effect upon the tanks.

Chemical Warfare. Concentrations of chemical agents interfere with the efficiency of tanks to a variable degree depending upon the circumstances. However, a tank affords some protection to the crew, and, if properly designed and equipped, will afford much. If it is necessary for the men to wear gas masks, their efficiency in their duties will be somewhat reduced.

Aerial Bombing. Generally speaking, tanks deployed for action make very poor targets from the air. However, if one side should build very large tanks with very thick armor, it is believed that the air force of the other side could be counted upon to bomb them extensively from low altitudes and with a considerable effect, although with some danger to themselves. Tanks entrained, in close formation upon roads, at fords or concentrated in parks make excellent targets for hostile aircraft.

Small Arms. Ordinary rifle and machine gun fire has little effect upon properly armored tanks. Such fire is not altogether impotent since all tanks have spots that are vulnerable to it. Tanks, by their rapid advance upon opposing foot troops, have a demoralizing effect which reduces the effectiveness of such fire. There is, however, a threat that small calibers may become highly efficient antitank weapons. H. Gerlich, an engineer of Kiel, Germany, reports that, in experimental firing against hardened chrome-nickel armor plate, one-half an inch in thickness, with his Halger-Ultra cal. .28 magazine rifle, having the Mauser bolt action, and his Halger-Ultra ammunition, with a muzzle velocity of between 4500 and 5000 foot seconds, at a range of about 60 yards, the bullets blasted holes 5/8 of an inch in diameter through the armor. This report is very interesting but the practicability of such a rifle as an antitank weapon is yet to be demonstrated, so far as we are concerned. Steels and bore surfaces have, of course, been improved in recent years. By the use of a molybdenum steel barrel, a chromium plated bore, and a more effective propellant, it seems undoubtedly true that a material increase over our present velocities could be attained and probably without excessive erosion. The increased recoil would, however, present a new problem in small arms of the ordinary rifle type. The report of Mr. Gerlich is perhaps not so alarming as it appears, but it does indicate that efforts should constantly be made to improve the quality of armor, to find practicable methods of using thicker armor without adversely affecting speed and other desirable qualities in tanks, and to increase further the protection afforded by placing the armor at favorable angles.

Hand Grenades: Special hand grenades of suitable power constitute a possible means of defense against tanks provided the grenadier can conceal himself at a point near which the tank will pass and at such a place that the tank must move slowly. The effectiveness of such efforts at defense is quite conjectural and the outlook is less favorable as the speed of tanks increases. The ordinary hand grenade, which might be of some use against armored cars, is not of sufficient power to be of any use against tanks, and even bunched grenades, which have been used to good advantage against the tracks, might not be very effective against tracks made of armor plate.

Antitank Mines. Antitank mines destroy tanks effectively, but only when it is possible to place them at points where the hostile tanks will pass. Due to the wide expanses of the battlefield the prospect of successfully placing the mines is, in most instances, rather slight. There is also a very considerable danger that planted mines may cause damage to friendly, rather than hostile, forces. The advisability of attempting to use such mines depends wholly upon the circumstances. While they may be successful under some circumstances, their use in most cases is believed to be inadvisable.

ANTITANK DEFENSE METHODS.

Tank vs. Tank.

In considering the problem of antitank defense, perhaps the first thought that occurs is, "What is the best means for accomplishing the purpose?" The second thought, suggested by articles read, may be, "The best antitank weapon is the tank." This idea has been expressed many times in these words and in others, and there is a great deal of truth in it. If we try to give our antitank weapon a maximum of desirable qualities we soon make a tank of it. We start with a good antitank gun. Many good reasons are advanced why the weapon should have a high degree of mobility—so we place it on a speedy motorized mount. Many good reasons are advanced why it must travel across country—so we make its mount similar to the chassis of a tank. Then someone advances some very good reasons why the crew must have some protection against hostile small arms fire—so we install armor. And then what have we? A tank, of course.

The foregoing line of reasoning seems sound. But this is a very important matter. It is entitled to consideration from all angles. For the sake of discussion let us admit, what is probably true, that the tank *is* the best antitank weapon. Even so, there are also other true statements:

The tank is the *most expensive* antitank weapon.

The tank is at a disadvantage in comparison with the simple (practically immobile) antitank weapon in that the former is *much less readily concealed* than the latter. (It must be admitted, however, that the armor of the tank offsets this to a considerable extent.)

The tank, which might otherwise serve as an antitank weapon, is often urgently needed in all available numbers for quite different purposes.

Thus we see that the tanks, with their high general-purpose value, must become the "card in the hole" as far as antitank defense is concerned. Tanks should not be used primarily for antitank purposes but they should be so used when the need is urgent and it is apparent that other antitank measures are inadequate, when such use will favor the inauguration of offensive action by the main force, or when a favorable opportunity arises to strike and cripple or destroy the hostile tank force.

Principal Defense Considerations.

Being now somewhat oriented as to the antitank use of tanks, we are in a better position to consider the general scheme of antitank defense. This difficult problem must be approached with a desire for economy and effectiveness. Economy demands that natural obstacles, which are the most economical of the efficacious agencies, receive first consideration. Hence the general of the future in planning defense will give consideration, first of all, to formidable natural tank obstacles. The lines will be so laid out as to take the fullest possible advantage of these features. By this procedure, in many cases, the areas available for the operation of the hostile tanks will be very materially curtailed. The remaining task is to supplement this skeleton of obstacles, through the placing of antitank guns, the planning of artillery concentrations, and the construction of artificial obstacles, so that, through the most economical means, the maximum damage may be done to hostile tanks when they attack. When practicable, matters are arranged so that the hostile tanks are directed by natural and artificial obstacles to points where they may be destroyed by antitank guns, mines and artillery concentrations. Due to the almost universal use of light armor for tank floors, small logs or earth works that require the tanks to assume positions such as to expose the bottom armor to antitank gunfire may contribute materially to the defense. Locations are planned for friendly tanks to occupy, when not otherwise engaged, so that they may in an emergency render the best assistance to the defense by making the final counter attack against the hostile tanks.

German World War Methods.

Germany was more concerned with defense against tanks during the World War than was any other combatant, and the methods developed by the German army, toward the last of the war, were becoming increasingly effective. Believing that the British tanks were armored against all machine gun fire, that they could operate only in daylight, and that they were, to a great extent, dependent upon roads, the Germans at first confined their antitank measures to various road obstacles such as pits or holes blown in roads and the use of indirect fire from all types of artillery.

Pill Boxes. Learning by experience that this distant defense was not effective, they began to use special antitank guns in concrete emplacements, called "pill boxes." These were effective in some cases if properly sited and camouflaged, but many of them were destroyed by artillery fire prior to the attack, and, since they were stationary, those that escaped the artillery were useful only against such tanks as came within their fields of fire. Since they were generally unable to combat tanks which approached from the rear, many of them were destroyed by tanks in this manner.

Armor Piercing Bullets. Small arms armor piercing bullets were used and did considerable damage so long as the Mark I tanks were employed but the Germans did not know how effective these bullets were until after the Battle of Bullécourt on April 11, 1917, when they captured two tanks. The British had discovered the fact that the Mark I armor was not proof against the armor piercing bullets and used thicker armor on the Mark IV and succeeding tanks.

Plate 72.
1. **H A I H A, 0.85 Inch, Dutch.**
(From *Royal Tank Corps Journal.*)
2. **Madsen, 0.79 Inch, Danish.**
(From *Royal Tank Corps Journal.*)
3. **Hotchkiss 0.52 Inch, French.**
(From *Taschenbuch der Tanks* by Heigl.)

Bullet Splash. Another feature of the use of small arms ammunition against tanks, which, at first, the Germans did not realize, was the great amount of hot lead or bullet splash that entered the tank through the cracks and other openings in the armor. Bullets striking armor will shatter, and, if the point of impact is within two inches of the opening, portions of the bullet will enter. Bullet splash was very bad on all World War tanks that used the open eye slits and had poor closures at the armor joints, doors, pistol ports, turret races, etc. Men who were burned by the particles of hot lead when observing through the open slits were loath to use them again, especially when the eye slit was under fire. Many men were blinded and some were killed by bullet fragments entering the tank in this manner. The use of small arms ammunition against open eye slits and poorly made or warped doors severely handicaps the crew and is a worth-while means of defense against tanks.

Hand Grenades. Hand grenades were employed but were ineffective if used singly. Bundles of four or five were quite effective when thrown under the tank tracks. Later five grenades were wired together and used as a bomb. The best two men of the squad were designated to attack the tanks by throwing these bombs on top of them.

Antitank Rifle. The Germans produced a 13.1 mm (.52 cal.) single shot antitank rifle, which was issued at first at the rate of two per infantry regiment and later in larger numbers. The armor piercing bullet fired by this gun penetrated a 12 mm (.472 inch) plate of armor with direct impact at 120 yards. At 60 yards and striking at an angle of 45 degrees, it failed to pierce a 7 mm (.275 inch) plate. The gun was mounted on a bipod one foot in height. It was sighted up to 500 meters. While it was effective against all British and French tanks, the recoil was such that few men would fire it. Hundreds of these guns were captured by the British, French and U. S. tank troops. The French reported that no more than 20 shots could be fired in succession on account of the heating of the barrel. This gun weighed 37 pounds and was 5.5 feet long. The gun was served by two men, a rifleman and a reserve rifleman. The former carried the gun and 20 cartridges, the latter 112 cartridges.

Antitank Machine Gun. Toward the end of the war the Germans brought out an antitank machine gun along the lines of the 13 mm antitank rifle. The machine gun was also intended to be used against aircraft.

Light Trench Mortars. The Germans adopted a new carriage for their light trench mortar, which permitted flatter trajectory fire and greater accuracy. The range was from 150 to 1100 meters, the fire being accurate up to 500 meters. The translation of a German document dated Aug. 21, 1918, which appears below, deals with this subject.

In two divisions, light trench mortars employed for antitank defense have done excellent work. In one of these divisions the results obtained do not permit of exact analysis because the fire of the artillery, trench mortars and machine guns was employed simultaneously. The other division (the 192d) made the following report:

(1) Tanks of which the armor had been pierced, took fire. It seems that neither the plates on the sides nor those of the turret were reinforced.

(2) The fire was accurate up to 500 meters. Up to that distance it is certain that the tank will be destroyed. At longer ranges (500 to 800 meters) the fire of light trench mortars was very effective: the tanks were forced to turn back.

(3) Fire was carried out from the flat trajectory carriage only. It has been recognized frequently, in the course of the many changes of position made necessary by the fighting, that this carriage is very mobile and practical.

(4) It has been demonstrated that the best and most effective method of using trench mortars is to use them as mobile pieces, and not as position artillery set in prepared emplacements. At a signal from the front line, the light trench mortars held in reserve near the battalion posts of command have been successfully used against advancing tanks. It is necessary to adopt distribution to a great depth for the trench mortars. The resting battalions must keep their light trench mortars with them, in order to have an arm against tanks at their disposal as soon as they come into line.

(5) It is important that the issue of regulation ammunition wagons (to carry 44 projectiles) to units be completed as soon as possible. (The regiment has only wagons carrying a maximum of 24 rounds. The scarcity of horses for the transport of munitions and the withdrawal of the guns has considerably decreased our efficiency. In order to maintain the fire effectiveness of the trench mortars for the longest possible time, it will be well to install a number of advanced ammunition dumps, distributed over the terrain.

(6) Measures taken by the enemy against our light trench mortars: Trench mortars in action have been discovered and machine-gunned by tanks at a distance of about 800 meters; probably as a result of the vibration of the moving tank, the machine gun fire has been too short; most of the time it hit the ground immediately in front of the tank. It is only at between 300 and 100 meters that the machine guns of the tanks have obtained some hits.

German Automobile Antitank Gun. Captured documents disclosed the fact that the Germans started to organize sections of antitank guns mounted upon trucks. These sections were assigned to the artillery.

Tank Barriers. In some cases tank barriers were constructed in streets. In one case these consisted of a kind of palisade made of small iron girders, one end of which was placed in a block of concrete which was buried

Plate 73.
1. Antitank Rifle, T-Rifle, German.
2. H A I H A, 0.79 Inch, Dutch.
    (From *Taschenbuch der Tanks* by Heigl.)
3. Oerlikon, 0.79 Inch, in Firing Position, British.
4. Oerlikon, 0.79 Inch, Antitank Unit, Towed by a Carden Loyd, British.
    (From *Royal Tank Corps Journal.*)

several feet in the ground. The barrier was extended at an angle toward the enemy, the upper end of the palisade being about two meters above ground. The girders were spaced about 20 inches apart. In other cases heavy chains were hung from large concrete posts as a tank barrier.

Wide Trenches. Following the Battle of Cambrai, November 20, 1917, the German Caudry Group issued an order dealing with antitank defense, in which trenches at least 13 to 16½ feet wide and 10 feet deep were advocated as tank obstacles. Wide trenches were used to very good advantage in opposing the French Schneider and St. Chamond tanks. It was necessary that special troops be assigned to help these tanks cross the German trenches.

Mines. Large stones were put in the road so that the tank would change its course and move to either side. Mines were well concealed by sod after being put into position on each side of the stones. Each mine weighed about 12 pounds, consisted of 20 powder charges of 200 grams each, placed in boxes approximately 14 x 16 x 2 inches and the boxes were concealed about 10 inches under the surface of the ground. Detonation was caused by a hand grenade placed inside and against one of the walls of the case so that its primer passed through the wall. It could function automatically as the tank passed over the ground above the mine, or at the will of the watchers nearby. A 900 pound roller run over the mine was sufficient to fire it.

Minenwerfer. Minenwerfer were utilized as a means for defense against tanks, as shown by a report of inspection dated September 16, 1918:

> One division placed its minenwerfer for antitank defense so near the main line of resistance that they were only in a position to engage tanks which might appear *in front of* the main line of resistance. Another division retains half its available minenwerfer in a mobile condition in the vicinity of the support battalion. This method allows for the absolutely indispensable distribution in depth of minenwerfer. It is especially necessary that minenwerfer should also be able to engage effectively tanks which have broken through.
> Wherever smoke clouds arise in conjunction with a hostile attack, they conceal danger and fire must therefore be directed into those clouds.
> *Passive antitank defense.* There is still little use made of tank barricades in sunken roads and in villages at points in the streets where the tank cannot make a detour.
> *General organization and supervision of antitank defenses.* This must, both in corps and divisions, be in the hands of an officer specially fitted and trained for the task, and not too young. As the most important element of antitank defense consists in the active engagement of tanks by artillery and minenwerfer, an artillery officer is the most suitable.

Defense in Depth. With reference to deep penetration of the German lines by tanks, a captured German document dated September 3, 1918, reported:

> The farther they push forward into our lines, however, the more surely do they become a prey to our deeply organized defenses, namely, artillery (direct fire), trench mortars with flat trajectory, machine guns (armor-piercing ammunition), tank rifles, and infantry rifles with armor-piercing ammunition, and the less are they able to do any harm to our infantry. The confidence of the infantry in its own weapons and in the sure effect of the means established in the rear for combatting tanks must be strengthened. The sure effect of these means has lately been almost daily mentioned in the summaries of information from General Headquarters. We must strive with all emphasis to stamp out the feeling on the part of the infantry that they are surrounded as soon as enemy tanks have broken through their lines. This feeling is entirely unjustified. These tanks are cut off just the same as, for example, machine gun crews that have been dropped in our rear by tanks (as has often happened); they are either destroyed by our fire or put out of action by our prompt counter attack.

The following quotation from German documents shows some of the methods worked out by them for combatting tanks:

> According to a communication to the IId and XVIIIth Armies in the course of the recent fighting, tanks have broken through in great numbers along narrow fronts and continuing straight ahead, have immediately attacked the battery posi-

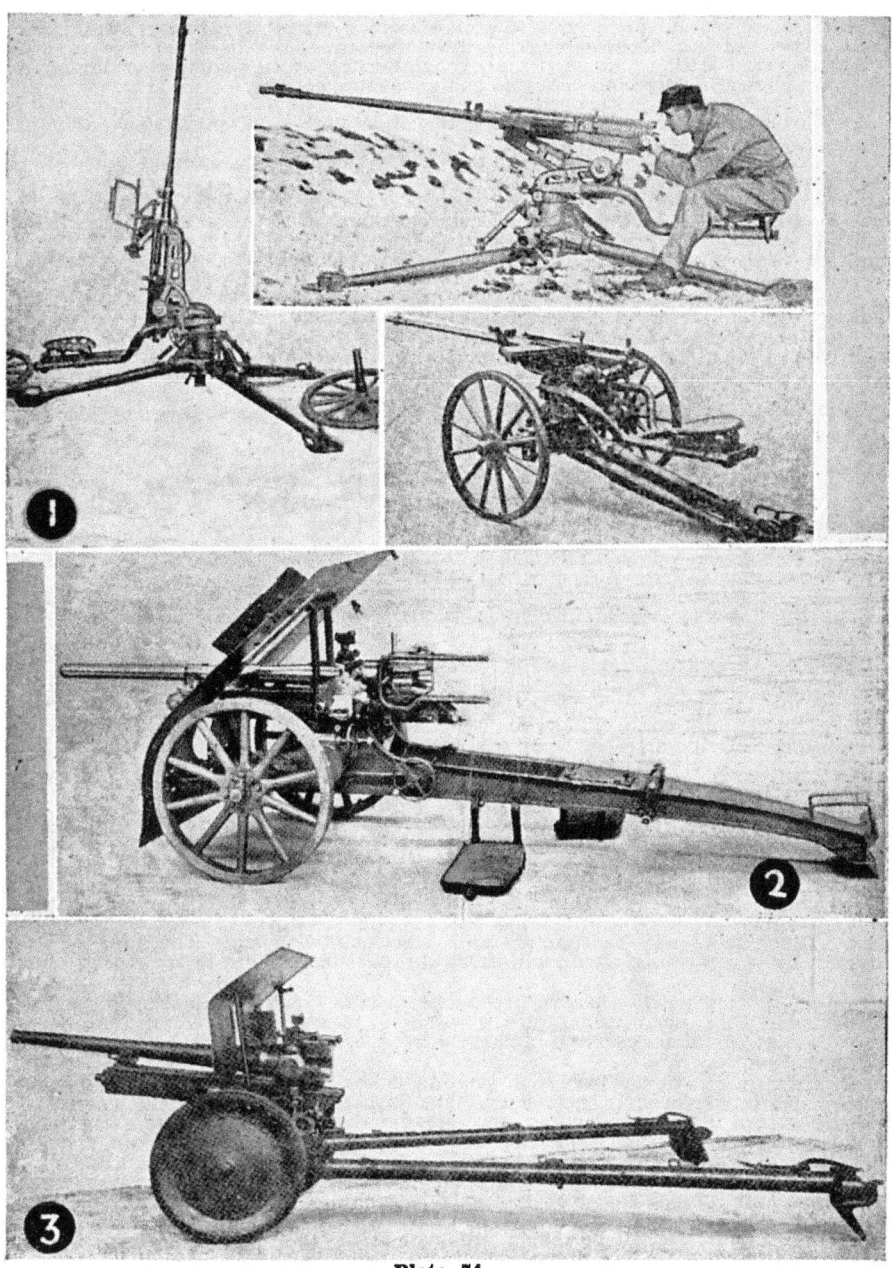

Plate 74.
1. **Solothurn S 5-100, Swiss.**
(From *Infantry Journal*, U. S. A.)
2. **Bofors 37 mm, Swedish.**
(From *Taschenbuch der Tanks* by Heigl.)
3. **HAIHA 47 mm, Dutch.**
(From *Taschenbuch der Tanks* by Heigl.)

tions and division headquarters. At several points, it seems to have been impossible to organize a defense in time owing to the fact that the batteries, having their guns dug in were not mobile enough and were not ready to fire in time to defend themselves quickly against the tanks which were attacking from all directions. The antitank defense must be developed upon the basis of these experiences. For the artillery in particular, regulations must be provided analogous to those which are in force for defense against cavalry attacks.

For this purpose it is necessary:

(1) To establish a direct antitank observation post functioning day and night in the vicinity of each battery.

(2) To avoid digging the guns in too deeply. It is sufficient that the men be protected.

(3) To dispose all guns, at least those of small and medium caliber, so that they may be able to withdraw from their positions in a few seconds and without difficulty execute direct fire in all directions, including to the rear. The guns must be ready to fire at all times. For the protection of heavy batteries and medium batteries in position special antitank sections should be designated.

(4) To increase the number of mobile antitank sections, to employ the mobile artillery antitank reserves by bringing them up quickly, placing them in position in the open and opening fire without delay; by making speedy arrangements which will permit the teams of all the batteries to be quickly brought up.

(5) To provide all the machine gun batteries with armor-piercing cartridges.

The captured German documents that follow show German views regarding the use of various weapons, in antitank defense.

Organization of German Antitank Defense.

In the organization of the antitank defense, the fact should be taken into consideration that water courses, stretches of marshy ground, and woods constitute obstacles for tanks. The distribution of the defensive arms of the service should be regulated in accordance with this consideration.

It is absolutely necessary even when the situation is quiet that all the arms suitable for antitank defense be distributed in depth and always ready to engage in action. The logical cooperation of the various arms must be assured.

I. *Artillery*.

It is well to prepare the antitank defense in such a manner that the greatest possible number of guns, distributed in depth, may participate in it. A distinction must be made between:

(1) *The antitank guns*, which, being distributed in depth between the main line of resistance and the artillery zone, are used for antitank defense only in the zone through which the tanks may advance with direct fire. Their number will depend upon this condition. The importance of antitank combat will offset the withdrawal of these guns from barrage fire and other missions.

An ammunition supply of 250 to 500 rounds will be sufficient for these guns. A shortage of steel-pointed shells may be made up by 150 mm shells; other shells may likewise be used in case of necessity.

(2) *Artillery sections assigned to antitank defense.* Each division designates an antitank section from the field artillery out of the line and at rest. This antitank section will be provided with horses and held in readiness in case of an urgent alert. The limbers will be supplied with steel-pointed shells. The guns will be followed by two caissons, which will carry shrapnel in addition to the steel-pointed shells.

Numerous firing positions, with a view to bringing these sections into position, even including the advanced infantry zone, should be reconnoitered. It is unnecessary to establish ammunition dumps at these positions, as ammunition caissons will be brought up.

(3) *Guns in position.* They will go into action, using direct fire, against the tanks which break through (field guns, light howitzers and heavy field howitzers). The use of a rather large number of these guns should be provided for. When they cannot fulfill their mission from their positions, steps will be taken to withdraw them promptly from their shelters. Frequent exercises will be held in order to make sure that they will be ready to open fire rapidly against tanks which have broken through. The proper supply of ammunition will be kept in readiness.

(4) *Heavy batteries* whose opportunities for observation are especially favorable for combatting tanks which have penetrated our lines. For these batteries, the terrain will be divided into zones for the purpose of antitank combat. The range to certain important points will be determined in advance and communicated to the artillerymen.

II. *Heavy machine guns*.

As a rule, all heavy machine guns, even those held in reserve, will be supplied with armor-piercing cartridges. The primary mission of each heavy machine gun is to combat tanks which penetrate into its zone, and it must be fully aware of this responsibility. Machine guns placed at points especially favorable to penetration by tanks must know that they have a very special responsibility. For this reason they will be designated by the name "Antitank machine guns."

III. *Antitank rifles.*

The antitank rifles will generally be placed in the main line of resistance or a short distance behind this line. Their short range (maximum, 500 meters) will be taken into consideration.

IV. *Trench mortars.*

The light trench mortars, echeloned in depth and engaged in groups of at least two, will be distributed in the infantry zone according to the principles applying to machine guns. Antitank combat will also be their permanent mission. When they are assigned barrage missions, it will be necessary to reconcile this mission with the antitank defense. The light trench mortars placed at points favorable to penetration by tanks will be called "Antitank trench mortars."

The medium and heavy trench mortars may also frequently intervene effectively in antitank combat.

V. *Antitank mines.*

They complete the action of the arms in antitank defense. They are best placed first of all in the outpost zone and then in sufficient number in the strong points of the intermediate zone and in the support positions. The troops will be trained in their use and will be taught to employ concentrated charges in antitank combat. When the tanks attack, all the arms suitable for the purpose must consider antitank combat their only mission until the last tank has been destroyed.

If the tanks are destroyed by fire, the entire attack fails. This fact must be known by all the troops.

* * * * * * * * *

(1). ....

(2). In another army, the enemy is making use of tanks in night-patrol raids. Darkness prevents the artillery from combatting them in an absolutely effective manner.

It is necessary to ascertain whether the arms used for antitank defense at the disposal of the infantry are in a state of readiness to enter action.

* * * * * * * *

The enemy has changed his tactics in the use of tanks. He is now employing them in large numbers and very seldom individually. Our method of defense must be adapted to this procedure.

Terrain, marshes, watercourses, thick forests, and steep slopes are impracticable for tanks. Trenches, hollow roads which are not too deep, thin forests, forest paths, crater fields, obstacles, carriage roads, even those with a fairly steep grade or crossing valleys, do not constitute obstacles for tanks.

Artillery. Against tanks attacking in mass, it is doubtful whether the use of individual guns or sections, stationary or mobile, will result in a decisive success. The fire of individual guns scatters easily and is too weak. Mobile sections are frequently brought into action too late. We must oppose a powerful fire to masses of tanks. Antitank batteries and sections will be assigned. They will be placed in position as ambushed batteries, but in such a way that they will command the entire outpost zone with direct fire observed from the immediate vicinity of the battery, and that they will have as effective ranges as possible. They will fire only at tanks or infantry in close range, defensive fighting. In addition to antitank batteries, every battery will take part in the antitank defense in the zone to which it is assigned. In the advance of the squadrons of tanks is observed early enough, they will be taken under the concentrated annihilating fire of as great a force of artillery as possible. The depth of antitank defense will include the entire zone of action of the divisions in line. The artillery must not be afraid to open fire on tanks which have broken through, even if they have already gained the rear of our infantry. If the fire is executed at short range with good observation, there is no fear of seriously endangering our infantry.

Infantry. For the same reasons as apply to the artillery, the scattered engagement of trench mortars, machine guns, and antitank rifles will be avoided. They must be assembled in groups of two or more; the antitank rifles preferably in groups of from six to eight. Each group is under the orders of an energetic leader. Armor-piercing cartridges (S.M.K.) are only really effective at very short distances.

The employment of concentrated explosive charges composed of four or more hand grenades tied together, thrown at close range, and from the rear underneath the track, promises good results. Hand grenades used singly are ineffective unless they are thrown from above into the apertures. This method is possible if the tanks are climbed from the rear. The infantry must, moreover, keep cool and observe very strict discipline. Anyone who has been in a tank knows that the accuracy of the fire of the guns and machine guns in action within the moving tanks is extremely uncertain. The effect produced by tanks is generally overestimated, especially when the infantry has not yet become familiar with this new arm. The infantry must quickly take cover from the view of tanks which have broken through, but immediately after their passage it must fill up the gaps and fire on the hostile infantry which follows after them. Support troops will be brought up in small groups. Long files of infantrymen offer an excellent target to the tanks. All the reserve battalions will, as a rule, be alloted an accompanying artillery, which if the opportunity offers, will take under its fire the tanks which have penetrated deeply.

Pioneers. Dig large traps, concealed from the observation of the enemy, with

Plate 75.
1. **Beardmore 47 mm, British.**
(Photo by Wm. Beardmore Co.)
2. **Bofors L 33, 47 mm, Swedish.**
(From *Taschenbuch der Tanks* by Heigl.)
3. **M2-E1, 37 mm, U.S.**

solid, step walls. revetted with hurdles, if possible. These are best sited at road defiles and along forest paths. Create marshes by damming brooks, destroy strong bridges and wherever crossings are necessary, replace them by wooden or foot bridges which are not strong enough to bear the weight of the tanks. Establish strong barricades; make slashings of trees across roads; prepare antitank mine fields, taking care to indicate them to our infantry by means of placards, wire fences, or the like. The ground in front of the obstacles should be swept by frontal and flanking fire.

If there is the immediate possibility of an attack, the above mentioned measures will be put into effect without delay. A concise report of what has been done will be sent in.

* * * * * * * * * *

Notes on Antitank Defense (Small British Tank).
I. *Infantry.*

(1) Good targets on the tank for light trench mortars, 13 mm (.53 inch) antitank rifles, machine guns, and infantry rifles are as follows:

Position of tank.	For light trench mortars.	Only for 13 mm (.530″) antitank rifles or armor piercing ammunition.	For every type of infantry ammunition (13 mm (.530″), armor piercing or ordinary ammunition.)
Front.	Petrol tank; behind it the water tank and engine room; at the back the driver's cab.	Driver's seat and machine gunners.	Apertures for aiming and vision.
Sides.	The whole broadside, especially the driver's cab.	Driver's and machine gunner's seats; petrol tank.	Apertures for aiming and vision.
Rear.	Above—driver's cab. Below — caterpillar track.	Seats of the crew.	Apertures for aiming and vision.

(2) Hand grenades: Used singly they are ineffective. Concentrated charges (two grenades without the stick tied on to a third), thrown on the track or horizontal portions of the tank, destroy the track or the armor plating.

II. *Artillery and medium or heavy trench mortars.*

A destructive effect is obtained by:
(1) All direct hits by field artillery striking at a good angle.
(2) Armor-piercing shell of all calibers, even against the strongest armor plating.
(3) Direct hits by medium and heavy calibers. Usually even the burst of the shell near the tank is sufficient to put the crew out of action.

With reference to the use of an artillery barrage or indirect fire as a means for combatting tanks it was found and reported that "a rigid scheme of artillery defense by means of barrage fire must be discarded. It has no effect, the fire is very seldom accurate, it is too thin, is usually opened too late, expends a large amount of ammunition, and is a considerable danger to the infantry in mobile warfare."

Antitank Forts. Antitank forts or specially prepared areas were formed by the Germans toward the end of the war. An extract from an Army order to the 243rd Division gives the following directions concerning these areas:

In order to increase the efficiency of our antitank defense, the following is ordered: In the forward battle zone, antitank groups, under especially energetic leaders, are to be formed. These will consist of antitank guns, machine guns, antitank rifles, and trench mortars. The various weapons of these groups need not be close together, but they must be able to render mutual support within their group, except in the case of antitank rifles, which must be in groups of four to six. Groups are to be distributed in depth in the battle zone. The first duty of the field artillery is to keep off the enemy's tanks. All other duties must give way to this.

Captured documents give the following details concerning the armament and location of these forts:

Two field guns. Two to three light minenwerfer on flat-trajectory carriages. Two to three machine guns (apparently drawn from artillery). Three to four antitank rifles. Two portable searchlights.

Some forts had no field guns, or alternately, no light minenwerfer, some had no antitank rifles. All weapons in each fort were mutually supporting.

The officer in command of each fort was called the Tank Fort Commander (Kommandant), and in many cases was an artillery or machine gun officer.

The area of the tank fort varies according to conditions of ground, but may cover several hundred square yards. According to captured maps, the most advanced tank forts were close behind the main line of resistance, and the remainder were distributed to a depth of 3,000 to 4,000 yards. One map shows nine such forts in an area about 1,500 yards broad and 3,500 yards deep.

The following further particulars of the construction of the redoubt were secured from German documents:

For antitank rifles. Circular sharpshooters' trenches (about 1.50 meters diam.) with places for ammunition and rifle.

For machine guns. Circular emplacements, with places for the gun and ammunition.

For light trench mortars. Emplacements in the open, with places for the ammunition.

For guns. Emplacements arranged to permit fire over the parapet, with inclines in rear and front so that the pieces can be quickly moved to neighboring ground, when it is impossible to fire from the emplacement itself. Places will be provided for ammunition.

Concealment. All works will be constantly concealed with great care from aerial observation, even during construction, which will not be begun before a sufficient amount of camouflage material is available.

Dugouts. The dugouts will be arranged so that the personnel of several pieces of defensive equipment have room in a single dugout and are able to reach the position very quickly from it. During the construction, strict cover from aerial observation will be provided.

Mine fields. Mine fields will be placed principally at the points from which the tanks would have particularly effective action on the fort and where the tanks could temporarily be sheltered from the action of the fort.

Dummy forts. At suitable points, between or by the side of the forts, in rear or in front, dummy forts will be constructed at a distance of at least 300 meters from the nearest fort.

As a means for spurring the German troops to the use of all possible initiative in preparing and executing antitank measures, General Ludendorff, on August 16, 1918, issued an order upon the subject, an extract from which is given:

The Commander-in-Chief is prepared to mention in the official communique the names of units as well as the names of individual officers, noncommissioned officers and men who have specially distinguished themselves in action against tanks.

I have also to request that the initiative and self confidence of individuals and units may be encouraged and increased by awarding distinctions for specially successful antitank defense.

Upon several occasions German troops surrounded and fought individual tanks at very close range and in some cases individuals climbed on the tanks and attacked the crews with pistols through all available openings. Others came up and fired through openings in the hull. In one or two cases German infantrymen grasped the tank weapons and tried to pull them out of the tank.

Having finally recognized the fact that the artillery piece was the best of their antitank weapons, orders were issued following the Battle of Amiens on August 8, 1918, which greatly increased the number of guns per division for antitank defense. In place of three or four guns per division formerly assigned to antitank defense, an order dated August 10, 1918, prescribed the following increase in the number of guns to be assigned to antitank defense:

One battery to execute direct fire from advanced positions. One surveillance battery occupying positions in the line of artillery protection. Five additional surveillance batteries, including two heavy batteries. Two batteries, horsed reserve of the division, for which special positions would be reconnoitered and prepared. All other batteries, if necessary.

Plate 76.
1. **75 mm Gun, No. 1, Model 1925 E1 on T2 Carriage in Antitank Firing Position, U.S.**
   (Photo by Ordnance Dept., U.S.A.)
2. **Skoda 47 mm, Czechoslovakian.**
   (From *Royal Tank Corps Journal*.)
3. **75 mm Gun, No. 1, Model 1925 E1 on T2 Carriage in Traveling Position, U.S.**
   (Photo by Ordnance Dept., U.S.A.)
4. **3.7 Inch Howitzer on Tracked Trailer, British.**

In order that all available artillery might be used to combat tanks promptly an order was issued directing that: "Messages concerning tanks will have priority over all other messages or calls whatsoever."

Present German Doctrine.

The present German doctrine of antitank defense is thought to be about as follows:

(1) That natural and artificial obstacles should be used when they are available.

(2) That weapons, especially artillery, using *direct fire methods* are the most effective means of combatting tanks. (To this end, should the Versailles treaty be nullified, the Germans plan to include in their infantry regiment a six-gun 77 mm battery and also to include in their infantry division an additional portée battery for antitank use.)

(3) That small caliber weapons should engage immobilized tanks by placing fire on the eye slits. Also that such weapons firing armor piercing ammunition may pierce the floor as the tank noses up over an obstacle.

(4) That the organization of the defensive zone should be in depth.

(5) That antitank weapons should be used only for engaging tanks.

(6) That the air service will locate tank concentration areas which will then be bombarded with heavy artillery or gassed.

Should the Versailles treaty be nullified, it is believed that Germany plans to equip infantry units with a 20 mm (.79 in.) automatic antitank gun.

Present British Doctrine.

The British favor combatting tanks by the use of:

(1) Tanks. (2) Self-propelled guns affording protection to the crew (18 pdr.). (3) Self-propelled guns protecting the crew by a gun shield only (Birch gun). (4) Oerlikon 20 mm gun mounted upon a trailer. (5) Pack howitzers on motorized mounts.

They believe that antitank guns should preferably have self-propelled mounts.

The following from the *Royal Tank Corps Journal* summarizes typical British ideas regarding antitank methods:

If possible, the defensive position should be chosen with a view to utilizing favorable topographical features which will assist in breaking up hostile tank attacks, separating them from their own infantry and drawing them into areas favorable for their destruction. When time and material are available the defensive position may be additionally strengthened by the use of artificial means, these means being placed well out in front of the main line of resistance.

Antitank artillery should be provided and disposed in such a manner that it can bring fire to bear over *open sights* on the areas favorable for tank advances. The antitank artillery should be concealed from air and ground observations up to the last moment, so as to gain the maximum advantage of surprise effect and, having once completed the immediate task assigned to it, should withdraw to its position of readiness. On no account should it be used for any purpose except antitank work. Observation posts well to the front should be provided to give *ample* warning of the approach of tank attacks or the assembly of tanks in front of or to the flanks of the defensive position. As it is considered the best tactics for the tanks to attack now either early in the morning or late in the evening, or on foggy days and not under the protection of artillery preparation prior to the attack; the defense should be alert to these conditions and prepared to function against tanks at these hours and under these conditions.

French Doctrine.

The French are believed to favor an antitank weapon capable of penetrating heavy armor up to about three inches in thickness. They believe that such a weapon should also be capable of functioning as ordinary divisional artillery and should form a part thereof. Their 75 mm gun, M 1917, does not fulfill these requirements.

According to the French view, antitank guns should be deployed in depth throughout the defensive position and so sited as to bring *direct fire* on the attacking tanks. They should be placed well to the front to break up the attack as quickly as possible.

Japanese Doctrine.

The Japanese have given extensive and detailed study to the matter of antitank defense and contemplate, in addition to antitank guns, artillery and mines, the use of specially trained infantrymen as bombers.

The Future of Antitank Defense.

In the future the task of the general on the defensive will be made much more difficult by the existence of fast tanks in large numbers. Their unprecedented cross-country mobility, their own ability for quick offensive action, and cooperation with other troops in sudden offensive movements will make the life of the defenders a nightmare. To have a flank out in the open when defending will become a tragedy. It is difficult enough to defend a front, but when one must defend the front, the rear and a flank, and when hostile offensive agencies in large numbers move long distances at high speed, efficient observation, communication, generalship and cooperation between all combat agencies will be required in a high degree in order to meet the situation.

In connection with the ability of antitank gunners to hit moving tanks, the following table shows the time required to cover certain distances at different rates of speed:

Rate Miles per hour	Time, Minutes and Seconds				
	Distance, Yards				
	200	300	500	700	1000
1	6m—49s	10m—14s	17m— 3s	24m—52s	34m— 5s
2	3m—25s	5m— 7s	8m—31s	11m—56s	17m— 3s
3	2m—16s	3m—25s	5m—41s	7m—57s	11m—22s.
4	1m—42s	2m—33s	4m—16s	5m—58s	8m—31s
5	1m—22s	2m— 3s	3m—25s	4m—46s	6m—49s
6	1m— 8s	1m—42s	2m—50s	3m—59s	5m—41s
7	58s	1m—28s	2m—26s	3m—25s	4m—52s
10	41s	1m— 1s	1m—42s	2m—23s	3m—25s
12	34s	51s	1m—25s	1m—59s	2m—50s
15	27s	41s	1m— 8s	1m—35s	2m—16s
20	20s	31s	51s	1m—12s	1m—42s
25	16s	25s	41s	57s	1m—22s
30	14s	20s	34s	48s	1m— 8s

Late Developments in Antitank Weapons.

Antitank weapons are being developed in several countries but only a few of them have all of the essential characteristics of an efficient weapon for this purpose. The table that follows includes the commonly known antitank weapons and certain available information regarding each. In considering the armor piercing ability, it should be borne in mind that normal impact is exceptional and that at less favorable angles the armor piercing ability is very materially less.

ANTITANK WEAPONS—Group One, .47-.55 in. (12.5-14 mm) caliber Automatic

Name	Nationality	Caliber	Projectile, weight	Muzzle velocity	Weight with mount	A.P. ability, normal impact	Cooling
Breda	Italian	.55 in. (14 mm)	1.92 oz.	3250 ft. per sec.	220 lbs.	1.14 in. 1100 yds.	Air
Browning	U. S.	.50 in. (12.7 mm)	1.83 oz.	2725 ft. per sec.	163 lbs.	.78 in. 1200 yds.	Water or air
Farquhar	British	.50 in. (12.7 mm)	1.76 oz.	2738 ft. per sec.	38 lbs.	.6 in. 800 yds.	Air
Fiat	Italian	.49 in. (12.5 mm)	1.28 oz.	2925 ft. per sec.	486 lbs.		Water or air
		.47 in. (12 mm)	1.41 oz.	2350 ft. per sec.	1466 lbs. (air)		
					1492 lbs. (water)		
Hotchkiss	French	.52 in. (13.2 mm)	1.83 oz.	2625 ft. per sec.	356 lbs.	.79 in. 100 yds.	Air
T-rifle	German	.52 in. (13.1 mm)	1.83 oz.	2525 ft. per sec.	37 lbs.	.47 in. 120 yds.	Air
T U F	German	.52 in. (13.1 mm)	1.76 oz.	2625 ft. per sec.			Water
Type 8	Russian	.52 in. (13.1 mm)	1.8 oz.		207 lbs.		

Group Two, .79-1.0 in. (20-25.4 mm) caliber Automatic

Name	Nationality	Caliber	Projectile, weight	Muzzle velocity	Weight with mount	A.P. ability, normal impact	Cooling
Becker	French	.79 in. (20.1 mm)	4.65 oz.	1705 ft. per sec.	88 lbs.		Air
Fiat-Revelli	Italian	1.0 in. (25.4 mm)	6.4 oz.	1445 ft. per sec.	88 lbs.		Air
			7.06 oz.				
H. A. I. H. A.	Dutch (new)	.79 in. (20 mm)	4 oz.	2460 ft. per sec.	295 lbs.	.79 in. 250 yds.	Air
	(old)	.85 in.	15.02 oz.	1900 ft. per sec.			
Hotchkiss	French	.79 in. (20 mm)	4.53 oz.	1965 ft. per sec.	176 lbs.	.59 in. 400 yds.	Air
Hotchkiss	French	.98 in. (24.9 mm)	4.6 oz.	3300 ft. per sec.			
			9 oz.	2952 ft. per sec.			
Madsen	Danish	.79 in. (20 mm)	0.6 oz.	2460 ft. per sec.	220 lbs.	.5 in. 300 yds.	Air
			4.94 oz.	2870 ft. per sec.			
			5.82 oz.	2500 ft. per sec.			
Oerlikon	Swiss	.79 in. (20.1 mm)	4.1 oz.	2681 ft. per sec.	375 lbs.	1 in. 500 yds.	Air
Solothurn }	Swiss	.79 in. (20 mm)	4.5 oz.	2880 ft. per sec.	242 lbs.		Air
S 5-100 }			4.41 oz.	2780 ft. per sec.	452 lbs.		
					531 lbs.		
Vickers	British	.98 in. (24.9 mm)	4.76 oz.	1997 ft. per sec.	299 lbs.		Water
			8 oz.				

Group Three, Calibers Greater than 1.0 in.

Name	Nationality	Caliber	Projectile, weight	Muzzle velocity	Weight with mount	A.P. ability, normal impact	Cooling (Remarks)
Armstrong	British	1.46 in. (37 mm)	1.5 lbs.	1400 ft. per sec.	196 lbs.	1.15 in. 330 yds.	
Beardmore	British	1.57 in. (40 mm)	1.98 lbs.	1900 ft. per sec.	400 lbs.	1.18 in. 300 yds.	
Beardmore	British	1.85 in. (47 mm)	3.3 lbs.	1620 ft. per sec.	517 lbs.	.79 in. 1350 yds.	
Bofors "L37"	Swedish	1.46 in. (37 mm)	1.71 lbs.	1590 ft. per sec.	500 lbs.	.78 in. 840 yds.	10° traverse
			1.32 lbs.	2000 ft. per sec.		1.0 in. 500 yds.	
Bofors "L33"	Swedish	1.85 in. (47 mm)	3.3 lbs.	1837 ft. per sec.	682 lbs.	1.57 in. 1000 yds.	40° traverse
						1.18 in. 1100 yds.	
						1.57 in. 950 yds.	
H. A. I. H. A.	Dutch	1.85 in. (47 mm)	3.3 lbs.	1320 ft. per sec.	772 lbs.	.79 in. 1000 yds.	
Maklen	Russian	1.46 in. (37 mm)	1.06 lbs.	1720 ft. per sec.	726 lbs.		
M 1916	French	1.46 in. (37 mm)	1.25 lbs.	2120 ft. per sec.	231 lbs.		45° traverse
M 1922	Japanese	1.46 in. (37 mm)	15 lbs.	2175 ft. per sec.	3250 lbs.		
M 1	U. S.	2.95 in. (75 mm)	1.45 lbs.	1850 ft. per sec.		1.0 in. 500 yds.	
M 2 E 1	U. S.	1.46 in. (37 mm)	3.3 lbs.	1525 ft. per sec.	530 lbs.		
Pocztsk	Polish	1.85 in. (47 mm)	1.1 lbs.	1430 ft. per sec.	396 lbs.		
Rosenberg	Russian	1.46 in. (37 mm)	1.81 lbs.	1968 ft. per sec.	387 lbs.		360° traverse
Skoda	Czech	1.25 in. (31.8 mm)	1.1 lbs.		330 lbs.		
Skoda	Czech	1.46 in. (37 mm)	1.81 lbs.	1510 ft. per sec.	442 lbs.	.79 in. 1100 yds.	
Vickers M/15	British	1.61 in. (41 mm)	1.98 lbs.	970 ft. per sec.	209 lbs.	1.18 in. 325 yds.	
Vickers	British	1.85 in. (47 mm)	3.3 lbs.	1600 ft. per sec.	559 lbs.		

CHAPTER XXVII.

TANK ORGANIZATION.

MAJOR ASPECTS.

In the major aspects of tank organization, as in the closely related matter of tank employment, the nations of the world appear to be feeling their way, rather slowly and somewhat blindly. Many ideas are advanced, but few of them are accepted generally. Slow progress is being made in the comparative weighing of the merits of the various conflicting ideas. Although the qualities and mechanical capabilities of modern tanks are generally known and recognized, there is a great diversity of opinion as to how they should be organized for use. It seems impossible to doubt, however, that, in the course of the next decade or two, the military minds of the world will view the utility of the efficient tank and its natural place in military organization in a very much more uniform light than is the case today.

Great Britain.

Great Britain is apparently completely sold on her Royal Tank Corps, which is a separate arm organized in 1917, and includes not only tanks but armored cars (excepting those that form part of the cavalry) and associated vehicles. We have been unable to find any British arguments in favor of making the tanks a part of the Infantry, Cavalry or Artillery.

United States.

The United States had a tank corps for two and one-half years. After its disbandment as such in 1920, the tanks were regarded as a subsidiary of the Infantry. The present trend of events seems to indicate that the tanks are to be divided, a major portion of them remaining subsidiary to the Infantry, while a minor portion become subsidiary to the Cavalry. Ours is the only one of the great countries of the world whose infantry division includes any sort of tank organization. We have a light tank company as an integral part of each infantry division.

France.

During the War, the French Tank Corps, although it functioned in a large measure as a separate branch, was under the administration of the Artillery Section in the War Department. In 1920, this organization, including the various tank units and the Tank School, was transferred to the control of the Infantry Section. The general officer, Inspector of Tanks, who heads the tank organization, is now, therefore, an assistant to the General Inspector of Infantry.

There has been in France much controversy regarding the advisability of subordinating the tanks to the Infantry. Much has been written in favor of making them a separate arm. A few excerpts from the views of some prominent French officers strongly tend to confirm these statements and may be of general interest. Colonel Romain wrote in the *Revue De Paris* in 1922 as follows:

> For what reasons does the Infantry claim the custody of the tanks? First of all because they are intended to open the way for it. Specious reason! All of the arms are intended to open the way for the infantryman. It is their mission. It is even their reason for existing. The artillery fire, the engineering works, the reconnaissance and the bombardments of the aviation, and even the charges of the cavalry

would be without avail if their aim was not to facilitate the advance of the infantry. The latter, regardless of what is said, or what is done, will always remain the greatest arbiter of battles; but it cannot, for that reason, claim to absorb the auxiliary arms. In this case, as always, the work should be divided.

Colonel Chedeville wrote in the *Revue Militaire Francaise*:

The experience gained in the war and the study of the tank requirements of small infantry units has brought out the facts that these requirements are considerable. During the engagements of the year 1918, we were always short of tank units in our efforts to support properly the operations undertaken. Moreover, the total resources of tanks will always be relatively weak with reference to the requirements of the armies. This consideration, together with the consideration that not all parts of the terrain are adapted to the use of tanks, precludes the desirable conception of organically alloting tanks to large units. By so doing we would surely run the risk of being short of tanks wherever a prolonged effort might be necessary, while in other places the tank units would remain unemployed opposite to fronts impracticable for their operation.

Colonel Romain also wrote:

* * * * * And they cannot be employed everywhere. What good would they be to troops behind watercourses, or in thick woods, or in front of the heaviest networks of trenches?

It is up to the High Command to make them operate in mass, and, as much as possible, by surprise, wherever it has been decided to make a decisive attack.
* * * * *

Finally the Infantry, in spite of its good will, could not take care of the task of assuming the improvement of the materiel. To be called a "technical arm" is not sufficient to become one.

* * * * * And even if it did succeed, it would never see in the tanks anything but an infantry-accompanying auxiliary, and it is from this viewpoint alone that it would wish to orient improvements. The scope of the tank should be much greater.

And Colonel Velpry wrote in *La Revue d' Infanterie*:

It has been said that the tank is merely an armored machine gun on a motorized carriage which it would be sufficient to place at the disposition of small infantry units or even incorporate therein. On the whole, the tank is, in reality, nothing more than that; but it is precisely the fact of being motorized and armored that endows it with qualitites, defects, and special properties which render its employment absolutely different from that of an ordinary machine gun.

The airplane which descends close to the ground to fire upon the enemy during combat, as it did in the last campaign and as it will undoubtedly do again. is also nothing more than a flying machine gun operating in favor of the infantry. Nobody thinks that we would obtain the greatest efficiency by alloting one airplane to each infantry company.

That which determines the mode of employment of a weapon is the manner in which it is carried. The cavalryman and the infantryman have in all times employed similar weapons: sledge hammer, hatchet, sword, lance, and rifle—these are peculiar neither to the infantry or the cavalry. The horse is what made the difference in the method of employment.

And this is the reason why, even in distant periods where the cavalryman and the dismounted man fought side by side on the battlefield, in close liaison which was not without analogy to the liaison of the modern tank and the infantryman, the cavalryman's mission being to open in the ranks of the adversary a breach through which the infantry might pass, or to overthrow the hostile cavalry which the infantry came to stab, even in those periods, they said, the amalgamation of the two arms never appeared to be a practical and advantageous solution. It will be the same in the case of the tank and the infantry for much more peremptory reasons.

Notwithstanding the ideas of various individuals as to what is desirable, the mechanization policy of France is forcibly established, to a large degree, by economic considerations. She maintains a large army. She has more serviceable tanks than any other country in the world. Although she has made extensive efforts to improve the efficiency of her tanks and has built a few new ones, she has apparently done little to develop fast modern tanks or to determine a combat role for such vehicles. To launch a mechanization program of significant proportions would involve a staggering addition to the military budget. France is naturally very loath to admit that her large numbers of serviceable tanks, which she cannot afford to replace, are obsolescent or unsatisfactory. Apparently, the official view, influenced by these considerations, continues to pin its faith to the accompanying role exclusively—for which these many tanks are suitable. And this may explain, in part, why France classifies her tanks as a part of her Infantry.

Although the tank service is considered part of the Infantry, the officers and men in it, in general, remain there. There is relatively little transferring in and out. The armored car units of France are under Cavalry control.

Russia.

Russia places her tanks and armored cars on a par with her artillery, and places both under control of an Inspectorate of Artillery and Armored Forces. Since the present-day trend is to motorize artillery, and since tank crews operate cannons, the Russian view that these elements are similar is plainly not without a reasonable foundation.

Italy.

The Italian tank organization consists wholly, or chiefly, of a tank regiment that is located at the Tank Center in Rome. This regiment, the officers of which have been drawn from the various older combat branches, engages in a considerable amount of experimental work. It is not considered a part of any other arm and is, in a small way, a separate corps. There was formerly little interest in tanks in Italy; but, since 1927, interest has increased very materially. Armored cars are included in the tank organization.

Japan.

A few years ago the Japanese tank organization consisted of but 2 companies. It is understood that this organization has, more recently, been materially expanded. The automobile school of the army furnishes specially trained personnel for the tank units.

Various Countries.

The armies of Belgium and Poland have one tank regiment each. Czechoslovakia has several battalions. Of the numerous other countries having tank organizations nothing will be said due to the small number of tanks involved and the scarcity of information available.

The treaty signed at the close of the Great War does not permit Germany, Austria, or Bulgaria to build or utilize tanks. Germany, however, simulates tanks by carrying canvas representations on light powerful motor cars. By means of these dummy tanks the German theories as to tank tactics are, from time to time, tested in maneuvers.

DETAILS OF ORGANIZATION.

United States.

The present organization of U. S. tank personnel is as shown below:

Regular Army. *66th Infantry (Light Tanks).* This regiment, until recently, was known as the 1st Tank Regiment. All organizations of it are active. All are equipped with the Six Ton tank M 1917 except Company E (at Fort Benning) which is equipped with the Mark VIII tank. The regiment is stationed as follows:

At Fort George G. Meade, Maryland: Hq., Hq. Co., Serv. Co., 1st Bn., and Co. I.
At Fort Benning, Georgia: 2d Bn.
At Fort Devens, Massachusetts: 3d Bn., less Co. I.
At Fort Hayes, Columbus, Ohio: Band.

67th Infantry (Medium Tanks). This regiment, until recently, was known as the 2d Tank Regiment. It is now all inactive except Company F, which is stationed at Fort Benning, Georgia, and is equipped with Christie M 1931 tanks, light tanks of the T1 series, the Medium T1, and the Medium T2.

Divisional tank companies. There are seven active tank companies which are stationed as follows:

1ST TANK COMPANY: Miller Field, New York.
2D TANK COMPANY: Fort Sam Houston, Texas.
3D TANK COMPANY: Fort Lewis, Washington.
4TH TANK COMPANY: Fort McClellan, Alabama.
5TH TANK COMPANY: (Only 2 platoons active) Fort Benjamin Harrison, Indiana, and Jefferson Barracks, Missouri.
6TH TANK COMPANY: Fort Snelling, Minnesota.
11TH TANK COMPANY: Schofield Barracks, T. H.

The Tank School. The Tank School of the U. S. Army was under the Chief of Tank Corps prior to 1920. From 1920 to 1932, it was at Fort George G. Meade, Maryland, under the Chief of Infantry. In 1932, it was moved to Fort Benning, Georgia, where it became the Tank Section of the Infantry School.

The 1st Cavalry. The 1st Cavalry (mechanized) is stationed at Fort Knox, Kentucky, and is equipped with armored cars and combat cars (including the Christie M 1931). **Combat car** is the Cavalry's name for a tank.

National Guard. There are 15 divisional tank companies in the National Guard, each having a total personnel of about 65. They are equipped with from six to eight, each, of the Six Ton tank M 1917 and are stationed as follows:

26TH TANK COMPANY: Boston, Massachusetts.
27TH TANK COMPANY: New York, New York.
28TH TANK COMPANY: Norristown, Pennsylvania.
29TH TANK COMPANY: Danville, Virginia.
30TH TANK COMPANY: Forsythe, Georgia.
31ST TANK COMPANY: Ozark, Alabama.
32D TANK COMPANY: Janesville, Wisconsin.
33D TANK COMPANY: Maywood, Illinois.
35TH TANK COMPANY: St. Joseph, Missouri.
37TH TANK COMPANY: Port Clinton, Ohio.
38TH TANK COMPANY: Covington, Kentucky.
40TH TANK COMPANY: Salinas, California.
41ST TANK COMPANY: Centralia, Washington.
43D TANK COMPANY: Hartford, Connecticut.
45TH TANK COMPANY: Denver, Colorado.

Organized Reserves. There are seven tank regiments in the Organized Reserves. They are located as follows:

420TH INFANTRY (LIGHT TANKS): Chicago, Illinois.
421ST INFANTRY (LIGHT TANKS): Scranton, Pennsylvania.
422D INFANTRY (HEAVY TANKS): Columbus, Georgia.
423D INFANTRY (HEAVY TANKS): Chicago, Illinois.
424TH INFANTRY (LIGHT TANKS): New York, New York (Queens).
425TH INFANTRY (LIGHT TANKS): Baltimore, Maryland.
426TH INFANTRY (LIGHT TANKS): Akron, Ohio.

The concise tables that follow show the essentials of tank organization in the United States Army. P indicates *according to peace tables of organization,* W indicates *according to war tables.*

	Number of tanks		Other motor vehicles and trailers		Total personnel	
	P	W	P	W	P	W
Light tank company, Infantry div.	18	24	27	41	103	160
Headquarters platoon	3	9			64	118
3 tank platoons—each	5	5			13	14
Headquarters platoon						
Headquarters section					36	75
Maint. and repl. section (P)	3				28	
Maintenance section (W)						23
Replacement section (W)		9				20
Light tank regiment	162	223	283	433	1,266	1,887
Headquarters					6	9
Band					29	29
Headquarters company			8	13	32	53
Service company		4	16	51	136	194
3 tank battalions—each	54	73	85	123	349	534
Attached medical			4	12	16	57
Service company		4	16	51	136	194
Headquarters platoon			14	19	58	46
2 Maintenance platoons (P)—each			1		39	
4 Maintenance platoons (W)—each		1		8		37
Light tank battalion						
Headquarters					5	7
Headquarters company		1	10	17	50	73
3 tank companies—each	18	24	25	34	98	145
Attached medcal				4		19
Headquarters company (Battalion)						
Company headquarters		1	9	17	27	55
1 maintenance platoon			1		23	18
Tank company, light tank battalion						
Headquarters platoon	3	9	10	34	59	103
Headquarters section			6	34	36	67
Maint. and repl. section (P)	3		4		23	
Maintenance section (W)						16
Replacement section (W)		9				20
3 Tank platoons—each	5	5			13	14
Heavy tank regiment	90	135	130	297	1,771	3,062
Headquarters					6	9
Band					29	29
Headquarters company			8	13	32	53
Service company			16	47	136	190
3 tank battalions—each	30	45	34	79	517	927
Attached medical			4	18	17	72
Service company						
Headquarters platoon			14	19	58	46
2 Maintenance platoons (P)—each			1		39	
4 Maintenance platoons (W)—each				7		36
Heavy tank battalion						
Headquarters					5	7
Headquarters company			10	22	50	89
3 tank companies—each	10	15	8	17	153	269
Attached medical				6		24
Headquarters company (Battalion)						
Company headquarters			9	22	27	60
1 Maintenance platoon			1		23	29
Tank company, heavy tank battalion						
Headquarters platoon	1	6	8	17	55	161
Headquarters section			8	17	22	65
Maint. and repl. section (P)	1				33	
Maintenance section (W)						29
Replacement section (W)		6				67
3 Tank platoons—each	3	3			33	36

Foreign Tank Organization.

The details of the organization of foreign tank units are not available for publication.

CHAPTER XXVIII.

SOME TANK COMBAT PRINCIPLES.

Napoleon Bonaparte in his sixteenth maxim wrote:
"One consequence deducible from this principle is, never to attack a position in front which you can gain by turning."

In an article in the *Infantry Journal* under the title *On Mechanization*, Brigadier General LeRoy Eltinge wrote:

> The good old infantry division will have to do the work, both in attack and in defense. But what use is there of a defense of these invaluable spots and areas if a swiftly moving hard-hitting force (referring to a mobile mechanized force) 'can go around them and destroy their lines and supply and the places on which they depend for food, ammunition, and replacement of men and materiel?
>
> * * *
>
> New inventions in offensive and defensive power are constantly taking place, conditions of life are changing and new emotions are being brought out by the presentation of ideas, which, while they may not be new, are at least presented afresh after they have been in a period of eclipse, and thus they appear new. So it comes about that a new school of tactics develops every few years in conformity with changed conditions, and *any army that does not adapt itself to these changes will be hopelessly inefficient in war.* (The italics are his.)

In future combat we may expect to see tanks used in large numbers and to a large extent against the flank or rear of the hostile position rather than frontally, except in those situations that preclude the practicability of attack in any other way. Hence, in turning a hostile position, there appears to be no reason why both leading and accompanying tanks should not be sent against the hostile flank. An armored force may be employed either against the hostile flank or rear.

Leading Tanks.

Leading tanks, according to the *Infantry Field Manual* (U. S.), may be used to assist in the main effort of a general attack by making a breach through a strongly organized defensive line, with the ultimate mission of disrupting artillery in position, and strong local reserves available to the enemy for counterattacking and closing the gap. Tanks thus employed form the spearhead of a blow against the hostile front or flank. They are under corps or army control, are used as a single unit under one commander, and have no tactical reserve within this function. These tanks do not delay their advance for the purpose of keeping close to the friendly foot troops. Leading tanks are not intended to take the place of accompanying tanks nor of tanks in general reserve. The accompanying role and a reserve of reasonable size for exploitation may be expected to have priority. After these needs have been met, it may be that additional tanks in the leading tank role will increase the prospect of victory. The leading tank idea is of course unproved in war. It may or may not, in its original conception or otherwise, survive the test of time.

The French have the same general view from a somewhat different angle. In their contemplated employment of slow very heavy tanks, to which they give the tactical name "breaking-through tanks", they plan to use them against defenses that are especially strongly organized, and in advance of the lighter accompanying tanks. They contemplate, even in the case of these tanks, that the speed of advance will be retarded sufficiently to keep them only a short distance ahead of the attacking foot troops.

A point should be brought out regarding leading tanks and depth. Reserves generally are so necessary in war that we get to thinking of them as a matter of course. Reserves of tanks are necessary, but both French and American writings indicate a general belief in both countries that no tactical reserves whatever should be held out from those tank units assigned to the leading role. The leading tanks may be formed in waves but they have no depth otherwise. A tactical reserve of tanks for use in the later stages of the attack is something entirely separate.

Accompanying Tanks.

The accompanying role seems to be generally recognized among nations, although Great Britain appears now to attach relatively little importance to it. In this role the essence of the idea is that the infantry battalion commander has a unit of tanks attached to his battalion for the attack, the tanks being kept close to the foot troops and used in the reduction of formidable points of resistance that may develop to the front or flanks. The French having an infantry division that includes three infantry regiments and no tanks, have employed a "standard dosage" of one tank battalion (three companies) for attachment to a division. Since the United States infantry division includes four infantry regiments and one tank company, we have been contemplating the same result by attaching one tank battalion of three *additional* companies to the division. In either such case, the resulting ratio is one company per regiment, or one platoon (5 tanks—the French call it a section) per battalion. More recently, however, the French teaching has veered to the idea, that where tanks are especially needed and the success of the attack must be insured, the "dosage" should be one tank company (three sections) per infantry battalion.

Accompanying tanks are normally rather small and light. If they are slow, in the usual method of employment, they cross the line of departure just in advance of the foot troops and continue to precede them at a short distance (not more than 200 yards) except at such times as it seems necessary to go farther forward in order to neutralize hostile weapons and thus make possible the resumption of the advance of the foot troops.

If the accompanying tanks are fast (which is preferable) the platoon does not advance at a relatively steady rate in front of the foot troops but moves rapidly by bounds at those particular times when its services are needed. After completing one minor task, the tank platoon takes a predetermined concealed or defiladed position or retires to a predetermined point until its services are again required or until it is necessary to move forward to avoid becoming separated from the battalion that it is serving.

In the United States, the accompanying role is regarded as necessary and highly important to the Infantry.

Assembly and Reservicing.

Leading and accompanying tanks, after completing the missions assigned to them in attack orders, retire to or assemble at predetermined localities known as **assembly points.** A platoon of slow accompanying tanks normally proceeds to its assembly point only after the final objective has been consolidated by the infantry foot troops. At assembly points, such reorganization is effected as may be necessary and the capability of the units for further effort is investigated. Here orders are received to proceed to the **reservicing point,** or, if practicable and necessary, to engage in further combat.

Tank units are normally sent to reservicing points as soon as it can be seen that they will not be needed further in the attack, or whenever the need for reservicing is imperative. At the reservicing points as many as practicable of the following are accomplished: resupply as to fuel, oil, water, and ammunition; cleaning and lubrication; adjustments; repairs; food for the crews; rest for the crews. The tanks are here made again ready for action as soon as possible so that, if an urgent need should develop, they will be available. Reservicing points are located only a moderate distance to the rear, although farther to the rear than assembly points, and close to reasonably good roads.

Movements in Preparation for Combat.

The movement from the places where tank units are detrained to the places from which they start the attack is much more simple if the tanks are fast. All such movements are normally made at night if practicable. The attack of leading and accompanying tanks is normally launched early in the morning. If the tanks are slow they may be expected to move the night of detrainment from the detraining point to a tank park. On another night they will move from the tank park to an intermediate position. On the night preceding the attack, they move from the intermediate position to the assault positions. They leave the assault positions at the proper time, cross the line of departure a few minutes later, and shortly after that they normally find themselves engaged in action. These positions, particularly the tank park and the intermediate position, are normally so selected as to afford thorough concealment to the tanks during the daytime. There is need for precautionary measures to prevent the enemy from locating the tanks by daytime aerial observation due to visible tank tracks leading to the places of concealment. If the tanks are fast, the tank park can be considerably farther to the rear than otherwise, and, since the tanks can move readily during the night preceding the attack from the tank park to the assault position, there is no need for intermediate positions.

Cavalry.

Slow tanks carried in trucks are of relatively little use to cavalry, primarily because they can be used only in those cases in which the enemy to be attacked has established himself with some degree of permanence in a definite position. When such tanks are used they are detrucked as close as practicable to the scene of employment and are concentrated where the main blow is to be struck, their attack naturally being coordinated with that of the cavalry troops.

Fast modern tanks, not requiring carriers and being capable of extended movement, add much to the power and efficiency of cavalry. Cavalry is expected to use such tanks chiefly as a powerful striking force held closely in hand, either on the offensive or defensive. In attack, such tanks would be thrown in quickly to add punch to the main effort when the situation had developed sufficiently to make such action practicable. When defending, the tanks add power to the main blow of the counter attack. And cavalry defense normally is of the active rather than passive type.

In reconnaissance and screening, cavalry will use a minor portion of the fast tanks to strengthen the advance echelon of armored cars. The larger portion will operate directly with the horse units to be employed as previously mentioned.

Cavalry also looks to other units of fast tanks in combination with motorized infantry to reinforce cavalry units when the latter have seized points of strategic importance, and thus release the cavalry units for more mobile missions.

Armored and Mechanized Forces.

The method of operation of an armored force is largely a theoretical matter in the United States. The mechanized forces that have been organized in this country have been composed chiefly of slow, unsuitable types of vehicles. Hence our lessons from our own practical efforts have been severely limited in scope. We can add somewhat to our knowledge by studying the available information regarding the British thought and their maneuvers of the past several years.

It appears reasonably certain that an armored force will either operate under the supreme commander in the field or be attached to an army. Its attachment to a corps seems unlikely.

Fast tanks constitute the essential element of an armored force. Other elements are auxiliary. In her initial practical efforts along this line, Great Britain assembled an heterogeneous array of obsolescent equipment in which the tank did not stand out with its due prominence, and included therewith a fairly large ratio of motorized foot troops. By maneuvers, studies, reports, etc., the British officers in authority gradually became fully aware of the incongruous aspects and moved toward high and uniform mobility, with the auxiliaries minimized. The most important auxiliaries appear to be airplanes, armored cars, an element to facilitate stream crossings, and elements to furnish supporting fire. Of those that may furnish supporting fire, the most important appears to be one that can throw a smoke screen.

In its experiments with mechanized forces, the United States appears to be following a path very similar to that previously taken by the British. Our country however seems to be markedly slow in acquiring the fast tanks and auxiliaries that are required in order that the necessary experimental maneuvers with such equipment may proceed.

The uses for an armored force are enumerated in Chapter XVIII. One of the best uses is an attack against the hostile flank or rear in coordination with an attack by the main force; one of the poorest is reconnaissance and counterreconnaissance. It is generally unwise to send an armored force far ahead of other troops unless there is an urgent need to delay a hostile force or it is desired to strike hostile armored elements. A distant blow upon the enemy is most likely to be premature and not yield the maximum results. It is usually better to strike later so that other friendly troops may attack simultaneously or directly afterward.

An armored force in action might now be discussed in detail if the requisite information were available. An associated subject, *Tanks versus Tanks,* is discussed a few pages farther on. In view of the lack of experience to date on this side of the water, any extended discussion of the methods of an armored force in action appears to be too closely akin to unfounded, speculative theory to suit the purposes of this book. One who is especially interested in this matter, however, may get an insight as to the more experienced British thought along this line by reading a rather fascinating bit of fiction by H. E. Graham, entitled, *The Battle of Dora.*

Tanks in Reserve.

To give the final punch to victory, to facilitate exploitation, or to assist in meeting an adverse emergency, it is quite obvious that a suitably large unit of fast tanks in reserve is extremely desirable. At first thought it seems reasonable to state that such a reserve is a necessity. However we must heed the dictates of common sense. We know that in future warfare there will be a great demand for efficient tanks. It is altogether probable that the supply will not be anywhere nearly equal to the demand. It does not seem quite reasonable to expect high commanders to deprive themselves of the use of available tanks urgently needed for active combat in order to maintain a large tactical reserve of tanks. It cannot be too strongly emphasized that the large reserve is very valuable and much to be desired, but, according to honest expectations the reserve in most cases will be smaller than is desired.

In addition to the principal reserve or reserves, however, it is expected that there will be accompanying tanks in reserve, at least initially. This is due to the plan of alloting at least one platoon for each infantry battalion in the zone where tanks are to be employed. As some of the battalions will be initially in reserve, so also will be the tank platoons pertaining to those battalions. Of course such tank platoons should be used when and where required; not necessarily with predetermined infantry battalions.

It seems that slow tanks can be used to much better advantage in the accompanying role than in reserve. To be fully effective the tanks in principal reserve require high speed. Having it, they can strike quickly when an especially favorable opportunity arrives, they can pass quickly through a break in the hostile line created by others, they can quickly cut off hostile columns in retreat, they can quickly strike threatening hostile reserves, or they can meet whatever other special emergency arises that is of sufficient importance to justify their use.

Reconnaissance and Intelligence.

In order that tanks may be employed over favorable ground and that they may not be wasted in attempts to negotiate unsuitable terrain, much reconnaissance is necessary by personnel familiar with the capabilities of tanks. At general headquarters there should be maps with this information graphically recorded thereon. Tank reconnaissance personnel must not only gather information of this character, but must make special reconnaissance for routes and sites.

The intelligence personnel of tank units are greatly interested in aerial photographs. It is likely to be only through the study of such photographs that facts can be determined that will govern the selection of routes to be followed by tank units after the initial penetration of the hostile front.

In general the reconnaissance and intelligence personnel of tank units, assisted by the corresponding personnel of infantry units, by the air corps, and by others, endeavor to obtain in a minimum time a maximum amount of information useful to their units.

Approaching Resistance.

Tanks approach hostile resistance similarly to a squad of men, secretly when practicable, taking advantage of concealment and defilade so as to avoid hostile observation and fire, returning the hostile fire whenever that can be done to advantage, and closing finally with the enemy at high speed. When crossing open terrain and exposed to direct antitank fire, tanks may seek to decrease their vulnerability by zigagging or pursuing irregular courses. By the proper training of the tank crews, the firing, the speeds, and the changes of direction can be so coordinated as to insure reasonably accurate fire and a decrease in vulnerability without an excessive diminution in the rate of advance. Sky lines and open places, being especially subject to hostile observation and fire, are avoided when practicable. When they must be crossed the crossing is made rapidly. When approaching a diagonal ridge, a unit of tanks should adjust its alignment so that the tanks will pass over the crest simultaneously.

When to Stop for Firing.

As a general rule when tanks are on the battlefield and not concealed, they keep in motion and maintain as high a speed as practicable in order to reduce as much as possible their vulnerability to hostile fire. However, to fire with adequate effectiveness there are times when they must slow up or stop. If a tank is believed to be under dangerous direct fire from an unlocated point, it should continue at a high speed. If a tank encounters a hostile tank and there are believed to be no hostile antitank ground guns about, it may stop momentarily in order to deliver fire of great accuracy against the hostile tank. When crossing a trench a tank may stop momentarily so as to enfilade the trench with its fire. Each situation should be estimated on its own merits, bearing in mind the general rule that stops should be made only for cogent reasons and that they should be extremely short in duration.

In Various Tactical Situations.

Tanks near at hand are very valuable for opposing hostile counter attacks. Any tanks available are used for this purpose. In such a situation the tanks should act quickly and boldly or the opportunity for effective action will be lost.

Tanks, if available, are also used in counter attacks. They are normally so used in the accompanying role. So far as the tanks are concerned, this is merely a quickly initiated, coordinated attack with a limited objective.

When a pursuit is decided upon, all tanks that can be used are employed. The faster units are normally used for cutting off and delaying the hostile retreat. Others should be used, if available, for attacking the hostile covering force. Most of those already engaged will continue to cooperate in applying direct pressure.

Tanks are greatly needed in the advance guard especially when opposed by a delaying force on a mission requiring a rapid march. Fast tanks which do not need carriers, are much to be preferred for such missions.

Fast tanks are very useful in affording flank protection to marching columns.

Although tanks are of value in rear guard operations, they should not be so used if it is possible otherwise to cope with the situation adequately.

Tanks may assist in covering a daylight withdrawal by making a sudden attack upon the advancing hostile foot troops, preferably from a flank.

Except when adequately screened by other troops, tank units must provide for their own security. When actually in combat, correct tank formations furnish a considerable measure of security. When armored cars are organized with tanks, the former by their reconnaissance also protect the latter. When necessary, tanks may act as scouts or as security and reconnaissance patrols. It may be desirable to have special scout tanks for such purposes, after the practice of the British.

TANKS VERSUS TANKS.

The last war was brought to a close before the mobility of the tank could be much improved and, as the employment of the new arm was almost wholly a one-sided affair, the possibilities and the problems attending the use of tanks against tanks were little considered, notwithstanding the fact that upon two ocassions a few British and German tanks met, by accident, upon the battlefield. But since most countries now contemplate the use of tanks in their plans for national defense, it is logical to assmue that, in future wars, tank forces of the opposing armies will meet in combat. In such a form of combat there are no significant precedents for our guidance but there are certain factors, both in the preparation for and the conduct of tank against tank actions, which appear to be logical and well worth considering. Inconclusive though they may be, the somewhat detailed considerations that follow are presented in order to furnish food for thought.

Preparation.

Preparation as used here comprises: the procurement and utilization of information regarding the possible or actual enemy, his vehicles and his methods; the development and supply of suitable materiel for our own use; and special attention to organization, communication and training suited to this particular purpose.

In the case of each potential enemy, not only should data be assembled regarding the details of his tanks and other vehicles, organization, and tactical methods, but this information should be constantly so digested and studied that adequate plans may be made for taking advantage of the enemy's weaknesses in these items, while the defects in our own materiel, organization, tactics, and training are brought to light and corrected. Tank personnel must know, among other things, the range and capabilities of hostile tank weapons, the speed of the tanks and the resistance of their armor, and which parts of the tanks are especially vulnerable to the fire of certain weapons.

In planning a tank to fight tanks the critical problem of a proper balance between speed or thick armor is particularly acute. If the enemy's tanks are faster than ours, he can outmaneuver us or run away from us at will. On the other hand, if our armor is more vulnerable than his, he may be making telling hits while our fire is ineffective. Yet in spite of all this it should be borne in mind that speed does not necessarily result from lack of armor. By more efficent design we may secure both better armor and higher speed than the enemy. This question of armor or speed which is continually bobbing up, reveals that tank authorities are divided upon the question. Some favor moderate armor and high speed; others are confident that effective armor is essential and high speed merely desirable.

A tank that is intended to fight tanks must be properly armed. It must have at least one antitank gun in which confidence may be rightfully reposed. A high muzzle velocity seems to be especially desirable. The qualities of the gun are entitled to most careful consideration. If most of the projectiles are to bounce harmlessly from the hostile armor the money put into the tank may be worse than wasted.

Conduct of the Attack.

The mobility of a modern tank force being considerably greater than that of the forces with which past wars have been fought, it is apparent that the events leading up to action, and also the events on the battlefield itself will be greatly speeded up.

The successful conduct of an action of tanks versus tanks, exclusive of materiel considerations, is dependent upon reconnaissance, communication, intelligent maneuver, utilization of terrain features, accurate fire, training and morale. The ability to effect surprise and to deliver accurate fire are among the favorable factors of outstanding importance.

Reconnaissance must be continuous before, during, and after combat. With the development of improved methods of cooperation from the air, more valuable assistance in this line may be expected from friendly airplanes. However, information from the air is not sufficient. A force of highly mobile tanks must protect itself through the use of armored cars and other reconnaissance agencies in accordance with the situation. Otherwise it may run into a trap or be surprised by a hostile force from an unexpected quarter while in an awkward formation or position. On the other hand, if the tank force reconnaissance agencies are able to fulfill their missions, the tank commander may be able to initiate action upon terrain of his own selection, or he may be able to surprise the enemy.

The effective utilization of the terrain opens up the general fields of strategy and tactics. The commander of a tank force operating semi-independently should be adept in strategy. Aside from strategic considerations, the ground may be utilized so as to effect surprise, to provide complete or partial protection while shifting to a more favorable position from which to attack, to provide concealment or partial defilade when opening fire, or to facilitate retreat.

If the enemy force is attacked just after it has passed through a defile, or over a stream where the crossings are limited, its maneuverability will be restricted and the same will be true if there is an impassable obstacle on one of its flanks. If attacked while crossing a stream or while passing a defile, the enemy will not be able to bring all of his vehicles into action at the start of the fight.

The effect of slopes in decreasing the speed of the particular hostile vehicles is a factor of importance. Their speeds, both on the level and on slopes, should be known to our personnel. The effect of this factor is felt in the various phases of battle maneuvers and in the increased prospect of hitting a target that can move only slowly.

Although accuracy of fire is improved by stopping to fire and although this may facilitate the utilization of partial defilade or perhaps even successful firing from concealment for a short time, the dangers and disadvantages resulting from such stops should be appreciated. The increased danger of being hit by the gunner of the hostile tank being fired upon need not be considered, since the increased accuracy of the fire delivered from the stationary tank upon the hostile tank more than offsets such danger. But in many instances a tank that stops may be unknowingly under hostile fire from two or more directions. If such fires are from tanks, antitank guns or both, stopping to fire may be only immediate suicide. Another objection to a tank stopping to fire is that such a practice may add seriously to the difficulties of control by the leaders. Successful control being of extreme importance in contributing toward successful action, it follows that if such stopping seriously interferes with control it cannot be tolerated.

The formation for attacking hostile tanks should normally be one of depth, but not excessive depth. A fairly compact formation, with some depth, tends to insure a local success. A local success, even with equal casualties, tends to lower the hostile morale and disrupt the hostile plans. Depth provides supporting fire for the assaulting waves. Supporting fire, by mobile mortars firing smoke, may be of importance in assisting tanks to close with dangerous resistance.

The method of attack is not of course necessarily simple and frontal. If the enemy can be struck from two directions at the same time, or with a sufficiently limited time between blows, he will be at a great disadvantage. If the second blow strikes shortly after he has issued orders for and turned to meet the first, he may be thrown into great confusion. However, if too much time should elapse between the blows of the parts of the attacking force, the advantage would lie with the enemy; hence the danger of that contingency must be carefully considered. The ambuscade is a procedure of great effectiveness when circumstances permit it. A feigned withdrawal has interesting possibilities at times in drawing the enemy into an unfavorable situation, planned in advance.

We may safely speculate upon the value of reconnaissance, carefully considered formations, control, surprise, and marksmanship, but the details of combat methods for tanks against tanks can be worked out satisfactorily and thoroughly only in maneuvers. Theories arrived at on paper or by debate are prone to ignore some essential factor or factors, especially in a new field such as this. Maneuvers under various conditions as to materiel, terrain, weather, darkness, etc., will give the theories a fairly good test, permit their intelligent revision, and furnish information that should influence, not only methods, but design and organization.

Tank Actions Away from the Main Force.

Tank actions that are fought away from the main force, where the other branches, with the exception of the air service, will be of no assistance, may occur under a variety of circumstances and few if any rules can safely be made without practical experience to serve as a guide. Some principles may be worked out by maneuvers where the requisite tank forces are available. From a purely academic viewpoint, however, it appears logical to believe that where the reconnaissance agencies have performed their missions, and the tank force commander is aware of the proximity of the enemy force, he will be enabled to make his dispositions for accomplishing his mission, although he may have to halt to do this. From this time on, the results obtained will depend upon the preparation that has been made for such combat, as previously mentioned, an understanding of his plan by the unit commanders, their knowledge of the terrain, their control over their units, their ability to maneuver as a unit for position, and the training of the crews in driving and marksmanship. Naturally, the ability to carry out any part of these operations, excepting possibly driving and marksmanship, can be acquired only by active training with the vehicles over various types of terrain under simulated battle conditions.

While control will never be complete after the action starts, it should nevertheless exist to the extent that the units engaged will follow the plan outlined to the best of their ability. A general control may be exercised, in cases where radio is not efficient, by staff observers and messengers. Boundaries beyond which the tanks are not to go without additional orders may prevent the force from becoming separated in pursuit and help the force commander to regain control.

In a tank against tank action, individual tanks must not be permitted to dodge about and roam over the terrain at will. The force commander's plan can be carried out only if the small unit commanders maintain control of their tanks. This control can be better exercised if the platoon consists of three, rather than five, tanks.

If, for any reason, the reconnaissance agencies have failed to function and our force is surprised by an enemy force, the results obtained will largely depend upon the terrain, our formation at the time, the relative sizes of the two forces, the degree of our readiness for instant action, the control exercised and the orders issued by our commander, but, more than any other one thing, upon the training received by our small unit commanders and their drivers and gunners *for just such a contingency*. If the small unit commanders are well acquainted with the general mission and the plan of the commander, have some knowledge of the terrain, have control of their units, have confidence in their vehicles and in their own ability, all of which comes from training, they may frequently be able to act in accordance with a prearranged plan, even in a surprise engagement.

Tank Actions When With the Main Force.

When enemy tanks have been located, prior to the time they are encountered by our tanks, they will, no doubt, be subjected to artillery fire if within range and our tanks will thus receive some support from our artillery prior to the tank action. With the exception of the air service, little can usually be accomplished by any other branch to assist our tanks in the combat, either before, during or after the engagement, unless the enemy tanks break through our tank force. In this case, the infantry antitank weapons will go into action as in repelling any tank attack. Reconnaissance by all of our agencies will have to be continued during the action in search of the enemy tank reserves, portions of the enemy tank force which may have become separated from the main force, may have retreated, or may have been moved to the flanks on special missions.

If our tanks are leading the infantry when the tank action starts, the infantry antitank weapons will accomplish more if they are at once placed in position to repel the enemy tanks, should any of them get past our tanks, than if these weapons are moved forward with the expectation of taking a hand in the close combat of the tank forces. Moreover, these weapons are not prepared for firing promptly nor subject to the best control during their move forward, and they may, under such adverse circumstances, be surprised by the appearance of the hostile tanks. To afford the best protection to the infantry troops, they must be ready for prompt and effective action.

In case our tanks are in rear of the infantry when the enemy tanks attack, the infantry, with the exception of its antitank weapons, may well move back if it can, and make way for our tanks to engage the enemy tanks. Otherwise the infantry may be between two fires and may interfere with our tanks in the fight.

The Melee

Whether the tank against tank action occurs at a distance or within supporting range of the main force, our tanks, in order to destroy the enemy tanks, must engage them in close combat. If it is desired merely to check the advance of the hostile force by harassing them, or to divert them from their direction and mission, close combat is not indicated. If they must be stopped, however, our course is plain. While some time may be spent in maneuvering for position, depending upon the time available, the force encountered and the terrain, the time will come when the two forces will meet, either by our force lying in wait at a favorable point or by moving toward the enemy force.

When they meet, the methods used by our force may depend upon the formation adopted by the hostile force and upon the terrain. If they are in line, many small tank actions at close quarters are probable. If our attack has been organized in depth, with reduced distances between waves, our leading wave will receive much fire support from the following waves. In order to bring the maximum amount of fire to bear upon the enemy force, our leading waves should pass through the first enemy waves, striking as many of these tanks as possible in passing, and give the other waves in succession similar opportunities to strike at close quarters with a minimum of confusion. In case favorable results have been secured during the melée, from then on individual tanks and small groups of the hostile force should be promptly hunted out and destroyed or captured. There will be little time during close combat, or the work which follows, for individual tanks to maneuver for position which will give them defilade so they may stop to fire. Control in close combat can probably be exercised only by the commanders of the smallest units, and by them, only by some simple method, such as "follow the leader," requiring few or no instructions to be given at the time.

If the enemy is, by any fortunate circumstance, advancing in column at the moment of contact, our force may have an opportunity to act under advantageous conditions. A part of our force might block the hostile advance while the remainder attacked one or both flanks. Or if our force, in wave formation, could sweep directly and longitudinally upon the hostile column, concentrating its fire successively upon the few tanks that were nearest, the prospect of victory would seem to be very favorable.

The employment of highly mobile forces against similar forces in open warfare situations has been but briefly touched upon in these pages. It is a new subject and, the more one studies the possibilities of increased mobility and the problems incidental to its use, the more apparent it becomes that generalship is becoming more important and, at the same time, due to employment of mechanical vehicles, more intricate. Although the means to be used and the degree of mobility is vastly different from the means and the mobility available to the leaders of the past, it is nevertheless true that a careful reading of the records left by the great leaders who, like Genghis Khan, understood the value of mobility and used it, will help one to understand the changes in tactics which are to come.

Mobility, the quality for which the armies of the world have been striving for centuries, is coming to the battlefield. When the emergency arrives, increased mobility will aid or it will confound us, according as we have learned, or have failed to learn, how to use it.

CHAPTER XXIX.

COOPERATION BETWEEN TANKS AND OTHER ELEMENTS.

Importance of Cooperation. The tactical successes of an arm or weapon are dependent, not alone upon its own efforts, but also upon the cooperative effort of other arms. Tanks are no exception. They assist and they frequently require assistance. They assist others to advance by reducing hostile resistance. In turn, they require assistance in various ways from the various arms. Especially do they need assistance when they are under, or about to come under, the fire of hostile antitank guns or other direct-firing cannons. The aim of such hostile gunners should be disturbed by all possible means, such as smoke, artillery fire, airplane attack, machine gun fire, rifle fire, and fire from tanks. The necessary cooperative effort in its various phases can best be secured through adequate instruction and combined training. This training should be so conducted as to emphasize in a practical way the points and methods that will insure the attainment of the desired ends. The instruction should cover particularly the powers and limitations of the arm with which cooperative efforts are to be made; also the specific methods by which mutual assistance may be effected.

Infantry Foot Troops.

In his tactical orders to attached tanks, the infantry unit commander should give the tank unit commander a definite task, but should, in most cases, avoid specifying methods. The tank unit commander should be allowed the greatest initiative and latitude that can be permitted in the solution of his problem. Infantry unit commanders should avoid unnecessary delay in authorizing tank units to proceed to reservicing points.

To insure proper cooperation between tanks and foot troops they should be trained together. Joint maneuvers with emphasis upon cooperative features are necessary. Each must be familiarized with the signals with which they are mutually concerned.

In the combined attack executed by accompanying tanks and infantry, it is extremely important that the foot troops advance as rapidly as practicable after the tanks. If the foot troops are tardy about resuming the advance when the tanks have sufficiently reduced the important points of resistance, they thereby waste the efforts of the tanks and jeopardize the success of the attack.

Foot troops should support their tanks by fire at every opportunity. By so doing, they may be able to neutralize one or more hostile guns, thus saving their tanks and insuring their own success.

Foot troops should maintain constant contact with their accompanying tanks. Although the ground may be irregular and wooded, scouts should keep the tanks constantly in view. Troops should not gather too closely around or in rear of advancing tanks. The tanks will draw hostile fire and create for them an unnecessary hazard.

Artillery.

There are two cases of artillery in support of tanks. In one the tanks receive such support as may be had from the same artillery that supports the foot troops. In the other case special tank units receive support from such special artillery as accompanies them.

In a brief preparation fire for a general attack, the artillery may be expected to neutralize some of the hostile antitank guns. During the early part of the tank advance the timely smoking of carefully selected areas may be most beneficial. There is no supporting fire of any kind of more importance to tanks than timely and well-placed smoke. The efficiency of such support is dependent, however, in a large measure, upon favorable weather conditions and good observation. The smoke should of course be placed directly upon those areas where it is suspected hostile antitank guns are located and so placed just before the friendly tanks come into view. Under favorable circumstances the tanks may thus cross an open place of several hundred yards with good vision while approaching an antitank area that is thoroughly clouded with smoke.

The flexibility of artillery fire maneuver is well suited to cooperation with and support of tanks in action within range *so long as observation and fire control can be maintained effectively.* The methods in detail for such cooperation are not definitely established at present. Artillery should fire upon hostile tanks whenever possible. Such fire if effective may be the best possible cooperation with friendly tanks. When the situation is befogged by lack of information, counterbattery fire is a good method of assisting tanks.

In addition to what would be considered artillery support of tanks, methods of cooperation may be developed for routing and demoralizing the hostile foot troops. For example, the advance of tanks into an area occupied by such troops may cause them to congregate densely in thick woods or other places where tanks cannot readily operate. When this has occured the artillery shells the areas so occupied.

Air Corps. Progress in the development of faster and more efficient tanks increases very greatly the importance of airplane cooperation. Airplanes may assist tanks by aerial photography, by reconnaissance, by carrying tank staff officers, by carrying tank commanders so that they may command from the air, by guiding tanks along selected routes, by carrying messages to and from tanks, by reporting locations and actions of friendly tanks, by so directing artillery fire as to protect tanks, by providing smoke screens so as to facilitate the tank advance against resistance, by attacking antitank guns, artillery or hostile tanks, and by reducing hostile aircraft observation. In some emergency situations in open warfare, it might be practicable for airplanes to carry supplies to a tank unit.

Engineers. Engineer troops assist tanks in the crossing of streams and other obstacles, in demolitions, and in camouflage.

Chemical Warfare Service. If chemical warfare troops man weapons, they cooperate with tanks by the use of smoke and gas in a manner similar to artillery.

Armored Cars. Armored cars cooperate with tanks through reconnaissance, security, and messenger functions, and by assisting in pursuit or other exploitation.

CHAPTER XXX.

COMMUNICATION AND CONTROL.

In the fields of communication and control there lies immediately in front of us much room for development, both as to means and as to methods.

Communication.

In interior communication, the voice being often if not generally inadequate, reliance is placed chiefly upon touch signals and the intratank telephone of a type employing a larynx transmitter. Visual signals made either with or without the aid of special mechanical equipment have possibilities.

In close exterior communication, reliance is placed primarily upon visual signals by arm, flag, light or pyrotechnics, and radio telephony. Although in present peace-time experimental exercises radio telephony is very effective and satisfactory, its effectiveness in battle is somewhat problematical. The air may be jammed by a number of transmitters in operation on or near the same wave length in use, or the enemy may interfere seriously by artificial static. Moreover, reception cannot be entirely satisfactory in all localities and in all weather. Radio messages may of course readily be picked up by the enemy. The use of a code might give some protection for a limited time. A device for the radio transmission and visual reception of numbers by means of which messages may be sent in code has been successfully demonstrated. It has an advantage over wireless telegraphy in that skill is unnecessary for its operation and it is claimed to work satisfactory under atmospheric conditions that would preclude the use of radio telephony. Sound by mechanical means might be practicable for very limited use in close exterior communication.

It seems probable that radio telephony will be the chief initial means of exterior communication for tanks in battle. If that fails, visual signals will probably be the first alternative for close exterior communication.

The use of tank radio telephony is at present more extensive in the British than in the United States army. In the former, the tanks of the commanders of the tank brigades, battalions, companies and sections are so equipped.

In distant exterior communication radio telegraphy, radio telephony, messenger tanks, messenger airplanes, armored cars, and motorcycles constitute the principal means. Carrier pigeons have been used in the past and may be used in the future. The latter can be used only to carry messages to lofts that have been established for some weeks. Wire communication is suitable only for use in the areas behind the line.

The importance of exterior communication, both close and distant is so great that provision must be made for two or more independent methods.

Formations.

It is fairly obvious that better control may be exercised over any tanks, but particularly fast tanks, if the organization is based upon a three-tank rather than a five-tank platoon. The reasons are:

(1) The three-tank unit is inherently well adapted to combat and the extremities of the formation are closer to the leader. Guiding upon the leader is simplified and facilitated, thus reducing communication requirements in combat. (2) For a given number of tanks there will be a greater ratio of officers present with the combat tanks.

Tank Orders.

The stabilization of the general situation that existed in the World War, the slowness of the tanks of that period, and the practical absence of hostile tanks made available (at least, theoretically) time for the issuance of detailed orders for the initiation and partial control of tank action. As we look to the future, we see an entirely different situation. With each side in possession of tireless forces that move at a rate of from 15 to 40 miles per hour and with an open-warfare general situation (which seems probable), the process that involves observation and transmission of information, digestion of information, decision, preparation of orders, and transmission of orders will have to be speeded up to a very extreme degree. Minutes will become as precious as former hours; seconds must take the place of minutes. To fail to speed up the customary process will be nothing less than a presentation to the enemy of all of the initiative and all hope of victory. These considerations indicate the necessity of a closer form of aerial cooperation than has been developed heretofore and the desirability of a code by means of which a sufficiently complete order can be given or transmitted in a minimum of time. In order that the meaning of the suggestion that a code be employed may be clear an example will be given. The commander of a unit of fast tanks gives to his unit an attack order as follows:

Information of the enemy (or if necessary, merely mention the enemy, as "Enemy column of infantry" or "Enemy tanks.")
The company (battalion) attacks.
Formation, 6.
Axis, ...(point) to ....(point).
Time, 8:22 (or *at once*).

And the order is complete! The first place mentioned after the word *axis* locates the exact point where the line of departure intersects the axis of the attack. The second place mentioned locates an advanced point on the axis of the attack and also the final objective. The training takes care of all other necessary details. The use of such a system would not of course preclude the use of much longer and more detailed orders if the available time made such a procedure practicable. It should be understood that this code method is merely a suggestion that might be developed into an acceptable system.

In this connection, Napoleon's method of issuing orders hastily is of interest. He would have close at hand several staff officers who understood his policies and to whom he had explained his tentative plans. When he made a decision and announced it, these staff officers departed immediately and issued verbal orders to the several subordinate commanders. In the British maneuvers of 1931, use was made of this system. The aides-decamp and also the tank battalion commanders were mounted in small, fast light tanks. The latter normally accompanied the force commander until just prior to the entry of the force into action.

Control from the Air.

With the development of fast tanks and of improved means of communication between airplanes and tanks, the idea of commanding tank units from the air, under some circumstances, has had a number of advocates. This idea is founded upon the great increase in the mobility of tanks and armored cars. It is claimed that, with such rapid-moving combat elements in action, there will sometimes be but very little time for the process of observation, transmission of information, digestion of information, decision, orders, and transmission of orders. It is proposed, under some circumstances, to put the commander in the air with a view to shortening the time consumed in observation, transmission of information, digestion of information, and, to some extent perhaps, in the transmission of orders. It is not contemplated that the commander should make it his business to perform the aerial reconnaissance for his unit, nor that he should be in the air habitually, nor that he should go aloft alone except for his pilot, nor that his plane should be without pursuit protection.

Brigadier General H. Rowan-Robinson states in his *Further Aspects of Mechanization:*

> Ground officers will have to cultivate airmindedness. They should be able, if not to drive, then at least to observe and command in the air. The plane will be to them what the horse was to their predecessor. The foot-soldier did not normally have as fine a seat on a horse as the hussar; and in the same way the tank officer will not be expected to display the same efficiency in a plane as the air-man. He must, however, be trained to the task. The occasions on which his military duties will take him there will, indeed be comparatively rare, but they will be so important that his whole career must be a novitiate for them.
>
> At the present moment there are no officers in the army equal to commanding in the air, and no great number among the more senior with much air experience of any kind. In the air force, on the other hand, there are no officers with ground experience of mechanized troops. Yet if we allow ourselves to be surprised in war by the difficulty of commanding highly mobile troops. we shall have to accept the exercise of control either by a soldier on the ground or by an airman in the air. The former will be well acquainted with ground tactics, but will be unable to apply his knowledge; because, in the first place, he will be unable to see the situation for himself, and, in the second place, he will have no time in the rush of such a rapid battle to collect information and issue orders on it. The latter, on the other hand, will be able to see. but will have no knowledge of ground tactics, nor any experience in the command of ground troops. Apart from all of which, the suitable plane and the very necessary service of signal communication will not have been developed.

This scheme has not, by any means, been accepted generally. Lieutenant D. M. Reeves, Air Corps, U.S.A., in an article in the *Infantry Journal,* of August, 1928, condemns it for many reasons. Some of the important objections to this plan are:

(1) The high ranking ground officer in the air would not be able to see much due to his limited eyesight, his lack of training in aerial observation, air sickness, etc.

(2) His ability to observe might be lessened due to having been forced to a higher altitude by hostile antiaircraft fire; due also to fog, haze, smoke, dust and vegetation.

(3) The commander in the air might not be able to distinguish friend from foe on the ground; and he might not be able to form any true picture of the situation from what he could see from his plane.

(4) The commander in the air would be in grave danger of losing his life. He might be brought down by fire from the ground or by hostile pursuit planes.

(5) The loss of an important commander while in flight would have far-reaching detrimental effects.

(6) The staff on the ground might not know whether or not their commander had been lost, with resulting uncertainty, confusion and hesitation.

The authors of this book neither support nor discredit the idea of commanding from the air in future warfare.

CHAPTER XXXI.

CONCEALMENT AND CAMOUFLAGE.[1]

The concealment of tanks from ground and aerial observation is of vital importance during all stages of the movement into the general area in which they are to be employed, and during their approach march from the tank park to the assault positions.

All movements, including those by rail, will have to be made under cover of darkness and, where it is necessary for the tanks to remain in the tank park or in the intermediate position during the daytime, effective measures must be taken to conceal them, particularly from aerial observation.

The presence of tanks in an area is much more likely to be discovered through the tracks they have made than through the location of the tanks themselves. During the approach march, while the tanks are using the same roads as other vehicles the characteristic marks made by their tracks will ordinarily be destroyed by other vehicles but, when it becomes necessary for the tanks to move across open country to a place of concealment, it is vitally important that the track marks be obliterated. A troop of cavalry, horse drawn vehicles, or in some cases, brush or harrows which are towed behind the last tanks in the column will aid in obliterating the marks made by the tracks. If it is possible to select routes with overhead cover, in going to the place of concealment, this should be done so that it will not be necessary to wipe out the track marks. In case it is absolutely necessary to move tanks across an open field where it is difficult to obliterate the track marks, as over plowed ground, through crops, etc., the path should be made past the point where the tanks are to be concealed, preferably to another likely place of concealment, thus preventing accurate determination by the hostile observer of the actual location of the tanks. If practicable, it is better to move the tanks in single column along the edge of the field, instead of taking them across it. The tracks thus made will not be so noticeable on a photograph on account of the natural line that appears along the edge or boundary of the field.

It will not be possible to conceal tanks in woods at every stop, so other suitable locations which lend themselves to effective concealment must sometimes be selected. Buildings, ruins, orchards, hedges, brush, individual trees, ravines or rough terrain may be used to advantage. Of the locations mentioned, it is apparent that woods or the shady sides of buildings are the best because, in addition to providing the most effective concealment, such locations permit maintenance activities, such as lubrication, refueling and repairs, to be made in daytime.

However, if the cover available does not effectively conceal the tanks (and, except where thick woods or buildings are available, the cover will usually be inadequate), they must be made inconspicuous by some form of camouflage.

Military camouflage is not merely concealment. It is, or should be, counter-intelligence work intended to deceive the enemy concerning the existence, nature or location of materiel, troops, or military works and thus prevent his aircraft, captive balloon and ground reconnaissance agencies from gathering information concerning the objects concealed.

---

[1] Much of the material under this heading was obtained from a lecture delivered by Major L. E. Oliver, C.E.

Due to the marked development of aerial photography, this form of reconnaissance has become very effective and consequently, the concealment of tanks from aerial observation is more difficult. To prevent the enemy observer from noticing the tank, and to reduce the probability of his camera recording

Plate 77.
1. A Six Ton Tank Concealed by a Tree from Aerial Observation.
(Photo by Tank School, U.S.A.)
2. A Vickers Tank with Camouflage Painting Concealed by Branches and its Proximity to a Hedge.
(Wide World Photo.)

its position, an effort is made to make the tank inconspicuous by destroying its outline with paint, by placing it in shade of another object, or by covering it with materiel which will make it appear as a part of the natural surroundings.

Paint, alone, will not conceal the tank from the aerial photographer for its shadow may disclose its position, but if it is painted in colors to resemble a given background and in broken tones which will cause it, at ordinary distances of observation, to show up indistinctly, this method of camouflage will aid in preventing detection by ground observers. The identity of the vehicle may be destroyed and its form broken up, when operating where the background consists of patches of green foliage, by using large, bold patches of color such as green, cream and brown, with black definitions between each color. Grey, violet, indigo-blue and light brown are not visible at any great distance. Green is fairly visible. Orange is the most visible of all; red is next in visibility to orange.

The pattern should not be stopped at an edge of the tank but should be continued over it, and the various parts of the tank hull and outside mechanism should be split up into a number of disassociated pieces. Camoufleurs recommend different color schemes for spring, summer and fall and, for tanks which are required to operate in snow, a dull faintly bluish white paint is recommended.

An example of the need for camouflaging tanks operating on snow covered terrain occured on April 11, 1917 at Bullecourt. The contrast between the tanks and the snow upon this occasion caused the dark colored tanks to stand out so clearly as to form a perfect artillery target. All of the tanks used were quickly destroyed and the attack failed.

The shade of trees and buildings will conceal the tank from both aerial and ground observers for a limited time only due to the movement of the shadow with the rotation of the earth, so this method of concealment will only be available for temporary use, while a more permanent method is being provided in cases where the tanks must remain at such a location for a longer period of time than the shade will continue

Where the location of the tanks is such that camouflage is necessary, it is especially important, in determining the methods and the materials to be used, that the camoufleur know how objects on the ground will appear on an aerial photograph in order that he may prevent detection by the enemy aerial photographer.

A photograph is a permanent record and shows every detail in its exact relation to other details in the whole area which it covers and, by comparing under a magnifying glass photographs of the same area taken on different dates, a skilled photographic interpreter will secure much information which may have been unattainable in any other way. An aerial photograph records colors, ground forms, and everything else, in various shades of gray. Where there is shadow, or light is absorbed, the photograph will be dark. Where light is reflected, the photograph will be light. Where the ground is uniform, as in a wheat field, the photograph will be a monotone and any irregularity will be easily detected. Where the ground is broken and varied, the photograph will have an intricate pattern and changes will be more difficult to detect. The pattern on the photograph is made by form, shadow, texture of surfaces and color.

Form and spacing are very important. Everything in nature is irregular. Perfect squares or circles and regular spacing of like objects practically never occur naturally. Artificial forms are usually regular. We make houses rectangular, plant trees in rows, and build roads and railroads straight or with smooth curves. Since in camouflage we generally wish our work to blend with natural objects we should avoid regularity. A single tank imperfectly camouflaged may not be noticed. If there are four of them at regular intervals they will be noticed at once.

Shadows disclose the form and height of the objects casting them. Camouflage must be erected so that its shadows merge with existing shadows in such a way as to be indistinguishable.

Texture of surfaces is illustrated by a rug with long nap. When the hairs of the nap are erect each casts a shadow and the rug is dark. When someone steps on the rug and the nap is pressed down that spot appears brighter because the small shadows are partly suppressed. In a place where texture is uniform, as in a meadow or wheat field, camouflage is just about impossible because of the difficulty of matching the texture. A concrete or metal surface cannot be made to resemble grass by simply painting it green, because the texture of the grassy surface would still be lacking.

Color shows up in the photograph only in various shades of gray, and is not so important as form, shadow, or texture. If camouflage is approximately the color of the surroundings, a small difference in shade will not matter provided the other elements are right.

We have covered very briefly the fundamental principles governing camouflage. Let us now examine, also very briefly, the general methods of applying those principles.

First, we must select a position. Tactical requirements must be met first of all, but there will generally be an area meeting the tactical requirements within which there is considerable room for choice of suitable terrain. Perhaps we can find a position where there is adequate existing cover such as woods or buildings, and where no camouflage at all is required. On the other hand we may be forced to use a position in the middle of an open field where no amount of work will satisfactorily conceal us. Troops in campaign are usually very tired so the necessity for additional work should be avoided wherever possible. We can thus see that proper selection of a position is very important.

In a wooded country, positions which require little or no camouflage are easily found. In an open agricultural country a position may be found along a hedge fence, among a group of buildings and debris in a farm yard, in an orchard, or at any other place where there is something which causes, a blot, blur or irregularity on the photograph. Such an object can be tied into and extended if necessary with camouflage without as much danger of detection as there would be if the work were placed in the open by itself. One consideration for a position is that it can be entered or departed from without leaving tracks which make its location evident.

Having selected a suitable location which requires some camouflage, the next thing to consider is the materials for the work. We may use either natural or artificial materials. For purely temporary use, natural materials, such as branches, grass, weeds, vines, etc., are best because they exactly match the natural materials of the locality. They have to be renewed at frequent intervals, else they wither and fade and so disclose the position. Unless great care is used in effecting this renewal the vegetation about the position is trampled down. This shows up on a photograph as a light ring about the position and is a complete betrayal.

Artificial camouflage materials are furnished by the engineer supply service. They generally consist of fish nets or wire netting garnished with strips of burlap painted in colors to suit the locality. Wire netting is generally furnished for permanent works and fish nets for tanks and other mobile units. It is harder to make artificial material match the locality, but once suitably erected it requires much less work to maintain.

Having obtained the materials, the next thing is to erect them. Fish nets or wire netting are generally erected on a framework consisting of stakes about the size of two-by-fours which are driven into the ground and cut off at a uniform elevation with wires stretched taut over their tops, the outer stakes securely guyed. The nets should be parallel to the ground and as close thereto as the objects and activities to be concealed will permit. The pattern of the camouflage material should be irregular, and the burlap strips should be thinned out near the edges of the nets so that they will cast no definite shadows. Care must be taken not to trample down the vegetation around the position in the process of erection. The camouflage net or flat-top as it is usually called, should be extensive enough to more than cover the object to be concealed, together with all the activities necessary in connection with it. If the position is not immediately adjacent to a road, and must be used for an extended period, the entrance from the road should be covered with an extension of the flat-top to conceal the turn-out from the road to the position. Trip wires should be stretched about knee high so as to confine personnel to prescribed paths or areas under the flat-top.

Camouflage of natural materials must be placed in natural positions. Branches should be right side up. Weeds or briars should be upright—not lying down. Either of two methods may be used. Wire netting may be set up in two horizontal layers, one four inches or more above the other. In this case, the stems of the natural materials are thrust down through the two layers of netting. Or wires may be stretched in various directions and the branches and weeds suspended therefrom in upright positions.

Having erected the camouflage, it is necessary to maintain it and to avoid any activities which will disclose it to the enemy. This requires *camouflage discipline*. Men are permitted to use only authorized paths. Roads or paths should not end at the position but should lead beyond it to other possible positions and on to a junction with a used road. A used road shows brighter on a photograph then an unused one. Therefore the road or path should be *actually used* beyond the position. New material for renewal of natural camouflage must not be cut near the position. Refuse must be carefully disposed of. Next to the proper selection of the position, camouflage discipline is the most important item in camouflage practise.

The erection and maintenance of camouflage is, in general, the function of the troop unit concerned. That is to say, each tank company must erect and maintain its own camouflage. There are no troops normally assigned to the division or corps whose primary duty is the erection of camouflage. One camouflage battalion forms part of the engineer service of each army. This battalion has a total of sixteen field camouflage sections, each consisting of one officer and twelve men. These work sections can do no more than take care of camouflage erection in special cases and conduct schools and provide demonstrations for the benefit of the personnel of the various troop units.

The air service should photograph the terrain immediately before and immediately after the camouflage installations so that any defects may be noted and if possible corrected before the enemy has obtained a photograph. It is expected that, under service conditions, photographs over hostile territory will usually have to be taken at an elevation of at least 15,000 feet. Such a height will make the work of the camoufleur somewhat easier. The military camoufleur can feel that, for the present, he need have no fears that his work will be disclosed by the use of filters. However, when color photography has been developed so that it can be used with instantaneous exposure, or when the autogiro has been developed so that it can remain suspended and take a time exposure, the camoufleur's work will become more difficult.

An unusual method of concealment and camouflage of tanks would be to bury small tanks or partially cover them with dirt in ravines or shell holes and cover the dirt with brush or weeds to make it resemble adjacent terrain.

World War experience has indicated that, having camouflaged the tanks to the best of his ability with the means at hand, the tank commander should then search for alternate positions for his tanks and be prepared to move to them promptly in case he has reason later to believe that his first position has been discovered.

CHAPTER XXXII.

LANDINGS AND STREAM CROSSINGS.

Water obstacles are of prime importance from the tank viewpoint and, in order to overcome them efficiently, much preliminary study and practical preparation are required.

Landings.

One water obstacle to be overcome is encountered in the landing of tanks on hostile shores. Experiments with armored self-propelled lighters and amphibious tanks indicate that landing operations will be facilitated by further developments along these lines.

STREAM CROSSINGS.

Streams have always constituted one of the principal obstacles to the movements of armies. They are particularly serious obstacles to tanks. In the crossing of inland waters by tanks, there are three principal situations: (1) The water obstacle is such that it can be forded by tanks. (2) The water cannot be forded but can be bridged. (3) The water cannot be forded and cannot be bridged without excessive delay and effort. In the last case the tanks must be crossed by floating or ferrying. Since the enemy will, whenever possible, make special efforts to intercept the tanks at stream crossings, the preparation for and the execution of the crossing in a minimum of time and with a minimum exposure of personnel to hostile small arms fire is essential to success. In practical research pertaining to this subject, it is therefore important that the factors of speed and protection be given the utmost of careful consideration.

Bridges, Rafts and Ferries.

Our engineers are prepared to construct quickly bridges for general military use but we have not as yet developed the special means needed by tanks nor the methods by which tanks can help themselves in their crossing. In the British army there has been a variety of intelligent efforts within this field of practical research. Some of these will be mentioned.

During the war, in the British army, a number of types of bridges were designed that could be carried at the front of a tank in an elevated position. When the tank reached the stream or canal to be crossed, by operation from the interior of the tank, the bridge could be lowered quickly into position. The tank could then cross over the bridge. Other tanks and other troops could follow. The Mark V Double Star tank employed for this purpose is shown in Fig. 1, Plate 78.

In advancing to attack on October 20th, 1918, British tanks crossed the river Selle over a special bridge which had been built at night with its surface beneath the surface of the water. The German aviation had failed to detect the bridge. Such a method is well entitled to future consideration.

In 1918 there were organized in the British army three tank bridging battalions. The British also developed a special type of tank bridge that could be transported easily and erected quickly that would span a 100-foot stream and support tanks up to 35 tons in weight. A bridge capable of spanning 70 feet was developed that could be put in place by a tank without exposing any of the personnel to fire. A standard box girder bridge was devised

300   THE FIGHTING TANKS SINCE 1916

which was made up of sections eight feet in length. By this means bridges of varied length and capacity could be put together quickly. By the use of 4 girders made up of these sections, a bridge was obtained that would support heavy tanks over an 80-foot span. A 30-foot tank bridge was built and carried on a military tractor.

In 1926-27, a folding raft that was carried on a tank and used for the ferrying of very light tanks was developed in the British army. There was also developed a method of crossing tanks over streams of moderate depth by means of timber crates. (See Fig. 2, Plate 78). These were launched slightly upstream from the selected point and, under the control of ropes allowed to float into position. When the water was deeper than the height of the crate, the approaching tank would force the crate beneath the surface and wade through the excess depth. By this method at a demonstration on one

Plate 78.
1. **A Mark V Double Star Tank Laying a Bridge and Immediately Crossing Upon it.**
    (From *Royal Tank Corps Journal*.)
2. **Self-Propelled Mount for 18-Pounder M1927 (Birch Gun) Passing Over Timber Crates, Sometimes called "Stepping Stones."**
    (From *In the Wake of the Tank* by Martel.)
3. **Vickers Armstrongs Straussler Collapsible Ponton Boats and other Ponton Equipage Loaded for Transportation.**
    (Photo by Vickers Armstrongs, Ltd.)
4. **The Ponton Collapsed.**
    (Photo by Vickers Armstrongs, Ltd.)
5. **The Ponton Open.**
    (Photo by Vickers Armstrongs, Ltd.)
6. **Ponton Equipment Used as a Ferry.**
    (Photo by Vickers Armstrongs, Ltd.)

occasion, the necessary crates were launched, placed, and the first tank completed the crossing of a 50-foot stream all within the space of one minute.

In 1927-28, a sectionalized box girder bridge was developed that could be launched readily across a 60-foot span. This was accomplished by means of rollers and a relatively light 27-foot launching nose attached at each end of each girder during the launching process. Five 3-ton trucks carried a 4-girder bridge that would support 16 ton vehicles over a 60-foot span. This type of bridge was deemed especially suitable for use where ordinary highway bridges had been destroyed.

There has been developed for the British army a ponton bridging equipage that is light in weight in relation to its capacity due to the employment of collapsible boats. The six boats and other material forming the cargo of the two trailers as illustrated have a total weight of about 8250 lbs. This material will bridge a distance of 84 feet. The entire assemblage of tractor, trailers and cargo weighs only 8.6 tons. This sort of ponton equipage is easily handled and occupies a minimum of space when being transported. Rafts made from ponton equipage are also employed for the ferrying of tanks in various countries. The British regard the idea of ferrying tanks with much favor.

The idea has emanated from several sources, of floating tanks in a partly submerged position through the attachment of special bouyant equipment to the sides of the tank.

There are quite a number of schemes for the utilization of motor power in connection with the ferrying of tanks.

AMPHIBIOUS TANKS.

Amphibious tanks appear to constitute a practicable solution of the stream crossing problem except for the difficulty of combining adequate armor with the lightness necessary for "swimming". It is believed, however, that there has been too strong an inclination to condemn the idea on the theory of inadequate armor and without sufficient experimental effort. The Christie Amphibian was never purchased, and has been destroyed. Several have been made in Great Britain. The latest, a Vickers Armstrongs production in 1931, has given some startling performances of exceptional merit. It may be that, in the future, amphibious tanks will lead the way across streams and protect the crossing points while tanks with thicker armor are crossed by ferrying or some more efficient means.

British. Light Infantry D, Amphibious.

Produced in 1920 by Experimental Establishment (Governmental). Total production, 1.

Crew: 2 or 3.
Armament: None.
Armor: 0.2 in.
Maximum speed: 30 mph on tracks; about 3 mph in water.
Suspension: Cable at first, later replaced by chain.
Tracks: Steel plates mounted upon a cable thus giving flexibility to plates and track.
General arrangement: Driver in front; gunners in center, engine and final drive in rear.
Dimensions: Length 21 ft.
Weight: 9 tons.
Engine: Hall-Scott, 100 HP, forced water cooling.
Horsepower per ton: 11.1.

A Light D Star tank was built, having a 48 HP Overland engine and rubber tracks.

U. S. Christie Amphibious Tank, First Model.

Produced by Front Drive Motor Company. Completed in June, 1921. Total production, 1.

Plate 79.
1. Light Infantry D, Amphibious.
   (Photo by F. Mitchell.)
2. Christie Amphibious Tank, First Model.
3. Christie Amphibious Tank, First Model, Afloat.

Crew: 2 or 3.
Armament: One 75 mm (2.95 in.) gun.
Tracks: Flat steel plates; width 10 in., pitch 10 in.
Special features: Except as to those differences apparent in the photographs, this model was much the same as the final (third) model. This model swam across the Hudson river.

U. S. Christie Amphibious Tank, Second Model.

Produced by Front Drive Motor Company. Completed in June, 1922. Total production, 1. This was a modification of the first model.

Crew: 2 or 3.
Armament: One 75 mm (2.95 in.) gun.
Tracks: Flat steel plates; width 10 in., pitch 10 in.
Special features: Except as to those differences apparent in the photographs, this

Plate 80.
1. Christie Amphibious Tank, Third Model.
2. Christie Amphibious Tank, Third Model, Cruising.
3. Christie Amphibious Tank, Second Model
   (Photo by Ordnance Dept., U.S.A.)

model was much the same as the final (third) model. This model swam across the Hudson river in December, 1922. Regarding this crossing, Brigadier General S. D. Rockenbach wrote as follows:

"Yesterday, the 5th instant, I witnessed the demonstration of the Christie 75 mm gun mount. The demonstration or trial of the machine started at Hoboken, N. J., crossed the 23rd Street Ferry, thence to Broadway, then up Broadway to 74th Street, thence out Riverside Drive to 205th Street. With the assistance of a motorcycle police squad the streets were kept cleared and the machine ran on its rubber tired wheels at the rate of 30 miles per hour. At the Dyckman Street

Ferry, the machine was moved on the ferryboat and was transported across the Hudson, and ran from the ferry on to the very narrow cinder road bordering the river. Its tracks were put on with heavy grousers, and thus the machine turned from the road up the steep bank of the river between the Palisades. Its climbing ability was something remarkable, and it continued to ascend for 100 feet, where the earth ended against the precipice. It then descended the bank much easier than the spectators, who were slipping and sliding, reached the road, turned abruptly to the left, went along the river for some 30 feet, and then descended a 6 foot stone wall into the Hudson River. It crossed the Hudson under its own power, but instead of the driver directing the machine to its designated landing place on the east shore, he headed south of the same and on getting across the river, faced a sheer stone precipice. At the time of crossing the river, the tide was going out very rapidly and the very worst possible for the machine to encounter. The driver turned his machine up the river and buffeted a tremendous tide and current successfully for over a mile. He then headed for the shore, reached a 20 foot rock revetment of the New York Central Railway Lines, ascended that, flopped down on the railway tracks turned south on rails and ties, and continued along the railway line until he reached the crossing at 205th Street.

"While this was not a tank, at the same time it was the most remarkable performance that I have seen or heard of. The combined wheel and caterpillar principle was a complete success. The engine difficulties that we have encountered with the Christie tank, the difficulties of the transmission, difficulties from obstacles getting between the tread and the wheels had all been overcome. The control of this gun mount was the simplest and best that is on any vehicle of similar weight and multiple operation.

"While as stated above, this is a gun mount and not a tank, and a great many details would have to be worked out before it could be converted into a tank; and while the flotation feature would cost too much in weight on a tank, from the exhibition I am persuaded that if the Christie tank now in our possession had the refinements and improvements on it that this gun mount had, there would be no hesitancy on my part in recommending the construction of a sufficient number of the machines to equip a company."

U. S. Christie Amphibious Tank, Third Model.

Produced by Front Drive Motor Company. Completed in November, 1923. Total production, 1. This was a modification of the second model.

Crew: 2 or 3.
Armament: One 75 mm (2.95 in.) gun.
Armor: 0.25 in.
Maximum speed: 18.5 mph on tracks; 30 mph on wheels; 7.5 mph in water.
Suspension: Rubber tired wheels, four on each side; coil springs except at rear wheel.
Tracks: Flat steel plates; width 10 in., pitch 10 in.
General arrangement: Gunner in front; driver in center; engine and final drive in rear.
Dimensions: Length 16 ft. 8 in.; width 7 ft.; height 7 ft. 6 in.
Weight: 7 tons.
Engine: Christie, 6 cylinder, 90 HP, forced water cooling.
Horsepower per ton: 12.9.
Transmission: Sliding gear, 3 speeds forward, 1 reverse.
Obstacle ability: Trench 7 ft.
Fuel capacity: 50 gallons.
Special features: This vehicle could be operated on tracks, on wheels, or in the water. It had two propellers for operation in water. Steering in the water was accomplished either by varying the speed of the propellers or of the tracks. This vehicle swam across the Hudson and Potomac rivers. At the Calebra Islands, it swam from the U. S. S. Wyoming to one of the islands. This tank was not purchased by the U. S. Government and was later destroyed. Plans of it were sold to Japan.

British. Carden Loyd Light Amphibious.

Produced in 1931 by Vickers Armstrongs, Ltd. Total production, 1.

Crew: 2.
Armament: One machine gun.
Armor: 0.276 in. to 0.354 in.
Maximum speed: 40 mph on tracks; 6 mph in water.
Suspension: Carden Loyd type with rubber tired wheels and leaf springs.
Tracks: Same as Carden Loyd Mark VII.
General arrangement: Driver at left center; gunner in rear of driver; engine at the right; final drive of tracks at front; propeller at rear.
Dimensions: Length 13 ft., width 6 ft. 10 in., height 6 ft.
Weight: 3.1 tons.
Engine: 50 HP.
Horsepower per ton: 16.1.

Plate 81.
1. Carden Loyd Light Amphibious Tank.
2. Carden Loyd Afloat.
(Photo by Vickers Armstrongs, Ltd.)
3. Stern of Carden Loyd Amphibious Tank.

Obstacle ability: Trench 5 ft.; stream any depth; slope 45 degrees for short distances; vertical wall 20 in.

Fuel capacity: 30 gallons.

Special features: Front armor (0.354 in.) will stop small arms A P bullets at 500 feet and side armor (0.276 in.) at 800 feet. The tank has a single rudder and a single propeller at the rear. A bracket below these gives them some protection. The maximum draft when afloat is about 3 ft. (near stern). To give increased transverse stability afloat, the hull is provided with balsa wood side rails, about 8 in. by 10 in. in section, which are inclosed in sheet steel. This tank performs well on land and in water.

The following comment has appeared in the *Army, Navy and Air Force Gazette,* a British Service publication:

From a military point of view it (the amphibious tank) helps in the solution of a vital problem. Rivers have always been a barrier, temporary perhaps, but still a barrier. They have delayed, and delay is of first importance in war. Armored fighting vehicles have restored mobility, but in a well-watered country that power of mobility has been lessened. * * * Armed attack from the sea will be simplified. The writer was in the Gallipoli campaign in 1915 and he and his friends would have welcomed an amphibious tank which could have gone ashore at Sedd-el-Bahr or Helles Point ahead of the rows of troops. Such would have destroyed the submerged wire entanglements and would have taken the first fierce response of the Turkish defense with men well under protection and well armed for the purpose.

It is quite apparent that much of a preparatory nature can and should be done so that, in war, inland waters may be crossed expeditiously by tanks.

CHAPTER XXXIII.

TANK SUPPLY, MAINTENANCE AND SALVAGE.

The Problem of Supply.

The special nature and mechanical complexity of tanks, the hard uses to which they are put, the uncertainty as to the particular units in the line with which they will be employed, and other factors make it necessary that matters of supply, maintenance, and salvage receive special consideration. Not knowing whether the bulk of the tanks are going to be used with this or that corps (not to mention other units), it is clearly not practicable for the normal supply establishments in the line to be prepared at all times to provide the unusual quantities and types of supplies (spare parts, special tank ammunition, fuel, oil, etc.) that would be required if tanks in large numbers should be brought into the particular area for use. On the other hand, in the interest of harmony and efficiency, corps and division authorities must exercise a coordinating control over activities generally, within their respective areas, including such special supply and maintenance establishments and activities as may become necessary due to the presence of tanks. Foresight and special arrangements are therefore plainly necessary. The armies of different countries meet this situation in different ways depending chiefly upon their organization. In any case there are special elements of the tank organization that, by virtue of their knowledge, special training and effort, meet, in a large measure, these special needs of the moment. The dumping of supplies in forward areas is, in general, considered objectionable especially in the more open forms of warfare.

Tank Supply.

Supply in the British Army is centralized in one branch, the Royal Army Service Corps. This simplified functioning automatically eliminates numerous complexities and uncertainties with which we (U. S.) have to contend in our supply system.

In the French army, when a unit of tanks is attached to an infantry division, the officer in command of the tanks is responsible for their supply. For this purpose, he utilizes the truck train under his command and procures the necessary supplies partly from the tank park of the corps. This tank supply system includes all types of supplies needed by the tank units. Thus all lesser infantry units and, to a considerable extent, the infantry division are relieved of the responsibility and effort pertaining to the supply of tank units. In the French tank company, each tank in the supply and maintenance section (corresponding to our replacement section) is provided with a trailer. By this means it is contemplated that a combat section (or combat platoon) will often be resupplied at what we call the assembly point.

In the United States Army the supply system is probably more complex than elsewhere. It is our policy that corps and division commanders shall be as supreme and fully responsible within their own areas as is possible. Supply in general is a function of the staff and the several supply branches. The supply facilities of our attached tank units are to be utilized, but under the full coordinating control of the corps and division staffs. Tank units have their own field and combat trains, which are wholly motorized. Rolling reserves of supplies are thus carried.

Under the plan contemplated in our army the supply of attached units is effected so far as practicable through the supply officers of the units to which attached. This even applies, to some extent, in the case of units attached to subordinate elements of the infantry division.

Although our country so far has no large force of fast tanks, the problem of supply for such a force seems entitled to special consideration. When acting with the main force, supply will not be very difficult unless tank versus tank combat occurs after a long period on the march, when the fuel carried in the vehicles is getting low. Should an enemy tank force attack with little warning, there may not be time for refueling the vehicles. They cannot be refueled in action and they are practically valueless when out of fuel. Hence, to the space and time factors, fuel and oil are added in supply calculations for motor vehicles. When a tank force is acting alone in situations where tank against tank actions may occur, control points, designated in advance, will be especially necessary in order to facilitate supply and regain control. The vehicles may thus be refuelled from supplies carried on special vehicles, preferably in small containers. Thus it is seen that a tank force, when acting alone, is, like all other independent forces, considerably restricted in its operations by the supply requirement. On account of this limitation, it will seldom be feasible, so far as can be seen at this time, for such a force to take supplies for more than one or at most two refills. Otherwise it would be burdened with a large and cumbersome supply echelon requiring protection.

Tank Maintenance.

The maintenance of tanks includes servicing, adjustments, the replacement of parts and assemblies, and repair. This work requires the services of more than one echelon. In the tank organizations of all armies that have tanks in any considerable number, are found special units that are trained and equipped for maintenance work. The forward echelons perform those tasks that are relatively simple and easy but which are sufficient to restore the tank to operating condition. The rear echelons replace difficult assemblies and execute general overhaul and repair. The older tanks of less mechanical reliability require a larger ratio of maintenance personnel than the modern better built tanks. It is convenient to divide maintenance into two classes, tactical and technical. The former comprises the simpler operations that are performed in the forward areas; the latter the more difficult operations performed in the rear areas.

Tactical maintenance includes such simple operations as are performed by the crew, and the adjustments, replacement of assemblies, etc. that are done by special units of tank personnel. As a check on tactical maintenance and to insure against untimely mechanical break-downs, frequent, intelligent, and thorough mechanical inspections of tanks are a vital necessity.

Technical maintenance is performed farther to the rear by such ordnance or other technical units as may be provided for the purpose. In the United States army, the technical maintenance of tanks is a task of the Ordnance Department; the technical maintenance of other motor vehicles belonging to tank units is a task of the Quartermaster Corps.

In the British army the maintenance activities are divided into 3 echelons. Foremost are the mechanics of the tank companies (who are assisted in some circumstances by the personnel of the tank salvage companies). The second, or intermediate echelon comprises the ordnance mobile workshops. These operate in forward areas and do not attempt tasks of more than moderate difficulty. The 3d echelon comprises the ordnance tank repair establishments at the bases well to the rear.

In the French army the echelon company of the tank battalion is the principal agency for maintenance. This is the only unit for tactical maintenance, and, having fairly heavy shop equipment, it handles repairs of moderate difficulty. Tanks requiring repairs too extensive for the echelon company are transported by trailer or otherwise by the echelon company to the corps tank park, where they are repaired by special workmen trained for the purpose.

In the United States army there are, for tactical maintenance, the following echelons: a maintenance section in the tank company, a maintenance platoon in the headquarters company of the tank battalion and a service company in the tank regiment. It is the U. S. policy to keep shop equipment to the absolute minimum in the tactical maintenance echelons and to effect the more difficult tasks of tactical maintenance through unit replacement rather than by repair. In other words, if an assembly is out of order, instead of attempting to find and remedy the difficulty, the whole assembly is at once replaced by a serviceable one, the unserviceable one being sent well to the rear for repair. The supervision of tactical maintenance is normally centered in the tank battalion commander. The service company of the tank regiment is so organized that platoons may be readily attached to the battalions of the regiment if desired.

For the technical maintenace of U. S. tanks, an ordnance heavy maintenance company (specially equipped for the purpose) is allotted to each tank regiment. Quartermaster repair units furnish corresponding technical maintenance for other motor vehicles.

Tank Salvage.

Salvage is, in reality, a part of maintenance but is sometimes considered separately. In war, definite steps should be taken to insure that tanks that become disabled in forward areas receive suitable attention without undue delay. Many of them can be restored to serviceability with a very moderate amount of work.

In the British Royal Tanks Corps, it is contemplated that, in war, the salvage companies will locate, inspect, and repair or salvage damaged or abandoned vehicles. Tanks reparable but with difficulty would be moved by the salvage companies to the ordnance repair organizations. As stated elsewhere it is contemplated that the British salvage companies will, under suitable circumstances, assist the tank organizations in minor maintenance work.

In the French army, as previously stated, the echelon company transports seriously damaged tanks to the corps tank park for repair there by special personnel.

In the United States army it is contemplated that the service companies of tank regiments will engage in salvage activities as required when other maintenance activities do not take precedence.

CHAPTER XXXIV.
OUTSTANDING CONCLUSIONS.

In the field of military mechanization development, there is at present much turmoil. In time the situation will clarify. But at present the progressives and the reactionaries are at odds. Many conflicting statements are made and the whole subject is foggy and intricately complex. The interested reader is usually at a loss as to what ideas should be accepted and what rejected. If this book will assist him in forming a clear picture of some chief and indisputable fundamentals, it would appear that it will therein have served a useful purpose.

The invention and development of motorized road vehicles and tanks thrust upon the armies of the world many new problems. Among the important questions to be answered were the following:

To what military purposes could such vehicles best be applied?

What preferred types of such vehicles were required?

What changes and innovations in organization should be effected in order to secure the best results?

What should be the employment methods and the technique for the new vehicles?

What development efforts should be inaugurated, and how should they be conducted?

Some efforts have been made to answer these questions but, generally, each of them still confronts the armies. Clarity and concurrence are still far from being achieved.

Some of the fundamentals are clear. The armies require:

(1) *Tactical vehicles.*

Combat vehicles (In the words of the British, "armored fighting vehicles —A. F. V.s.").

Combat carriers (Vehicles for carrying men and weapons to combat localities for dismounted combat).

(2) *Transport vehicles.*
   Cross-country.
   Road.

(3) *Tractors and trailers,* for special purposes, when necessary.

(This classification is covered in greater detail in Chapter I.)

It is seldom now that anyone disputes the need and value of the increased combat and supply mobility, made possible by mechanization developments. Napoleon Bonaparte wrote:

"Maxim IX. The strength of an army, like the power in mechanics, is estimated by multiplying the mass by the rapidity; a rapid march augments the morale of an army, and increases its means of victory. Press on!"

Greatly increased speed in tanks, now being potentially available, is necessary for the following reasons:

(1) It effects increased demoralization among hostile troops, diminishing their opportunity for escape and increasing their prospect of being run over.

(2) It makes hostile fire upon the tanks much less accurate.

(3) It affords hostile gunners time for but few shots while the tanks are closing upon them.

(4) It affords greater freedom as to the selection of the point to be assaulted by the tanks.

(5) It may enable the tanks to reach an important point earlier than the enemy, thus saving a bridge from destruction or otherwise securing an advantage.

(6) It may deprive the enemy of time needed for the organization of his defense in accordance with his normal plans.

(7) It increases the possibility of surprise to an important degree.

(8) It greatly simplifies the movement of tanks from a remote point to a suitable area from which to launch an attack.

The chief limiting factors restricting the use of tanks are as follows:

(1) *Mechanical weakness in the vehicle.* This was a serious matter in the World War. It is different as to the tank of today and the tank of the future. Reliability in motor vehicles is achievable. That fact is becoming increasingly apparent in the case of the tank. Road vehicles of today are reliable; tanks of today are also reliable in large measure; tanks of tomorrow will be more so.

(2) *Obstacles.* They are avoided in two ways: first, by planning tank operations for areas where the obstacles are relatively few; second, by detouring around small obstacles. And, in many cases, by special effort, obstacles may be passed over.

(3) *Hostile fire.* This constitutes a serious danger to tanks. But what man, vehicle, weapon or other agency in the combat zone is not vulnerable to hostile activity under some circumstances? The primary purpose of the tank is not to give the weapon armor, but to give it mobility. The most important protection for tanks lies not in their armor, but in their proper employment. By their mobility they may be massed against weaker elements; by their mobility, their fire power, and their numbers they can overcome the resistance. The hostile defense against tanks cannot be unlimited as to width and depth. The tanks can go around it or break through it, after which they may create great havoc at the expense of but few casualties.

But of course the tank *is* armored. It is and will be armored against at least small arms weapons. By its mobility, power, and method of employment, with the assistance that it receives from others, and protected from small arms by its armor, it will ultimately overcome the forces that oppose it.

(4) *Control.* The control of tanks is difficult after the launching of an attack and this difficulty is increased somewhat as the speed of the tanks is increased. However, the developments of improved visual communication and of radio, and other improvements in communication are, it is believed, in process of alleviating this difficulty to a considerable degree. Moreover, if tanks are launched in attack in pursuance of well-conceived orders, and if they are equipped with the best available direction indicators it is believed that they can carry on with a considerable degree of satisfaction even though communication fails completely while the attack is in progress.

(5) *Human limitations.* We might build a tank that could be driven 47 hours out of 48. But where would we find a man to drive it? The interior of a tank is not a comfortable place. It is often very hot. The air is frequently bad. Riding in a steel shell through fields and woods is not a comfortable occupation. After several hours of fatiguing operation, the crew must be expected to go to work immediately upon their tank and prepare it for further operation. The human element should be carefully considered by higher commanders in assigning tasks to tanks.

(6) *Production.* Difficulties of rapid production in quantity constitute the most potent limiting factor of all. In time of peace we think we need

only a few tanks. The moment we find ourselves at war we decide that we need thousands of them. Much precious time is required to build them. Not only that, there are the casualties. The number of casualties among men in tanks is small in comparison to the number of lives of foot soldiers that they save, but the number of casualties among the tanks themselves is high at each operation. Roughly we may assume 25% tank casualties for each attack. Although many of these may be reconditioned and used again, replacements will be frequently required in considerable numbers. It is therefore extremely important that everything practicable be done to insure an early, rapid, and continuous production of the desired type of tank at the beginning of a war.

To attack modern defenses without tanks is a frightfully expensive undertaking from the standpoint of human lives. Ordinary machine guns account for much the slaughter. Both theoretical considerations and experience in the World War indicate that the number of lives expended in gaining one square mile is very large if tanks are not used and much less if they are used. General Fuller makes the following statement:[1]

> As regards casualties the comparisons are amazing. On the first day of the battle of the Somme, July 1, 1916, when no tanks were used, the British casualties were approximately 60,000. On the first day of the battle of Amiens, August 8, 1918, when 415 tanks were used, they were slightly under 1,000. Between July and November 1916, British casualties per square mile of battlefield gained were 5,300; during the same months in 1917, at the Third Battle of Ypres, they were 8,200; and during the same period in 1918 they were 83. In the third period alone were tanks used in numbers and efficiently.

It may be that some will think these figures tend to exaggerate the value of the tank, but it is believed conservative to say that 10 efficient tanks, as part of the whole, may save 200 lives in the capture of one square mile. In war, our government insures the life of practically every soldier for $10,000. Their enrollment, maintenance, training, etc., costs another $1,000 each. So, even if all ten of the tanks are destroyed in the engagement, they will have saved the government from a gross loss of $2,200,000, which is close to two million dollars, net. From a purely mercenary standpoint that is making or saving money rather fast.

Ignoring for a moment the value of increased power and efficiency in the means for national defense, let us consider the view held by some that mechanization is markedly expensive. The question at issue appears to be:

Is it expensive in the long run to substitute motors and machines for mules, horses and men in such a way that the fighting power will be slightly increased?

In the 34th Infantry, a very complete and detailed study (based upon the exact figures of experience) was made of the comparative costs of animalization and motorization as applied to an infantry regiment. It was found that the initial cost of providing motors was considerably more than the initial cost of providing mules and horses, but that the cost of maintenance was so much higher for animals than for motors, that, after three years the total cost, initial plus maintenance, was in favor of the motors. With a liberal allowance for the replacement of motor vehicles and a scant allowance for the replacement of animals there resulted a rather large and constant net saving in favor of the motors, the total personnel being the same in either case.

A similar result was found by an officer of the Ordnance Department[2] who, in a published study, compared the fighting power and the cost of a light tank battalion with that of an infantry regiment. He found that the

---
[1] *The Mechanization of War*, by Major General J. F. C. Fuller in *Army Ordnance*, (U. S.) January-February, 1930.
[2] Captain John K. Christmas in *Army Ordnance*, July-August, 1930.

tank battalion had 1.34 times the fire power of the infantry regiment; and that, per man risked in combat, the tank battalion had 7.85 times as much fire power as the infantry regiment. In the continuation of his study he found that, although the initial cost of equipment was much greater for the tank battalion than for the infantry regiment, this difference was more than wiped out by a single year of maintenance costs, due chiefly to the difference in numbers of personnel. He found that the annual maintenance cost was in favor of the tank battalion by more than two million dollars.

There is no possible room for an argument against the procurement and use of tanks unless it be founded upon a belief that tanks will not save the lives of foot soldiers in attack. It does not appear that any such belief is justified.

The discussion leads us to the following conclusions:

Armies must either take advantage in an intelligent and comprehensive way of the greatly increased mobility as to combat and supply, or fight their future wars under tremendous handicaps with a resulting great waste of lives, money and effort.

Although the limitations applicable to the tank are serious and must not be ignored, its advantages greatly outweigh its disadvantages. Modern, fast, well-built tanks are invaluable fighting instruments that should be provided in liberal numbers. Although expensive initially in dollars, they are cheap in relation to the lives they save, the victories that they so effectively help to insure, and the economies that their possession and use make possible. Our army should be equipped as soon as possible with a reasonable number of them. In addition to being invaluable in war, they are necessary for peacetime training and the development of tactics.

BIBLIOGRAPHY.

Most of the books here listed deal extensively with some phase of the subject of tanks. Some devote, thereto, only a chapter or so. *The latter have been marked with an asterisk.

Books and Pamphlets.

A Company of Tanks, Watson (Blackwood, London) 1920.
A Brief History of the Royal Tank Corps, Woolnough (Gale and Polden, Aldershot) 1923.
*Aerial Photography, Reeves.
*Airmen or Noahs, Sueter (Pitman and Sons, London) 1928.
A Night Guide for Tanks, Kennedy.
Armored Car Training, Vol, I, (H. M. Stationery Office, London) 1930.
Artillery Today and Tomorrow, Rowan-Robinson (Clowes and Sons, London) 1929.
*A Saga of the Sword, Austin (Macmillan Company, New York) 1929.
A Short History of the Royal Tank Corps, (Gale and Polden) 1930.
A Tank Driver's Experiences, Jenkins (Elliot Stock, London) 1920.
*Australian Victories in France, 1918, Monash.
Battle of Dora, The, Graham (Wm. Clowes and Son, London) 1931.
Boevoe Primenenie Tankov i Borba s Nimi, Heigl (Griz, Moscow) 1928.
Carden Loyd Light Amphibious Tank, The, (Vickers Armstrongs, Ltd., London) 1931.
Carden Loyd Light Armored Vehicle, The, (Vickers Armstrongs, Ltd., London) 1927.
Complete Guide to the Hotchkiss Machine Gun, (Gale and Polden, Aldershot).
Crew Drill for Medium Tanks, (H. M. Stationery Office, London) 1930.
Crew Drill for the Rolls-Royce Armored Car, (H. M. Stationery Office, London) 1930.
*Daily Mail Year Book 1921, (London *Daily Mail*).
Defense of Bowler Bridge, Graham (Clowes and Sons) 1930.
Der Kampf Gegen Tanks, Borchet (Mittler, Berlin) 1931.
Der Kampfwagen in der Heutigen Kriegführung, Volckheim (Mittler, Berlin) 1927.
*Der Stellungskrieg, (Mittler, Berlin) 1926.
Die Deutschen Kampfwagen im Weltkrieg, Volckheim (Mittler, Berlin) 1923.
Die Kampfwagen Fremder Heere, (Eisenschmidt, Berlin) 1926.
*Die Militärischen Lehren des Grossen Krieges, Schwarte.
Die Schweren Französischen Tanks, Die Italienischen Tanks, Heigl (Eisenschmidt, Berlin) 1925.
Dowodzenie i Walka Broni Polaczonych, Quirini (Glowna Ksiegarnia Wojskowa, Warsaw) 1926.
Employment of Tanks in Combat, (U. S. General Service Schools, Fort Leavenworth, Kansas) 1925.
Eye-Witness and the Origin of the Tanks, Swinton (Doubleday, Doran & Co., N. Y.) 1933.
Fighting Tanks, Wilson (Murray, London) 1930.
Foundation of the Science of War, The, Fuller.
Framework of the Science of Infantry Tactics, The, Liddell-Hart (Murray, London) 1927.

*From Chauffeur to Brigadier, Baker-Carr (Ernest Benn, London) 1930.
Further Aspects of Mechanization, Rowan-Robinson (Clowes and Sons) 1930.
Future of the British Army, The, Dening (H. F. and G. Witherby, London) 1928.
Handbook of the Six Ton Special Tractor, (U. S. Government Printing Office, Washington) 1918.
Handbook of the Mark VIII Tank, (U. S. Government Printing Office, Washington) 1918.
History of the 1st Tank Battalion, (Royal Tank Corps Journal).
History of the 13th Tank Battalion, (Royal Tank Corps Journal).
History of the 17th (Armored Car) Tank Battalion, (Royal Tank Corps Journal).
I Carri d'Assalto, Verse.
*Infanterie Begleitwaffen, Däniker, 1928.
Infantry Field Manual, Vol. II (Tank Units) (U. S. Government Printing Office, Washington) 1931.
Instrutsja po Boevomu Primeneniu Bronevomu Avto, Tajna (Griz) 1916.
Instrutsja po Boevomu Primeneniu Tankov R.K.K.A. (Griz) 1928.
Instr. Prov. sur l'Emploi et la Mar. des Unites Auto-Mitrall. de Cavallerie.
Instr. Prov. sur l'Emploi des Chars de Combat comme Engins d'Infanterie, 1929.
In the Wake of the Tank, Martel (Sifton Praed, London) 1931.
La Division legere Automobile, Boullaire.
La Motorization de l'Armee et la Maneuvre Strategique, Camon (Berger-Levrault, Paris) 1926.
l'Artillerie d'Assaut de 1916 à 1918, Lieutenant Colonel Lafitte, (Librairie Charles-Lavauzelle, 124 Boulevard Saint-Germain, Paris).
Lectures on F. S. R. III. (Operations Between Mechanized Forces), Fuller (Sifton Praed, London) 1932.
Les Chars d'Assaut, Captain L. Dutil (Imprimerie et Librairie Berger-Levrault, Nancy).
Lewis Machine Gun Mechanism Made Easy, (Gale and Polden, Aldershot).
Life in a Tank, Haigh (Houghton Mifflin, Boston) 1920.
*Long Road to Victory, The, ed. By Buchan (Nelson, London).
Maintenance for Full and Half Tracked Vehicles, (H. M. Stationery Office, London) 1927.
Manuel Pratique du Char Renault, Goutay (Fournier, Paris) 1923.
Materiale d Infanterie i d'Artiglieria, (Scuole Centr. di Fant., Civitavecchia) 1923.
Mechanization, (Vickers Armstrongs, Ltd., London) 1931.
Mechanization of War, Germains (Sifton Praed, London) 1926.
Men and Tanks, Macintosh (John Lane, London) 1919.
*Military Motor Transportation, (U. S. Coast Artillery School).
Motorization in the Armies of Tomorrow, Allehaut.
Narrative History of the 6th and 7th Tank Battalions, Woolnough (Gale and Polden, Aldershot) 1923.
Obrona Prizecienczolgowa, Korezynski i Kuszelewski (Glowna Ksiegarnia Wojskowa, Warsaw) 1926.
On Future Warfare, Fuller (Sifton Praed, London) 1929.
*Ordnance and Gunnery, McFarland (John Wiley and Sons, New York) 1930.
Panzerautomobile Gegen die Walachen, Siemer (Eisenschmidt, Berlin) 1924.
Paris, or the Future of War, Liddell-Hart (Murray, London) 1925.

Pociagi Pancerne, Frasunkiewicz (Glowna Ksiegarnia Wojskowa, Warsaw) 1928.
Pociagi Pancerne, Szawrow.
Reformation of War, The, Fuller (Hutchinson, London).
Remaking of Modern Armies, The, Liddell-Hart (Murray, London) 1926.
Reno Russkij, Fotjakov (Griz, Moscow).
Riddle of the Rhine, The, Lefebvre (American Chemical Foundation) 1922.
Samochody Pancerne, Zyrkiewicz (Glowna Ksiegarnia Wojskowa, Warsaw) 1928.
Some Aspects of Mechanization, Rowan-Robinson (Clowes and Sons) 1929.
Sous l'Armure, Lestringuez.
Spravotchnik po Bronevomu Delu, Dervtsov i Pushkin (Griz. Moscow) 1926.
Stand der Tankfrage, Heigl (Wojna i Mir, Berlin) 1924.
Strassenpanzerwagen: Die Sonderwagen der Schutzpolizei, Schmitt (Eisenschmidt, Berlin) 1925.
Taktika Avto-Bronevikh Chastej, Uvanov (Griz, Moscow) 1926.
Taktika Bronevomu Delu, Gladkow.
Tank and Armored Car Training, Vol. II, (H. M. Stationery Office, London) 1927.
Tank Corps Book of Honor, The, Ballantyne (Spottiswoode).
Tank Corps, The, Major C. Williams-Ellis and A. Williams-Ellis (*Country Life,* London) 1919.
Tank in Action, The, Browne (Blackwood) 1919.
Tank, The, Its Birth and Development, (Foster, Lincoln).
Tanki, Golowin.
Tankovike Vojska Inostrannikh Gosudarstv, Vishnev (Griz, Moscow) 1926.
Tanks, The, Swinton (Doubleday) 1919.
Tanks, Krüger (Richard Carl Schmidt, Berlin) 1921.
Tanks, (U. S. Infantry School, Fort Benning, Georgia) 1931.
Tanks 1914-1918: The Log Book of a Pioneer, Stern (Hodder and Stoughton, London).
Tanks and Tank Tactics, (U. S. Infantry School, Fort Benning, Georgia) 1926.
Tanks, Gas, Bombing, and Liquid Fire, Dion 1918.
Tanks in the Great War 1914-1918, Fuller (Murray, London) 1920.
Tank Tales, Tank Major and Eric Wood.
Tank Training, Vol. I, (H. M. Stationery Office, London) 1930.
Tanks und Tank Bekämpfung, Volckheim (Mittler, Berlin) 1927.
Taschenbuch der Tanks, Wesen, Erkennung, Bekämpfung, Heigl (Lehmann, Munich) 1926.
Taschenbuch der Tanks, Ergänzungsband 1927, Heigl (Lehmann, Munich).
Taschenbuch der Tanks, Ausgabe 1930, Heigl (Lehmann, Munich).
Technical Regulations 1320-39, Browning Tank Machine Gun Cal. .30, Model 1919, (War Dept.)
Technical Regulations 1320-75, 37 mm Gun and Tank Cradle, Model 1926, (War Dept.)
Technical Regulations 1325-A, Six Ton Tank M 1917, (War Dept.)
Technical Regulations 1325-B, Mark VIII Tank, (Ordnance Document No. 1977).
Technical Regulations 1350-2.24A, Ammunition for 2.24-inch Tank Gun Mark II (British gun), (War Dept.)

Technical Regulations 1350-37A, Ammunition for 37 mm Gun, M 1916, (War Dept.)
Technical Regulations 1425-A Instructions for Ordnance Maintenance Companies, Six Ton Tank M 1917, (War Dept.)
Technische Mitteilungen über Kampfwagen, Dörffer (Eisenschmidt, Berlin) 1922.
3 Pounder Q F Gun for Medium Tanks, The (H. M. Stationery Office, London) 1929.
.303 Vickers Machine Gun, The (H. M. Stationery Office, London) 1930.
*Training Regulations 195-40, Camouflage for All Arms, (War Dept.)
Training Regulations 425-90, Armored Car Marksmanship, (War Dept.)
Voivoj Ustav Bronevikh sil A.K.K.A. Kniga I: Tanki, The (Griz, Moscow) 1919.
*War in the Garden of Eden, The Roosevelt (Murray, London) 1920.
War History of the 3rd (Light) Tank Battalion, The (Royal Tank Corps Journal).
War History of the 6th Tank Battalion, The (Royal Tank Corps Journal).
War History of the 2nd Tank Brigade, The (Royal Tank Corps Journal).
War History of the 5th Tank Brigade, The (Royal Tank Corps Journal).
Zasady Uzycia Czolgow, Romszowski (Glowna Ksiegarnia Wojskowa, Warsaw) 1925.

Most of these books may be procured from the *Infantry Journal*, Washington, D. C.

Periodicals.

Army and Navy Journal	United States
Army and Navy Register	British
Army, Navy and Air Force Gazette	British
Army Ordnance	United States
Army Quarterly	British
Bulletin des Sciences Militaires	Belgian
Canadian Defense Quarterly	Canadian
Cavalry Journal	United States
Cavalry Journal	British
Chemical Warfare Journal	United States
Coast Artillery Journal	United States
Der Kraftzug in Wirtschaft und Heer	German
Field Artillery Journal	United States
Fighting Forces	British
Infantry Journal	United States
Irish Military Journal	Irish
Journal of the Royal Artillery	British
Journal of the Royal United Service Institute	British
La Revue d'Artillerie	French
La Revue d'Infanterie	French
La Revue Militaire Suisse	Swiss
Militär-Wochenblatt	German
Military Engineer, The	United States
Royal Engineers Journal	British
Royal Tank Corps Journal	British
Service Motor Transport Gazette	United States
S. A. E. Journal	United States
Signal Corps Bulletin	United States
Tank Notes (discontinued)	United States
Technische Mitteilungen u. Milit.-Wissensch.	Austrian

INDEX

Numbers refer to pages.

Accessibility 202
Accessories 203-217
Accompanying tanks 279
Aerial control 292, 293
Air cleaners 216
Airplanes and tanks (See *Cooperation between Tanks and Air Corps*)
Ammonia, use of 49, 214
Ammunition racks 206
Amphibious tanks 182, 301-306
(See also *Medium D. pp. 117, 118, 119*)
Antiaircraft weapons for tank units 185, 249, 250
Antitank defense (See *Defense against tanks*)
Antitank weapons 272, 256, 259-272
Armament 184, 249, 250
Armor 62, 191, 192
Armored cars 1, 35, 36, 39, 42, 49, 50, 53, 90, 225-229
(See also *Half track—track combat vehicles*)
Armored force 1, 281
Armor piercing ammunition 11, 12, 18, 40, 62, 257, 258
Arrangement (See *General arrangement*)
Arrangement of items, this book 115
(See also *Table of Contents*)
Artillery, self-propelled 230-237
Artillery, use of against tanks 22, 37, 47, 57, 73, 88, 256, 265, 270
Assembly point 279
Attacking tanks with iron bars 93
Barriers, tank 261, 262
Battles, Engagements, Campaigns, Villages, Rivers, etc. (See *List of localities p. 321*)
Belt loading machines 206
Bibliography 314-317
Bombs, hand (See *Grenades, hand*)
Bridges, 299, 300, 301
Bridges, submerged 51, 299
Bullet splash 23, 47, 49, 62, 192, 207, 208, 209, 259, 260
Burned tanks 7, 39, 42, 57, 58, 59, 65, 76
Camouflage (See *Concealment and camouflage*)
Camouflage nets 211
Campaigns (See *Battles*)
Cargo carriers (See *Cross-country cargo carriers*)
Cargo carriers converted from tanks 17, 31, 52, 243, 244
Casualties reduced by mechanization 105, 312
Cavalry and tanks 21, 37, 280
Center of gravity of tank 197
Chief of Tank Corps, U. S. A. 107
Christie, Mr. J. Walter 168, 169

Classification of tanks 2, 3, 180-182, 184, 310
Classification of vehicles 2
Climbing ability of tanks 198
Close support tanks 182
Combat cars 171, 276
Combat principles, use of tanks 278-288
(See also *Principles of tank employment*)
Comfort of tank crew 193, 194
Command and signal tanks 218-222
Command post vehicles 222
Communication 203-206, 291
Compasses 51, 52, 214
Concealment and camouflage 294-298
Conclusions 310-313
Control 291, 292, 293
Control from the air 292, 293
Cooling methods for engines 194
Cooperation between tanks and:
 Air Corps 32, 35, 36, 38, 41, 46, 290, 292, 293
 Armored cars 290
 Artillery 289, 290
 Chemical warfare 290
 Engineers 290
 Infantry foot troops 21, 69, 289
Cost, relative, of mechanization 2, 312, 313
Cross-country cargo carriers 244, 245
(See also *Cargo carriers converted from tanks, Power carts, and Military tractors*)
Cross-country traction aids 241, 242
Defense against tanks 257, 258, 271
 British doctrine 270
 French doctrine 270, 271
 German doctrine 270
 German World War methods 24, 258-270
 Japanese doctrine 271
Definitions 1
Deserts 15
Design of tanks 183-202
Design of cross-country cargo carriers 201
Diagonal ridge, crossing of 23
Direction indicators 214
(See also *Compasses*)
Dismounted fighting by tank crews 28
Employment of tanks (See *Principles of tank employment*)
Engagements (See *List of Localities, p. 321*)
Equipment 203-217
Farms (See *List of Localities, p. 321*)
Fascines 20, 215
Ferries 300, 301
Fire, protection against 193, 211
First aid 214
Flags 205

319

Flame thrower in tank 153
Flares 44
Flash hiders 250
Fording ability 198
Formations 291
Forts, antitank 267, 268
Fuel distance 202
Fuel tanks 202
Fumes, protection against 36, 39, 41, 193
Gallons, kind of 3
Gas, poisonous:
   Operations in 17, 44, 52, 79, 103
   Protection against 193, 211, 212, 256
Gauges 216
General arrangement of compartments, engine, etc., cross-country cargo carrier 201
General arrangement of compartments, engine, etc., tank 198-201
Glass, laminated, protective use of 207, 208
Goggles 211
Governor 216
Grapnels 21, 217
Grenades, hand 24, 257, 260, 267
Ground pressure 195
Grousers 196, 215
Guns (See *Weapons*)
Gunnery 251, 252, 253
Gyroscopes 214
Half-track combat vehicles 222-224
Halger-Ultra rifle and ammunition 256, 257
Heat, protection against 36, 39, 41, 193
Heavy tanks 182
Hipkins traction device 241, 242
History of tanks in action (See *Table of Contents*; also *List of localities*, p. 321)
Horsepower requirements 186, 194
Infantry carrying tanks (See *Troop carrying tanks*)
Intelligence 282
Landings on hostile shores 299
Leading tanks 184, 278, 279
Lights 202, 205, 206, 214
Limitations of tanks 176, 177, 311, 312
Lives saved by mechanization 105, 312
Localities referred to in text 321
Locations of tank units U.S.A. 275, 276
Lubrication 202
Maintenance 308, 309
Maintenance vehicles 248
Maneuverability 202
Map boards 214
Mechanization 1, 2
Mechanized force 1, 281
Melee of tank vs. tank actions 287, 288
Military tractors 237-240
Mines, antitank 43, 47, 87, 88, 112, 213, 257, 262
Motorized force 1
Mounting of weapons 184, 185, 250
Mufflers 202
Night operations 30, 37, 40, 51, 52
Noise for concealing tank movements 32, 35, 88
Obstacles 254, 255, 256
Obstacles, methods of crossing 20, 45, 196, 197, 198
Oil cooling 216

Orders for fast tank units 292
Organization of tank units
   British 5, 34, 273
   French 55, 89, 273, 274, 275
   German 96, 105
   In World War 5, 34, 55, 96, 105
   Italian 275
   Japanese 275
   Major aspects of 273, 274, 275
   Russian 275
   U.S. 273, 275, 276, 277
   Various countries 275
Periscopes 210
Pershing, General, letter from 114
Personal experiences:
   A British officer opposed by German tanks 99, 100
   Haimbaugh, Lt., U.S.A. 49
   Marrison, L. A., member British tank crew 39, 44
   Spremberg, Lt., German Inf. 24
Phosphorus bombs 41
Pigeons 206
Pillboxes 19, 20, 258
Pistol ports 30, 68, 192
Platoons, tank, size of 277, 291
Ponton equipment 300, 301
Power carts 244, 245, 246
Power requirements 186, 194
Powers of tanks 174, 175, 176
Priming, engine 215
Principles of tank employment 25, 26, 174-182
   (See also *Combat principles*)
Pumps, water expelling 214
Pyrotechnics 205
Radio 23, 48, 49, 206, 218-222
Reconnaissance 282
Reserve, tanks in 281
Reservicing point 279
River crossings (See *Stream crossings*)
Rivers (See *List of Localities*, p. 321)
Role of the tank 177
Rockenbach, S. D., Brig. Gen. 107
Rubber tracks, use of in action 93, 94
Salvage 309
Self-propelled artillery 230-237
Semaphores 44, 205
Shell bags 206
Shock absorbers 189
Sights 250, 251
Signal tanks (See *Command and signal tanks*)
Silence in operation 202
Small arms fire upon tanks 256, 257
   (See also *Bullet splash*)
Smoke 20, 33, 47, 79, 212, 213
Snow 12
Space as a factor in tank design 193, 194
Spare parts 217
Speed as a factor in tank design 185, 186
Speedometers 214
Sprockets 196
Starting, engine 215, 216
Stations of tank units U.S.A. 275, 276
Stopping to fire 283
Stream crossings 51, 299-306
Stroboscopes 210, 211
Submergence 195
Substitutes for tanks (German) 141
Superstructure 201, 202

Supply 307, 308
Supply tanks 17, 31, 52, 243, 244
Suspensions of tanks 186-191
Suspension of Christie tank 190
Swamps, operations in 11, 18, 19
Table of antitank weapons 272
Table of distance, speed, and time 271
Tachometers 216
Tactics (See *Combat principles and Principles of tank employment*)
Tailpiece 195
Tank carriers 246, 247, 248
Tank Corps, U.S. Army 107, 113, 114
Tank, origin of name 1
Tanks, classification of 2, 3
Tanks vs. tanks 48, 99, 257, 258, 283-288
TANKS (See also *Amphibious tanks, Command and signal tanks, and Half-track combat vehicles*)
  By countries:
    British 5, 6, 9, 10, 16, 17, 28, 29, 30, 31, 32, 34, 35, 115-131
    Czechoslovakian 150, 151, 152
    French 55, 56, 59, 60, 63, 64, 132-138
    German 95, 96, 139-142
    Italian 142-146
    Japanese 147, 148, 149, 150
    Polish 152
    Russian 147, 148
    Spanish 151, 152
    Swedish 150, 151
    U.S. 153-171
  Classification of 2, 3, 180-182, 184, 310
  List of, all countries 172, 173
  Radio controlled 149, 150
  Shapes and relative sizes 173
  By name of tank—Photographs, drawings descriptions, and lists of battles in which engaged:
    A7V (German) 95, 96
    A7VU (German) 139, 140
    Ansaldo (Italian) 144, 145
    Big Willie (British) 115, 116
    Bridging Tank, R.E. (British) 119, 120, 121, 213
    Carden Loyd, One Man, Mark III (British) 122, 123
    Carden Loyd, Mark V (British) 122, 123
    Carden Loyd, Mark VI (Machine Gun Carrier) (British) 122, 123, 124
    Carden Loyd, Mark VII (British) 123, 124
    (Carden Loyd) Light, Mark 1A (British) 124
    (Carden Loyd) Light, Mark II (British) 218, 219
    Cardosowitz (Polish) 152
    Char Moyen (French) 136
    Char 1A (French) 136, 137
    Char 1C (French) 136, 137
    Char 2C (French) 136, 137
    Char 3C (French) 136, 137, 138
    Chenilette St. Chamond, M1921 (French) 134, 135
    Chenilette St. Chamond, M1924 (French) 134, 135
    Chenilette St. Chamond, M1926 (French) 134, 135
    Christie M1919 (U.S.) 166, 167
    Christie M1921 (U.S.) 166, 167, 168
    Christie Chassis M1928 (U.S.) 167, 168
    Christie M1931 (U.S.) 168, 169, 170
    Christie Light M1932 (U.S.) 168, 169, 171
    Christie suspension (U.S.) 190
    Combat Car, T2 (U.S.) 171
    Convertible Medium, T3 (U.S.) 168, 169, 170
    Delaunay-Belleville (French) 135, 136
    Eighty Ton (Russian) 147
    Fiat, Type 3000 (Italian) 142, 143
    Fiat, Type 3000 B (Italian) 142, 143
    Fiat, Type 2000 (Italian) 144, 145
    Flying Elephant (British) 115, 116, 117
    Ford, Three Ton (U.S.) 154, 155
    Gas-Electric (U.S.) 153, 154
    G.L. 4 (Italian) 146
    Heavy Pavesi (Italian) 145, 146
    Hornet (British) 117, 118
    Independent Mark I and Mark II (British) 128, 129
    Infantry Carrier (Mark IX) (British) 120, 121, 122
    Italian Light (Carden Loyd type) (British) 146
    KH 50-60-70 (Kola Housenka) (Czech) 151, 152
    K Vehicle (German) 140, 141, 142
    Landskrona (Swedish) 150
    Light Dragon Machine Gun Carrier (British) 127, 128
    Light Tank (Czech) 150, 152
    Light Tank (Italian) (Carden Loyd type) 146
    Light Tank (Russian) 147, 148
    Light Tank (U.S.) T1 162, 163
    Light Tank (U.S.) T1-E1 163, 164
    Light Tank (U.S.) T1-E2 164, 165
    Light Tank (U.S.) T1-E3 164, 165
    Light Tank (U.S.) T1-E4 164, 165, 166
    Little Willie (British) 115, 116
    L.K.I (German) 139, 140
    L.K.II (German) 139, 140
    L.K.III (German) 139, 140
    M21 (Swedish) 150, 151
    Mark I (British) 5, 6
    Mark I, Three Man (U.S.) 155, 156
    Mark II (British) 9, 10
    Mark III (British) 9, 10
    Mark IV (British) 16, 17
    Mark IV, Lengthened Tail (British) 118, 119
    Mark IV, Modified (British) 119
    Mark V, Experimental (British) 118, 119
    Mark V (British) 31
    Mark V, Sprung Tracks (British) 119, 120
    Mark V Star (British) 34, 35
    Mark V Double Star (British) 119, 120, 121, 213
    Mark VI (British) 120, 121
    Mark VII (British) 120, 121
    Mark VIII (British) 121
    Mark VIII (U.S.) 156, 157
    Mark IX (Infantry Carrier) (British) 120, 121, 122
    Medium A (British) 28, 29, 30

Medium A, Modified (British) 117, 118
Medium A, M1921 (U.S.) 158, 159, 160
Medium B (British) 117, 118
Medium C (British) 117, 118
Medium D (British) 117, 118, 119
Medium M1922 (U.S.) 160, 161
Medium T1 (U.S.) 160, 161
Medium T2 (U.S.) 160, 161, 162
Mother (British) 115, 116
One Man Tank, Experimental (U.S.) 162, 163
Patrol, Vickers Carden Loyd (British) 125, 126
Pavesi, Heavy (Italian) 145, 146
Pavesi, M1925 (Italian) 142, 143, 144
Pavesi, M1926 (Italian) 143, 144
Pavesi Tank Destroyer (Italian) 144, 145
Peugeot (French) 134, 135
Radio Controlled (Japanese) 149, 150
Renault (French) 63, 64
Renault, Model BS (French) 132, 133
Renault, M1923 (French) 132, 133
Renault with Rubber Tracks (French) 132, 133
Renault, NC, M1927 (French) 134, 135
Russian Renault (Russian) 147, 148
Schneider (French) 55, 56
Schneider (Supertank) (French) 138
Six Ton, M1917 (U.S.) 157, 158
Six Ton, M1917, A1, Pilot Model (U.S.) 157, 158
Six Ton, M1917, A1 (U.S.) 158, 159
Sixteen Ton (British) 128, 129
Skeleton (U.S.) 154, 155
St. Chamond (French) 59, 60
Steam, Three Wheeled (U.S.) 153, 154, 155
Steam, Track Laying (U.S.) 153, 154
Substitutes for Tanks (German) 141
Track Development Chassis, T1 (U.S.) 162, 163

Trubia (Spanish) 151, 152
Vickers Carden Loyd Patrol (British) 125, 126
Vickers, Convertible (British) 129, 130
Vickers, Mark I (British) 126, 127
Vickers, Mark IA (British) 126
Vickers, Mark II (British) 126, 127
Vickers, Mark IIA (British) 126, 127
Vickers, Medium C (Japanese) 147, 148, 149, 150
Vickers Armstrongs Six Ton, Alternative A (British) 130, 131
Vickers Armstrongs Six Ton, Alternative B (British) 130, 131
Vickers Wheel and Track (British) 127, 128
Whippet (British) 28, 29
Target designation equipment 207
Telephones 204, 205
Tons, kinds of 3
Tools 216, 217
Tracks for tanks 195, 196
Traction aids 241, 242
Tractors, military 237-240
Troop-carrying tanks 35, 43, 44
Turrets 201, 202
Types of tanks and their uses 180, 181, 182
Unditching gear 43, 213, 215
Unusual uses of tanks 89
Uses of tanks 177, 178, 180, 181, 182
Vehicles
  Classification of 2
  For maintenance 248
Villages (See *List of localities, p. 321*)
Vision protection 207-211
Weapons
  Antiaircraft 185, 249, 250
  Antitank 256, 259-272
  Tank 184, 249, 250
Woods (See *List of localities, p. 321*)

LIST OF LOCALITIES MENTIONED IN THE TEXT

B—battle; C—campaign; F—farm; R—river; V—village or town; W—wood or forest.

Achiet le Grand (V) 40, 41
Achiet le Petit (V) 39
Adelpare (BF) 62
Ailette Canal 78
Aire (R) 110
Aisne-Marne Canal 101
Aisne (R) 57
Albert (V) 39
Alberta (V) 19
Albert-Arras R.R. 38
Amervalles (V) 51
Amiens (B) 34
Ancre (R) 38
Andon (R) 111
Andon (V) 110
Anneux (V) 23
Apremont (V) 110, 111
Aquennes (W) 98
Ardoye (V) 83
Argonne (B)
Argonne (W) 111

Arrachis (W) 33
Arras (1st B) 10
Arras (2d B) 41
Avillers (V) 108
Bancourt (V) 42
Battles of U.S. Tank Corps 113
Battoir 81
Bapaume (B) 38
Baulny (V) 110
Beauchamp Ridge 45
Beaumont-Hamel (B) 9
Beaurevoir (V) 48
Beer (V) 84
Bellecroix (VW) 61
Bellicourt 47, 103
Belloy (B) 67
Belloy (V) 68
Beney (V) 106, 107
Berteaucourt-Thennes Road 35

Besme (V) 78
Benges (W) 110
Beugnatre (V) 42
Beugneux (B) 76
Beugny (V) 28, 42
Beveren (V) 83
Biefvillers (V) 41
Bieuxy (V) 77, 78
Bihucourt (V) 40
Blanzy (V) 74
Blecourt (V) 48
Blerancourdelle 78
Bony (V) 47
Bouc (W) 81
Bouchoir (V) 37
Bourlon (V) 23, 46
Bourlon (W) 23, 45, 46
Bousseoux Ravine 61
Bouzencourt (V) 30
Brancourt (B) 48
Brancourt (V) 48

Index

Bray (V) 41
Bray sur Somme (V) 39
Brie Bridge 28
Bucquoy (V) 39
Bucquoy Night Raid 30
Bucy le Long 79
Bullecourt (B) 12
Bullin (BF) 76
Buvignies (V) 53
Cachy (V) 97, 99
Calvaire (V) 64
Cambrai (1st B) 20, 105
Cambrai (2d B) 45
Cambrai-St. Quentin (B) 45
Cambrai, North of 104
Camelin (V) 78
Camouflage (W) 41
Camp de Cesar (V) 77
Cantigny (B) 62
Cantaing (V) 23
Castel (BV) 62
Catelet-Bony (B) 46
Catillon (V) 52
Celles les Conde (V) 71
Chafosse (B) 68
Chalelet (W) 75
Champagne (B) 80
Charantigny (V) 73, 74
Charleroi (V) 105
Charpentry (V) 110
Chaudron (F) 110
Chaudun (V) 72
Chavignon (F) 61
Chavignon Plateau 61
Chavigny (F) 66
Chazelle Ravine 72
Chazelle (V) 65
Chemin des Dames (B) 56
Chene Sec (W) 111
Cheneviere (F) 75
Cheppy (V) 109
Chipilly Ridge 36
Chipilly (W) 37
Chuignolles (V) 36, 40
Cierges (V) 110
Coeuvres (B) 68
Cointicourt (V) 73
Colincamps (V) 30
Cologne (R) 28
Colombe (BF) 80
Conchy les Pots (V) 103
Concroix 73
Conde 88
Conde (W) 71
Conde en Brie (W) 71
Connantreuil (B) 76
Connigis (V) 71
Coolscamp (V) 83
Corcy (B) 65
Corcy (V) 66
Corey (V) 66
Courcelles (V) 39, 40, 67
Courchamps (V) 73
Courjumelles (V) 87
Couronne (W) 79
Cravancon (F) 65
Crecy au Mont (B) 78
Crinchon Valley 11

Croix Muzart 81
Crouy (B) 78
Cuffies (V) 78, 79
Cuisy (W) 109
Cunel (V) 110, 111
Cutry (B) 69
Cutry (V) 70
Cuvillers (V) 48
Cuvilly (V) 67, 103
Dammard (V) 73
Deligny Hill 46
Demi-Lieue (V) 51
Demuin (V) 36
Dermicourt (V) 42
Derni Lune (F)
Dix Hommes (BW) 76
Doignies (V) 27
Dommiers (V) 72
Doon (W) 48
Drocourt-Queant Line 11, 42
Du Gros-Hetre (BW) 62
Du Nord Canal 45
Echelles Ravine 72, 73
Eecke (V) 88
Ecoust 42
Eclisfontaine 110
El Arish Redoubt 14
Emilie (V) 27
Epayelles (V) 67
Epehy (B) 43
Epehy (V) 28
Epinoy (V) 46
Escaille (V) 84
Escaut (BR) 88
Espilly (B) 76
Essey (V) 106
Etinehem Spur 37
Exermont 110
Fanny's Farm 17
Faucon (W) 79
Faverolles (B) 65
Faverolles (V) 66
Fendu (W) 77
Fermicourt 104
Flanders (C) 83
Flers 7
Flesquieres (V) 23, 45
Fleury (BW) 76
Fonsomme Line 48
Fontaine (V) 24
Fontaine Notre Dame (V) 23, 46
Forges (R) 109
Fort Pompelle 101, 102
Fraicourt (W) 48
Framerville (V) 37
Fremicourt (V) 42
Fresnoy (V) 43, 77
Fresnoy le Petit (V) 43
Gand (V) 83
Gaza (2d B) 14
Gaza (3d B) 14
Geite St. Joseph (V) 83
Gesnes (V) 111
Gird Trench 9
Gitsberg (V) 83
Glaux (F)
Gomiecourt (V) 40

Gonnelieu (V) 46
Goyencourt (V) 77
Graincourt (V) 23
Grand (W) 81
Grand Ravine 20
Grand Rozoy (V) 76
Grand Thiolet (V) 86
Grange (F) 77
Grevillers (V) 41
Grille (F) 65
Grivesnes Park (B) 62
Grougis Mill 85
Guemappe (V) 42
Guery (V) 78
Gueudecourt (V) 9
Guillaucourt (V) 36
Guillemont (F) 43, 46, 47
Guise, South of (B) 87
Gros Hetre (BW) 62
Hagrand (W) 103
Hamel (B) 31
Hamel (W) 32
Hamelincourt (V) 40
Hangest-en-Santerre (B) 76
Haplincourt (V) 42
Happeharbes (V) 52
Happy Valley 41
Harbonnieres (V) 36
Hargicourt (V) 43
Harp, The (fortification) 11
Harpon (W) 33
Hartennes (V) 73
Hatton Chatel (V) 108
Havrincourt (V) 21
Haynecourt (V) 46
Hebuterne 30
Heine Selve Plateau 87
Hendecourt 12
Heninel (V) 13, 40
Henin sur Cojeul (V) 41
Hermies (V) 42
Hervilly (V) 28
Hill 132 85
Hill 205 (B) 76
Hille (V) 83
Hindenburg Line 12, 13, 20, 43, 46, 47
Hoinets (W) 61
Hommes (W) 76
Honnechy (V) 49, 50
Hooglede (B) 83
Hooilhoek (V) 84
Hundung Stellung (B) 88
Jardin 72
Javelle (F) 78
Jolimetz (V) 53
Jonville (V) 108
Juvigny (V) 78, 79
Kirbet El Sihan (redoubt) 14
Knoll (F) 43, 46, 47
La Dormoise 81
l'Araignée (W) 81
Laffaux Mill (B) 60
Lagnicourt (V) 42
La Grille (F) 66, 67
La Jonquiere (F) 78

Landrecies (V) 52
Landres et St. George (V) 113
La Neuville (V) 77
Langemarck-St. Julien (B) 19
La Pince Heights 81
La Targette 48
Lataule (B) 67
La Taule Park 77
Lateau (W) 23
Latilly (W) 75
Laucourt (V) 77
Le Banc de Pierre 79
Le Cateau (V) 50, 51
Le Fresne (B) 78
Le Madine (R) 106
Le Plessier Hulen (V) 74
Le Quesnoy (V) 53
l'Escaut Canal 20
Lessart (F) 73
Les Trois Peupliers 69
Leury (V) 79
Lihons (V) 37
Limonval (F) 79
Loges (BF) 70
Lombray (B) 78
Longatte 42
Longpoint (V) 65
Louport (W) 41
Luce River 35
Machet Mill 77
Macogny (V) 73
Madeleine (W) 110
Magny (V) 47
Mairie (V) 22
Maizerais (V) 108
Malancourt (W) 109
Malmaison (B) 61
Manicamp (V) 78
Manloy (W) 73
Manre (V) 81
Manre Tunnel 81
Many (F) 61
Marchavenne (F) 85
Marcoing (V) 21, 23
Maretz (V) 49
Marfaux (B) 76
Marine (W) 111
Marizy (V) 73
Marne, South of the (B) 71
Marquion Line 46
Masnieres (V) 21
Matz (B) 102
Matz Valley 67
Maubeuge (B) 52
Maubry (V) 75
Maurois (V) 50
Meaulte (V) 39
Medeah (F) 82
Meharicourt (V) 37
Mennejean (BF) 80
Mennevret (V) 86
Menuet (W) 75
Mery (B) 67
Messines (B) 16
Meuse-Argonne (B) 109
Mezieres (V) 36
Missy aux Bois Ravine 66
Monchy (B) 13

Monchy (V) 11
Monchy le Preux (V) 42
Mongival (V) 62
Monnes (V) 73, 74
Mont Blanc 82
Montbrehain (V) 48
Mont de Courmelles (F) 73
Mont de Leuilly 79
Montfaucon (V W) 109, 110
Montgobert (B)
Montmirail (V) 71
Montparnasse Heights 61
Montron (V) 73
Mont Rouge 28
Morchies (V) 42
Moreuil (B) 33
Moreuil Switch 42
Moreuil (V) 42
Mormal (W) 53
Morocco (C) 1925 89
Mortemer (W) 67
Mory (W) 41
Mory (V) 41
Nampcel (B) 78
Nampcel Plateau 78
Nampcel Ravine 78
Nantillois (V) 110
Nauroy (V) 47
Nazareth (V) 88
Neufoy (V) 68
Neuilly (V) 74
Neuville Vitasse (B) 13
Neuvilly (V) 51
Niergnies 48
Nizy le Comte (V) 88
Nonsard (V) 106
Notre Dame des Champs 82
Nouvron-Vingre (B) 77, 78
Oeuilly (B) 71
Ognons (W) 110, 111
Oiseaux Ravine 81
Oosttaverne Line 17, 18
Orfeuil Ridge 82
Orme de Grand Rozoy 75
Orme Signal 74
Orme (W) 73
Orvillers (V) 102, 103
Outpost Hill 14
Palestine (C) 1917 13
Pannes (V) 106, 107
Parcy (V) 74
Parvillers (V) 37
Passy en Valois (V) 73
Pernant Ravine 66
Peronne (V) 28
Perriere (F) 79
Petit Thiolet (V) 86
Petit Verley Region (B) 84
Pine (W) 82
Ploissy (V) 65
Ploissy-Chazelle (B) 64
Pont St. Mard 78
Porte (BF) 70
Premont (V) 50
Premy Chapel 45
Proyart (V) 36, 37
Py (R) 82
Quadrilateral (fortification) 43, 44

Quart de Reserve (W) 106
Quennemont (F) 43, 46, 47
Rafa Redoubt 14
Raillencourt (V) 46
Rainecourt (V) 37
Ramicourt (V) 48
Rassy (V) 74
Rate (W) 106
Recouvrance 88
Reims (B) 102
Reims (BW) 76
Reincourt 12
Ressons Station 77
Ressons sur Matz (B) 77
Ressons (W) 103
Ressons (V) 77
Retz (BW) 66
Retz (W) 64
Reygerie (V) 83
Richecourt 106
Riguerval (W) 50
Roisel (V) 28
Romagne (V) 111
Ronssoy (W) 27
Ronssoy (V) 43
Rosieres (V) 37
Roulers (V) 83
Rouvroy (V) 37
Roye (B) 77
Roye-Chaulnes Line 35
Rupt de Mad (R) 106
Russia (C) 1918 53
Saconin Missy aux Bois Ravine 72
Samson Ridge 14
Saniere (F) 87
Sapignies (V) 40, 41
Sauchy-Lestree (V) 46
Sauvillers (B) 33
Sauvillers (V) 33, 62
Savarts (BF) 76
Scarpe (R) 11
Seboncourt (B) 84
Sechelles (W) 103
Selency (V) 43
Selle (B) 50
Senecat (BW) 62
Septsarges (V) 110
Sequehart (V) 48
Serain (V) 48
Shanghai, 1932 150
Sheikh Ajlin 15
Sheikh Hassan 15
Signal d'Origny 87
Silesia (C) 1921 54
Silkem Chapel 46
Soissons (B) 72, 100
Somme (1st B) 5
Somme (2nd B) 26
Somme (R) 37
Sommelans (V) 74
Somme Py Heights 81
Sorny (V) 79
Staden (V) 83
St. Benoit (V) 108
Steenbeck (R) 19
St. Front (V) 74
St. Genevieve (V) 73
St. George (V) 113

INDEX

St. Julien (V) 19
St. Leger (V) 41
St. Mard (V) 77
St. Marie a Py Station 82
St. Maur (B) 67
St. Maur (V) 68
St. Maurice (V) 108
St. Mihiel (B) 106
St. Olle (V) 46
St. Paul (F) 66, 78
St. Pierre-Aigle (B) 69
St. Pierre-Aigle (V) 64
St. Pierre-Aigle Brook 68
St. Quentin (B) 96
St. Quentin Canal 46, 47
St. Quentin (V) 84, 86
St. Quentin le Petit (V) 88
St. Remy (V) 74
St. Ribert (W) 33
St. Ribert (V) 33
Synghem (V) 88
Syria (C) 1920 94
Tara Hill 40
Taux (V) 73
Telegraph Hill 11

Terny (W) 79
Terny (V) 79
Thiaucourt (W) 109
Thielt (B) 83
Thiepval (VB) 9
Tigny (V) 73, 74
Tillaloy (B) 77
Tilloy (V) 48
Tilloy les Mafflaines 11
Translon 72
Troesnes Road 66
Tournelle (W) 75
Tourtelle (W) 81
Usna Hill 40
Vailly (V) 80
Vaire (W) 32
Vareille (F) 75
Valènciennes (V) 51
Varennes (V) 109
Vaudesson (V) 61
Vauvillers (V) 37
Vaux Andigny (V) 50
Vaux Castile (V) 72, 73
Velu (W) 42
Vertefeuille (F) 65, 66, 72

Very (V) 109, 110
Viermont (V) 103
Vierzy (V) 72
Vigneulles (V) 108
Villar au Flos 42
Villemontoire (V) 74
Villers Bretonneux (B) 97
Villers Cotterets (F) 65
Villers Guislain (V) 46
Villers Helon (V) 72
Villers Outreaux (V) 48
Vimy Heights 10
Visen Artois (V) 11
Vracourt 42
Vergny Plateau 79
Vouty (V) 65, 66
Voux-Vracourt 42
Wacquemoulin (V) 67
Wancourt (V) 13, 42
Wilpré Spring Ravine 111
Woel (V) 108
Wood Switch 46
Wytschaete (V) 17
Ypres (3d B) 18
Zeswege (V) 83

Coachwhip Publications
CoachwhipBooks.com

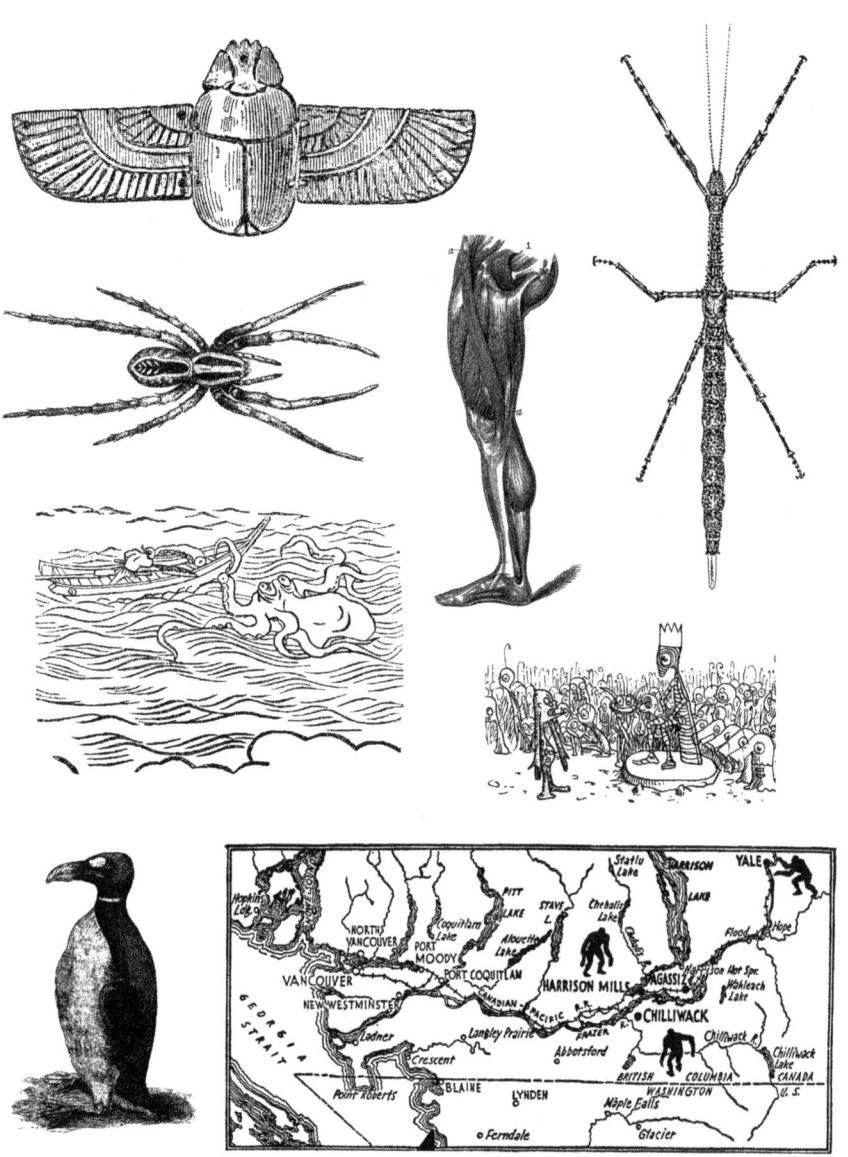

Coachwhip Publications
Also Available

TANKS
and How to Draw Them

Terence T. Cuneo

Tanks and How to Draw Them
ISBN 1-61646-021-0

Coachwhip Publications
Also Available

Tank-Fighter Team
ISBN 1-61646-023-7

Coachwhip Publications
Also Available

Bastogne
ISBN 1-61646-062-8

Coachwhip Publications

Also Available

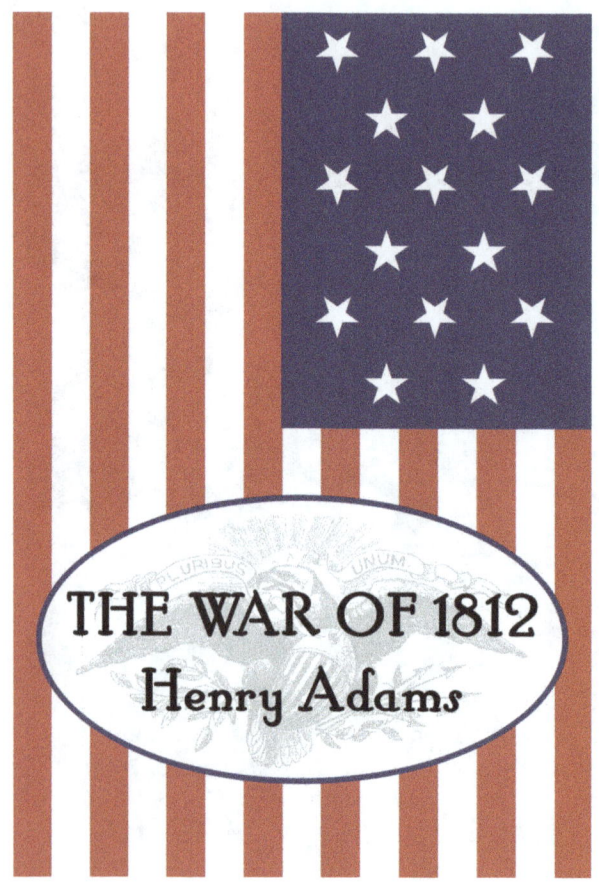

The War of 1812
ISBN 1-61646-065-2

Coachwhip Publications
Also Available

YORKTOWN

THE STRATEGY, PEOPLE, AND EVENTS
SURROUNDING THE FINAL BATTLE IN THE
AMERICAN WAR OF INDEPENDENCE

Yorktown
ISBN 1-61646-101-2

www.ingramcontent.com/pod-product-compliance
Lightning Source LLC
Chambersburg PA
CBHW080834230426
43665CB00021B/2842